Y0-BDP-821

Advances in Industrial Control

Other titles published in this series:

Digital Controller Implementation and Fragility
Robert S.H. Istepanian and James F. Whidborne (Eds.)

Optimisation of Industrial Processes at Supervisory Level
Doris Sáez, Aldo Cipriano and Andrzej W. Ordys

Robust Control of Diesel Ship Propulsion
Nikolaos Xiros

Hydraulic Servo-systems
Mohieddine Jelali and Andreas Kroll

Model-based Fault Diagnosis in Dynamic Systems Using Identification Techniques
Silvio Simani, Cesare Fantuzzi and Ron J. Patton

Strategies for Feedback Linearisation
Freddy Garces, Victor M. Becerra, Chandrasekhar Kambhampati and Kevin Warwick

Robust Autonomous Guidance
Alberto Isidori, Lorenzo Marconi and Andrea Serrani

Dynamic Modelling of Gas Turbines
Gennady G. Kulikov and Haydn A. Thompson (Eds.)

Control of Fuel Cell Power Systems
Jay T. Pukrushpan, Anna G. Stefanopoulou and Huei Peng

Fuzzy Logic, Identification and Predictive Control
Jairo Espinosa, Joos Vandewalle and Vincent Wertz

Optimal Real-time Control of Sewer Networks
Magdalene Marinaki and Markos Papageorgiou

Process Modelling for Control
Benoît Codrons

Computational Intelligence in Time Series Forecasting
Ajoy K. Palit and Dobrivoje Popovic

Modelling and Control of Mini-Flying Machines
Pedro Castillo, Rogelio Lozano and Alejandro Dzul

Ship Motion Control
Tristan Perez

Hard Disk Drive Servo Systems (2nd Ed.)
Ben M. Chen, Tong H. Lee, Kemao Peng and Venkatakrishnan Venkataramanan

Measurement, Control, and Communication Using IEEE 1588
John C. Eidson

Piezoelectric Transducers for Vibration Control and Damping
S.O. Reza Moheimani and Andrew J. Fleming

Manufacturing Systems Control Design
Stjepan Bogdan, Frank L. Lewis, Zdenko Kovačić and José Mireles Jr.

Windup in Control
Peter Hippe

Nonlinear H_2/H_∞ Constrained Feedback Control
Murad Abu-Khalaf, Jie Huang and Frank L. Lewis

Practical Grey-box Process Identification
Torsten Bohlin

Control of Traffic Systems in Buildings
Sandor Markon, Hajime Kita, Hiroshi Kise and Thomas Bartz-Beielstein

Wind Turbine Control Systems
Fernando D. Bianchi, Hernán De Battista and Ricardo J. Mantz

Advanced Fuzzy Logic Technologies in Industrial Applications
Ying Bai, Hanqi Zhuang and Dali Wang (Eds.)

Practical PID Control
Antonio Visioli

Tan Kok Kiong • Lee Tong Heng • Huang Sunan

Precision Motion Control

Design and Implementation

Second Edition

Tan Kok Kiong, PhD
Lee Tong Heng, PhD
Huang Sunan, PhD

Department of Electrical and Computer Engineering
National University of Singapore
4 Engineering Drive 3
Singapore 117576
Singapore

ISBN 978-1-84800-020-9 e-ISBN 978-1-84800-021-6

DOI 10.1007/978-1-84800-021-6

Advances in Industrial Control series ISSN 1430-9491

British Library Cataloguing in Publication Data
Tan, Kok Kiong, 1967-
Precision motion control : design and implementation. - 2nd ed. - (Advances in industrial control)
1. Motion control devices 2. Automatic control
I. Title II. Lee, Tong Heng, 1958- III. Huang, Sunan, 1962-
629.8
ISBN-13: 9781848000209

Library of Congress Control Number: 2007939805

Cover design: eStudio Calamar S.L., Girona, Spain

Printed on acid-free paper

9 8 7 6 5 4 3 2 1

springer.com

Advances in Industrial Control

Series Editors' Foreword

The series *Advances in Industrial Control* aims to report and encourage technology transfer in control engineering. The rapid development of control technology has an impact on all areas of the control discipline. New theory, new controllers, actuators, sensors, new industrial processes, computer methods, new applications, new philosophies..., new challenges. Much of this development work resides in industrial reports, feasibility study papers and the reports of advanced collaborative projects. The series offers an opportunity for researchers to present an extended exposition of such new work in all aspects of industrial control for wider and rapid dissemination.

A striking development in the *Advances in Industrial Control* series that has occurred over the last few years have been the appearance in the series of highly authoritative volumes that are more comprehensive than the usual monograph for a particular technical area in industrial control. The Editors believe these volumes have set new standards for the presentation of knowledge and industrial control research in their specific fields. Typical examples are: *Hydraulic Servo-systems* by Mohieddine Jelali and Andreas Kroll, *Control of Fuel Cell Power Systems* by Jay Pukrushpan, Anna Stephanopoulou and Huei Peng, *Hard Disk Drive Servo Systems* (now in its second edition) by Ben Chen, Tong Heng Lee, Kemao Peng and Venkatakrishnan Venkataramanan, *Piezoelectric Transducers for Vibration Control and Damping* by Reza Moheimani and Andrew Fleming, and finally *Wind Turbine Control Systems* by Fernando Bianchi, Hernán De Battista and Ricardo Mantz. These and other volumes like them in the series all seemed to capture the "spirit of the age" in the field of individual control in the new millennium. To these volumes, the Editors are very pleased to add a second edition of *Precision Motion Control* by Kok Kiong Tan, Tong Heng Lee and Sunan Huang; this is a revision of a volume that was first published in the *Advances in Industrial Control* series in 2001.

The new volume presents a revised and systematic coverage of many of the theoretical and practical aspects of precision motion control. A strong feature of the volume is its presentation and integration of industrial control methods with new advanced control solutions in this field. This is often illustrated by presenting experimental results that span the full range of hardware implementations, from classical control through to new control solutions advanced by the authors. The authors often approach the control engineering problems by starting with standard industrial control solutions (usually PID) and go on to show how system performance can be enhanced by the addition of advanced control features. In some cases completely new advanced control approaches are proposed. Models, too, are sufficiently complex to capture important nonlinear system effects like friction and ripple.

An example of the authors' approach can be found in their treatment of the control for permanent magnet linear motors. The system model is basically linear in structure with additional nonlinear loops to represent the debilitating physical effects of friction and force ripple. The control system design begins from classical PID control which is then augmented by feed-forward control and adaptive radial-basis-function (RBF) compensation to achieve enhanced system performance in the presence of nonlinear disturbances.

As well as being a compendium of the technology used in state-of-the-art precision motion control – and indeed there are many excellent descriptions of equipment, hardware, software and techniques used in this field – the volume has interesting reports of new approaches to real problems in precision motion control. Three examples illustrate the types of approach taken. In Chapter 3, new applications are described for the relay experiment; in this particular case, it is used to identify a friction model to enhance the control design. In Chapter 4, the co-ordinated control of a gantry system is investigated, and the performance of PID compared with that of an adaptive controller; however, real performance enhancements are shown to accrue to those control strategies that overcome the multivariable interactions that are present in the system. Finally, Chapter 7 reports on a mechatronics approach to control design where the structural design of the machine is seamlessly integrated with the control system design.

This second edition of *Precision Motion Control* is likely to become a source book for a very wide range of readers. It has industrial perspectives, current state-of-the-art hardware descriptions, academic perspectives and advanced control system solutions often explained from initial conception right through to results from laboratory rigs and prototype tests. Thus, the volume makes a very welcome and appropriate contribution to the *Advances in Industrial Control* series.

Industrial Control Centre
Glasgow
Scotland, UK
2007

M.J. Grimble
M.A. Johnson

Preface

Precision manufacturing has been steadily gathering momentum and attention over the last century in terms of research, development, and application to product innovation. The driving force in this development appears to arise from requirements for much higher performance of products, higher reliability, longer life, lower cost, and miniaturisation. This development is also widely known as precision engineering and, today, it can be generally defined as manufacturing to tolerances which are better than one part in 10^5.

The historical roots of precision engineering are arguably in the field of horology, the development of chronometers, watches and optics, *e.g.*, the manufacture of mirrors and lenses for telescopes and microscopes. Major contributions were made to the development of high-precision machine tools and instruments in the late 1800s and early 1900s by ruling engines for the manufacture of scales, reticules and spectrographic diffraction gratings. Today, ultra-precision machine tools under computer control can position the tool relative to the workpiece to a resolution and positioning accuracy in an order better than micrometers. It must be noted that achievable "machining" accuracy includes the use of not only machine tools and abrasive techniques, but also energy beam processes such as ion beam and electron beam machining, as well as scanning probe systems for surface measurement and pick-and-place types of manipulation.

In the new millenium, ultra-precision manufacture is poised to progress further and to enter the nanometer scale regime (nanotechnology). Increasing packing density on integrated circuits and sustained breakthrough in minimum feature dimensions of semiconductors set the pace in the electronics industry. Emerging technologies such as Micro-electro-mechanical Systems (MEMS), otherwise known as Micro-systems Technology (MST) in Europe further expands the scope of miniaturisation and integration of electrical and mechanical components.

This book is focused on the enabling technologies in the realisation of precision motion positioning systems. It is a compilation of the major results and publications from projects set out to develop state-of-the-art high-speed, ultra-precision robotic systems. A comprehensive and thorough treatment of the subject matter is provided in a manner amenable to a broad base of readers, ranging from academics to practitioners, by providing detailed experimental verifications of the developed materials.

The book begins with an introduction to precision engineering, and provides a brief survey of its development and applications. Chapter 2 addresses the control system technology to achieve high-precision motion control in motion systems. Intelligent control schemes are presented which can yield high performance in terms of tracking accuracy. These control schemes use different combinations of advanced control theory and artificial intelligence according to the information available and the nature of operations. These include an adaptive control scheme, a composite control scheme comprising linear and non-linear control components, an adaptive ripple compensation scheme, a disturbance observer and compensation scheme, and a learning control strategy. Experimental results are duly provided for comparison and verification of the performance and improvement achievable over standard controllers. The use of a high grade accelerometer in providing direct acceleration measurements and an illustration of the possible enhancement in tracking performance achievable with additional state feedback are clearly elaborated. While the materials are applied to the subject matter, they are sufficiently generic to interest general control specialists and practitioners.

Chapter 3 presents relay feedback configurations and techniques which are suitable to produce nominal models for the motion systems, based on sustained small amplitude oscillations induced in the closed-loop. In this way, the control systems as presented in Chapter 2 can be automatically tuned and commissioned, and yet satisfactory performance can be achieved. A variation of the basic configuration to facilitate the automatic modelling of the frictional effects is also given. These models can be used to commission feedforward and feedback controllers, and they are also useful for the initialisation of adaptive control. A scheme is provided for optimal features extraction from possibly noisy relay oscillations.

Chapter 4 addresses a popular configuration of precision Cartesian robotic systems, the moving gantry stage, which is frequently employed in wafer steppers and fine resolution assembly machines. Apart from individual servo tracking requirements, it is also necessary that the parallel servo systems move in tandem to minimise the inter-axis offsets. Different control configurations are presented and compared in terms of their performance. These include control schemes used in existing industrial control systems, as well as more recent developments.

Chapter 5 presents a comprehensive treatment of the topic of geometrical error calibration and compensation. The sources of geometrical errors, the calibration equipment used in their measurement, treatment and modelling from the raw data set to the final compensation *via* the control system are among the topics which will be delivered systematically in this chapter. Recent and refreshing advances in geometrical calibration and compensation are also presented in the chapter, which include the use of *Artificial Intelligence* (AI) approaches in geometrical error modelling. Possible probabilistic approaches, formulated to reduce the influence of random errors from affecting the systematic error compensation, are also presented in the chapter.

Chapter 6 addresses explicitly the measurement system. Precision motion control can only be possible with precision motion measurements. Encoder interpolation is a cost effective way to derive fine resolution position measurements using only devices and instruments at moderate costs. Techniques are presented to correct for imperfections in encoder signals and to derive fractional resolution from the corrected signals to fulfil high-resolution requirements in the input signals for the control system.

Chapter 7 will touch on the topic of vibration monitoring and control. Three approaches are presented. The first focuses on a proper mechanical design, based on the determinacy of machine structure, to reduce the mechanical vibration to a minimum. The second approach is based on the notch filter and its application as part of the control system to suppress frequencies which may excite undesirable mechanical resonance. An adaptive technique based on *Fast Fourier Transform* (FFT) tracks the resonant frequency and adapts the filter accordingly. The third approach uses a technique based on sensor fusion to monitor and analyse the vibration of precision machines. A DSP device is used to learn and capture the vibration signature of the machine under normal operational circumstances. When the machine deviates from its normal operational condition, the device can detect the abnormality and activates appropriate fault diagnostic and maintenance measures.

Finally, in Chapter 8, other important engineering aspects behind the construction of a high-precision motion control system are discussed. These include the considerations behind selection of components, hardware architecture, software development platform, user interface design, evaluation tests which are crucial in determining the final success of the system, and digital communication protocols.

This book provides extensive and up-to-date coverage of the methodology and algorithms of precision motion control considered mainly in the context of control engineering and soft computing.

Compared to the first edition, the new edition has incorporated a series of modifications, updates and extensions. Some six years after publication of the first edition, precision engineering has remained an important area in control engineering and new results have emerged. The first edition has been updated with new contents, including piezo actuator modelling and control (Chapter 2), adaptive co-ordinated control scheme (Chapter 4), parametric model for interpolation (Chapter 6), mechanical design to minimise vibration (Chapter 7), and digital communication protocols (Chapter 8). The introductory chapter has been substantially revised to reflect the state-of-the-art of precision motion control.

This book would not be possible without the generous assistance of the following colleagues and friends: Dr Lim Ser Yong, Mr Andi Sudjana, Mr

Teo Chek Sing, Dr Tang Kok Zuea, Dr Zhou Huixing and Mr Jiang Xi. The authors would like to express their sincere appreciation of their kind assistance provided in the writing of the book. They would also like to thank the National University of Singapore (NUS) and Singapore Institute of Manufacturing Technology for co-funding the projects from which most of the information and results reported in the book have originated. The authors also acknowledge the kind permission from Hewlett Packard for the reproduction of figures relating to laser measurement systems.

Finally, the authors would like to dedicate the book to our families for their love and support.

Singapore,
May 2007

Kok Kiong Tan
Tong Heng Lee
Sunan Huang

Contents

1

Introduction

Precision control is one of the core requirements to be met by ultra precision machines. A well chosen control strategy will enable a comprehensive complete control and compensation of the mechanical system to achieve precise positioning. The field of high-precision motion control is now an interesting subject of research. Precision control technology will be discussed here with respect to the following broad fields:

- Precision engineering
- Micromanufacturing
- Biotechnology
- Nanotechnology

1.1 Fields Requiring Precision Control

1.1.1 Precision Engineering

Machining is an essential process in the manufacturing industry concerned with removing excess or unwanted material by the use of machine tools, such as cutting, grinding, and finishing. Conventional machining is executed *via* turning machines, drilling machines, milling machines, *etc.* While they are still in use, the development of machining processes to provide high precision components has introduced new and non-conventional machining *via* laser cutting, hydrodynamic fluids, chemical substances, *etc.* Nowadays, there has been a trend towards non-contact machining as opposed to contact machining.

Ultra-precision Spindles

Ultra-precision spindles are used to drive loads at high speed, and moderate or low torque. They are implemented in high-precision manufacturing devices such as high speed turning and milling machines, as well as non-manufacturing devices such as high performance magnetic memory disk file systems, high definition large scale projection televisions, and video cassette recorders. These applications call for highly precise positioning, which poses a challenge since it is also to be accomplished at high speed.

To achieve the required specifications, air-bearing is typically employed. The characteristic of interest in air-bearing is its low asynchronous error motion making it possible to achieve high rotational accuracy. The disadvantage, however, is its low stiffness and damping ability. Figure 1.1 shows the working diagram of an air-bearing, where pressurized air is used to maintain the gap between the rotating and the static parts of the machine (*e.g.* spindles).

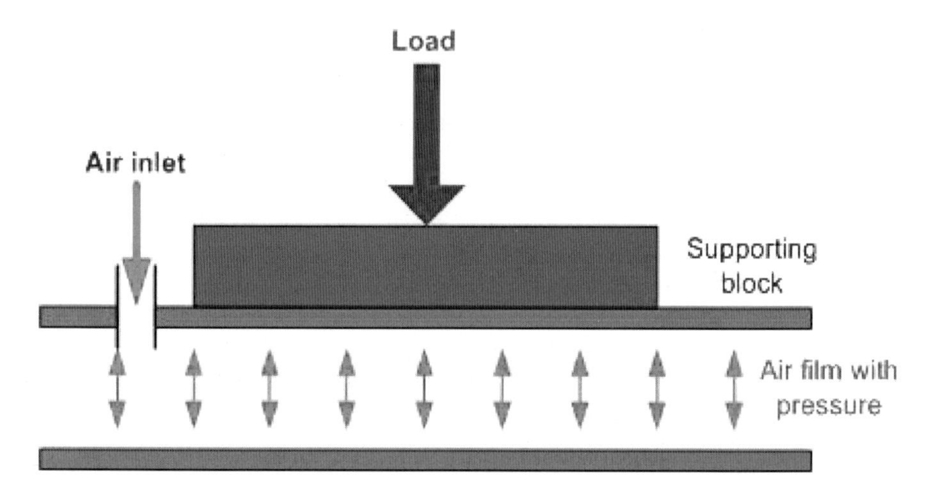

Fig. 1.1. Air bearing

To achieve high stiffness, a hybrid solution involving the integration of air-bearing with conventional oil bearing has also been developed.

Excimer Laser Micromachining

Lasers, in particular excimer lasers, are today widely used for micromachining of different kinds of materials due to their unique pulsed ultra violet (UV) emission. They have been used of in research laboratories since 1977 and about 10 years later they were successfully introduced into industrial processing and manufacturing. Excimer lasers have been used for the highly precise marking of glass (such as in eyeglasses) and ceramics, especially in surface mounted devices (SMD). In microelectronics production lines, drilling into printed-circuit boards can be performed with this technique. In semiconductor processes, it can be used as a direct writing tool to replace photomasks, as a micro-drill for multilayer chip, and as an ablation tool for non-chemical etching and repair in semiconductor processes.

The excimer laser is excited by a rare gas halide or rare gas metal vapor, often employing noble gases due to their stability. Controlling the flow and pressure of the gas is necessary in order to maintain precision. Excimer laser control includes controlling gas exhaust filters, vacuum pumps, and gas mixers.

Precision Metrology and Test

The measurement precision associated with Co-ordinate Measuring Machines (CMM) has been continually increasing over the years. When these machines are fitted with precision tools, such as a probe, vision device, or a microscope, special applications can be set up in the area of metrology and tests. One of the applications is Scanning Probe Microscopy (SPM).

The first generation of SPM is Scanning Electron Microscopy (SEM), where an electron beam is focused into a small spot on the object and electromagnetically raster scanned across it. Images can be formed by collecting the secondary electrons generated by the impact of the impinging electron beam, by detecting the backscattered electrons, or by detecting the X-rays generated. In this way, several different aspects of the object can be characterized, including morphology, average atomic number and composition.

In another technique, Scanning Tunneling Microscopy (STM), the evanescent wave is an electron wave function with an intrinsic wavelength of about 1 nm which extends beyond the surface of a sharp metal tip. If a conducting surface is brought to within about 1 nm of the tip and a potential difference is applied between them, then a tunneling current will be induced. The magnitude of this current is an exponentially-decaying function of distance and is also dependent upon the difference between the work-functions of the two materials. Thus, information can be derived on both the topography of the surface and its chemical composition. The limitation of STM is that it can only work with conducting surfaces.

Scanning Force Microscopy (SFM), usually referred to as the Atomic Force Microscopy (AFM), has been developed to overcome the limitation of STM. The initial instrument used a diamond stylus on a gold foil cantilever scanned lightly across the surface of the specimen, with the repulsion being detected using a tunneling tip. The change in cantilever resonance frequency is sensed as the tip approaches the sample surface and is affected by the van der Waals attraction. This type of microscopy has been used for a very wide range of surface characterization, including imaging and topography.

1.1.2 Micromanufacturing

One of the major inventions in the twentieth century is microelectronics, the science of micro-devices. The design of a micro-device is not a trivial task as it leads to integrated circuits which are fabricated in submicron size. Microfabrication covers a range of manufacturing processes that produce patterns or layers of material to form microstructures. Lithography and MEMS (or

MST) are common examples of micro-fabrication processes. Micro-assembly is another important process of precision engineering.

Lithography

The semiconductor and microelectronics industries have led to the development and application of photo- and electron beam lithography techniques which are expected to serve as the main basis for continuing miniaturization in large scale production in the future. Features and dimensions are printed on silicon chips using a process called photolithography, in which UV light from a mercury vapor lamp is shone through a mask containing the features of the chip and projected onto the surface of the silicon wafers in a machine known as a photolithographic "stepper" — so called because it prints an image of one chip and then "steps" to the next location on the wafer to print the pattern for the next and so on. For feature size of smaller than 0.1 μm, shorter wavelength radiation in the form of electron beams or X-rays can be used.

Ultra large scale integration (ULSI) chips will be the harvest of precision lithography. These are fast becoming smaller, faster, cheaper and come equipped with more memory. They are expected to bring further massive improvements to the performance of microprocessors and computers, and will, in turn, lead to direct benefits for telecommunications, domestic, automotive, and medical products and services. Figure 1.2 illustrates the lithography process, showing mainly the control configuration of the process. In this process, a piezoelectric actuator is employed. A piezoelectric substance is a material, usually made from ceramic, with a capability of transforming an electrical signal (voltage) into a motion of the order of nanometer level. The piezoelectric tube actuator drives the cantilever, which in turn drive a silicon probe tip. This tip is the source of electrons that will develop the pattern on the working object. Therefore, the performance of the overall system is very much dictated by the motion precision of the piezoelectric actuator. The application of piezoelectric material in lithography is an example of how material science will also enhance and influence the developments of motion systems.

Micro-electro-mechanical Systems (MEMS)

MEMS is the integration of mechanical and electronic elements, including sensors and actuators a common substrate, usually silicon. MEMS components are fabricated using micromachining processes that selectively etch away parts of the silicon wafer or add new structural layers to form the intended structures. With MEMS it is possible to develop a system-on-chip; a term commonly used to refer to a multifunctional chip.

From the early examples of accelerometers and gyroscopes, MEMS products with micro-mechanical features such as specialized sensors, arrays of sensors, and actuators fully integrated into the same silicon chip, are already

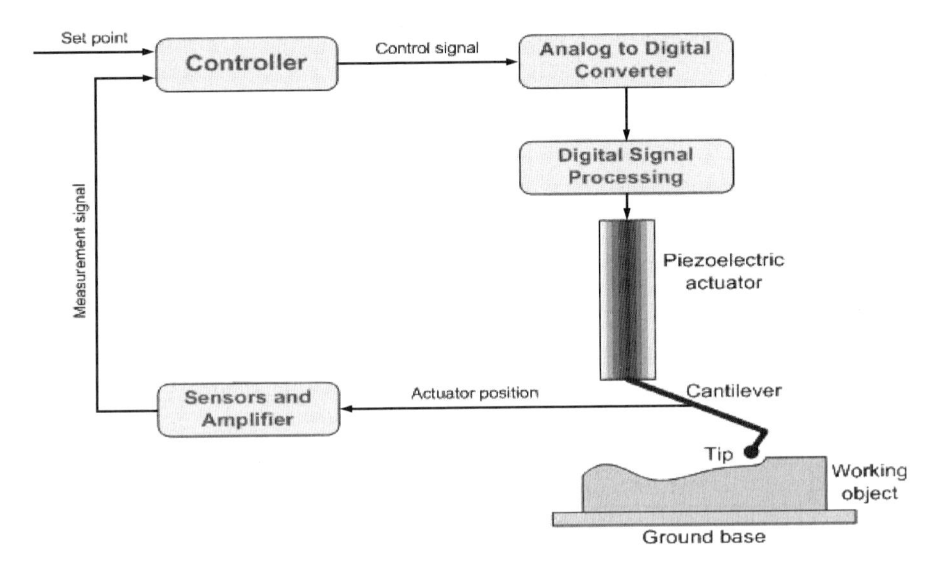

Fig. 1.2. The lithography process

burgeoning; applications are expected to expand in the navigational, automotive, biomedical and pharmaceutical industries.

Micro-assembly

Another process involving high precision is in the area of pick-and-place micro-assembly. One example of micro-assembly process is a flip chip assembly.

A flip chip is a chip mounted on the substrate with various interconnect materials and methods, such as tape-automated bonding, flux-less solder bumps, wire interconnects, isotropic and anisotropic conductive adhesives, metal bumps, compliant bumps and pressure contacts, as long as the chip surface (active area or I/O side) is facing the substrate.

One of the earliest flip chip technologies was solder-bumped flip chip technology, as a possible replacement for the expensive, unreliable, low productivity, and manually operated face-up wire-bonding technology. Bumps are formed by injecting molten solder into etched cavities in a glass mold plate across a wafer. The mold plate is heated to just below melting point of the solder. The injector includes a slightly pressurized reservoir of molten solder of any composition. Figure 1.3 illustrates the process of solder bump deposition.

The use of flip chip technologies in the manufacture of IC devices has increased tremendously in recent years. As the size of devices gets smaller, the precision required to align the solder bumps on the chip to the pads on the substrate becomes more crucial.

Besides flip chip assembly, high-precision robots are also used to assemble micro-electronic and mechanical components.

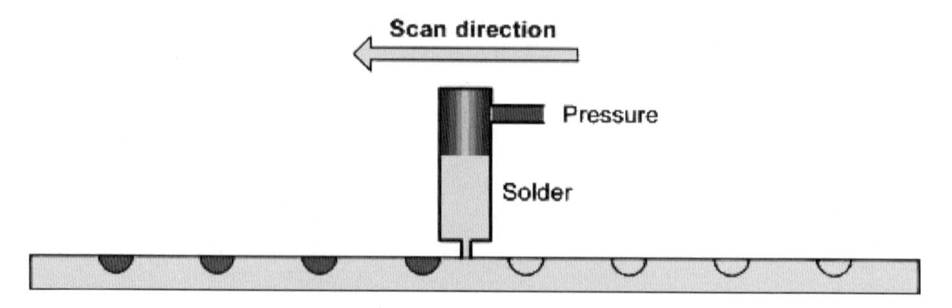

Fig. 1.3. Solder bump deposition

1.1.3 Biotechnology

Biotechnology is the technology to manipulate the structure and function of biological systems, especially when used in food science, agriculture, and medicine. Modern biotechnology is often related to genetic alteration of living materials, such as microorganisms, plants, and animals.

Examples of biotechnology applications include the following:

- Minimally invasive surgery, assisted by remotely operated surgical instruments and diagnostic tools, *e.g.*, micro-catheters down to 100μm diameter incorporating optical fibers for delivery and retrieval of light images for high-resolution cameras; nano-scale sensors for measuring blood chemistry; and tip-mounted micro-turbine rotary cutters for arterial plaque removal.
- Intracytoplasmic sperm injection (ICSI), a method to help fertilization by the injection of sperm to an egg cell which requires high-precision actuator to minimize the damage imposed to the cell.
- Accurate and efficient drug targeting and delivery by nano-particle technology, acting as medicinal bullets.
- Replacement of damaged nerves by artificial equivalents.
- Improved adhesion growth of living tissue cells on to prosthetic implants by micro- and nano-surface patterning of implant materials.

Another biomedical application of nanotechnology is in the fabrication of tiny biochips, a technology sometimes referred to as "laboratories on a chip". Brief information on biochips is given in the following sub-section.

Biochips and High-density Sensor Arrays

Many examples exist where large numbers of individual biological analyses, *i.e.*, biological assays, commonly 10^3 to 10^6, need to be performed and include the screening of libraries of potential pharmaceutical compounds and various protocols for the screening and sequencing of genetic material. Such large number dictate the parallel processing of assays to enable completion in reasonable timescales and the common availability of only small sample quantity dictates small size. Thus, microfabricated high-density arrays of biosensor-like sensor elements have been investigated where the size of individual elements approaches the nanotechnology domain. Such approaches are often termed "biochips", generally meaning an integration of biology with microchip type technologies. For example, devices are being developed for genetic screening that contain two dimensional arrays with greater than 1×10^5 elements each comprising a differing DNA sequence and where each element is optically examined for specific interaction with complementary genetic material.

1.1.4 Nanotechnology

Nanotechnology is a group of generic technologies that are becoming significantly important to many industrial applications and it is poised to revolutionalise new trends in technological advancement. Following McKeown (1996), nanotechnology may be defined as the study, development and processing of materials, devices and systems in which structure on a dimension of less than 100nm is essential to obtain the required functional performance. It covers nano-fabrication processes, the design, behaviour and modelling of nano structures, methods of measurement and characterisation at the nanometre scale. Nanotechnology may be deemed as a natural next step to precision engineering as ultra-precision manufacturing progresses through micrometre accuracy capability to enter the nanometre scale regime.

Nanotechnology creates opportunities for the international business community which arise from the science and engineering research base in microsystem technologies, nano science and nanotechnology. The main driving forces in this broad field from micro to nano systems are:

- New products that can work only on a very small scale or by virtue of ultra-precision tolerances,
- Higher systems performance,
- Miniaturisation, motivated by "smaller, faster, cheaper",
- Higher reliability, and
- Lower cost.

The term nanotechnology was coined by Professor Norio Taniguchi, formerly of Tokyo Science University, in 1974 at the International Conference on Production Engineering in Tokyo. Taniguchi has used the term to relate specifically to precision machining — the processing of a material to nano scale precision using primarily ultra-sonic machining. Professor Taniguchi was subsequently deeply involved in the research and application of electron beam processes for nano-fabrication.

Although Taniguchi was the first to coin the term, the concept of nanotechnology was arguably first enunciated by the American physicist Dr Richard Feynman in a visionary lecture delivered to the annual meeting of the American Physical Society in 1959. His talk entitled, "There's plenty of room at the bottom" questioned the traditional concept of space. At the outset he asked,

"Why cannot we write the entire 24 volumes of the Encyclopaedia Britannica on the head of pin?". He reasoned that if the head of the pin can be magnified by 25 thousand times, the area would then be sufficient to contain all the pages of the Encyclopaedia Britannica. This magnification will be equivalent to reducing the size of all the writing by the same 25 thousand times. He also predicted then that the Scanning Electron Microscope (SEM) could be improved in resolution and stability to be able to resolve atoms and went on further to predict the possibility of direct atom arrangement to build tiny structures leading to molecular or atomic synthesis of materials. In hindsight, his foresights and predictions have been very accurate. He did not explicitly use the term nanotechnology as such, but has accurately predicted its potential and applications.

There can be no doubt that many new and interesting developments and products will arise from today's nanoscience and nanotechnology R&D work. Waves of product miniaturisation to follow will see existing macro products replaced by Microsystem Technologies (MST) and nanotechnology products, produced by new nanotechnology-based manufacturing facilities. Nanotechnology is a major new technological force that will have substantial socio-economic effects throughout the world, and many benefits in standards of living and quality of life can be confidently expected.

1.2 Precision Machines and Tools

In order to implement and use the advanced technology processes described in the previous sections, ultra-precision machines and instruments are needed to control the three-dimensional (3D) spatial relationship of the "tool" to the workpiece to accuracies in the order of less than 0.1nm. The tools can be:

- Solid tools for cutting, abrasive or chemico-mechanical action,
- Energy beam tools,
- Scanning probe tools such as STM, AFM, magnetic, thermal or chemical-reactive probes *etc.*

Ultra-precision machine systems generally fall into three main classifications:

- Computer Numerical Control (CNC) macro-machines for measuring, shaping or forming conventional macro-sized component parts; today, this can mean working to nano tolerances on macro-components,
- Instruments for metrological applications to macro- and micro-components,
- Very small specialised machines ranging in size from a few millimetres down to micrometre dimensions for specific applications.

1.3 Applications of Precision Motion Control Systems

A summary of the relevant industries and applications of precision motion control systems are briefly summarised and outlined in the following subsections.

1.3.1 Semiconductor

- Microlithography
- Substrate coating
- Memory repair
- Laser direct writing
- Microscope XY inspection
- Wafer probing
- Wire bonding

1.3.2 Magnetic and Optical Memory Manufacturing

- Disk drive read/write head machining
- Disk and head inspection
- Air bearing spin stands
- Optical disk mastering
- Precision grinding, dicing, and slicing
- Tape head machining
- Flying height testing

1.3.3 Optical Manufacturing

- Lens and mirror diamond turning machines
- Optical grinding machines
- Precision rotary scanning
- Diamond fly cutting machines
- Contact lens lathes
- Encoder and grating ruling engines

1.3.4 High-resolution Imaging

- Flat panel displays
- Internal and external drum plotters
- Constant velocity motion

1.3.5 Precision Metrology

- Sub-micron coordinate measuring machines
- Flatness and roundness measuring systems
- Vision and optical inspection
- Automotive, medical, electronics, optical components

1.4 Scope of the Book

This book will be mainly focused on the control systems and instrumentation technologies for the realisation of precision motion positioning systems. It will address several important challenges to the design of precision motion control systems, including motion control algorithms, geometrical error compensation, encoder interpolation, mechanical vibration monitoring and control, and other related engineering aspects. It is a compilation of the major results and publications from projects set out to develop a state-of-the-art high-speed, ultra-precision robotic system. A comprehensive and thorough treatment of the subject matter is provided in a manner which is amenable to a broad base of readers, ranging from the academics to the practitioners, by providing detailed experimental verifications of the developed materials. Engineering aspects relating to precision control system design which is crucially important in ensuring the final success of the overall system are also provided in the book for general interest.

2

Precision Tracking Motion Control

In this chapters, two precision actuation systems using piezoelectric actuators and permanent magnet linear motors will be presented. Common configurations of these systems, their mathematical model as well as control schemes will be included.

2.1 Piezoelectric Actuators

In the area of micro- and nano-scale systems, the piezoelectric actuator (PA) has become an increasingly popular candidate as a precise actuator, due to its ability to achieve high precision and its versatility to be implemented in a wide range of applications. More specifically, the PA can provide very precise positioning (of the order of nanometers) and generate high forces (up to few thousands newton). The increasingly widespread industrial applications of the PA in various optical fibre alignment, mask alignment, and medical micro-manipulation systems are self-evident testimonies of the effectiveness of the PA in these application domains.

In spite of the benefits of a PA in these application domains, there are challenges in the design and control of these devices. In the following sections, issues pertaining to PA design and control will be discussed in detail.

2.1.1 Types of Piezoelectric Actuator Configuration

The PA offers unique and compelling advantages in nanometer resolution and high-speed applications. To derive maximum performance from PAs, a variety of configuration can be designed to adapt to various requirements.

Stack Design

The most common design for PAs is a stack of ceramic layers (see Figure 2.1). Such devices are capable of achieving high displacements and holding forces. Standard designs which can withstand pressures of up to 100kN are available in commercial products (*e.g.*, products from Physik Instrumente), and preloaded actuators can also be operated in a push-pull mode.

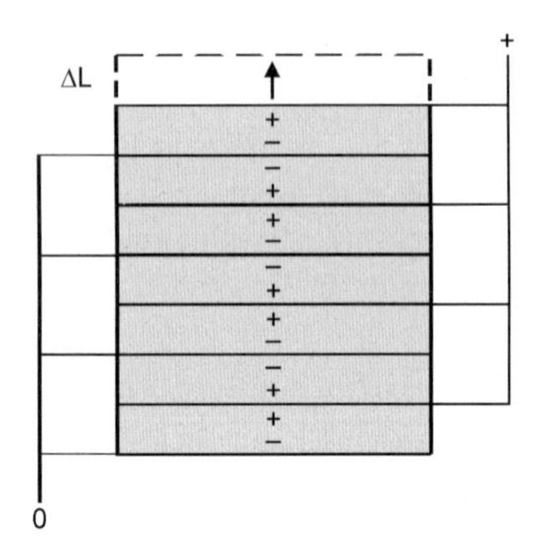

Fig. 2.1. Construction of a stack actuator

Laminar Design

This design uses thin laminated ceramic sheets (see Figure 2.2). When a voltage is applied to the device, the actuator sheet contracts. The displacement in the device is caused by the contraction in the material being perpendicular to the direction of polarization and electric field application. The maximum travel of the laminar actuators is a function of the length of the sheets, while the number of sheets arranged in parallel will determine the stiffness and force generation of the ceramic element. Laminar actuators are easily integrated in conventional composite layers.

Tubular Design

The monolithic ceramic tube is yet another form of piezo actuator. Figure 2.3 shows a design structure. The surface of a tube is partitioned into four regions and they are connected along with one end of the tube to electrodes. Thus, it becomes possible to apply voltages to the tube to initiate motion in various directions. For example, when an electric voltage is applied between the outer and inner diameter of a thin-walled tube, the tube contracts axially and radially. A variety of chemical and materials processing applications use ceramic tubes. Ceramic tubes are also used to fabricate electrical parts for high voltage or power applications such as insulators, igniters or heating elements.

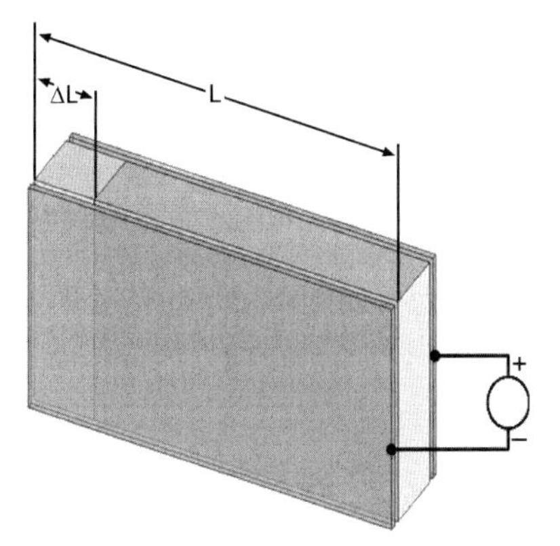

Fig. 2.2. Laminar design

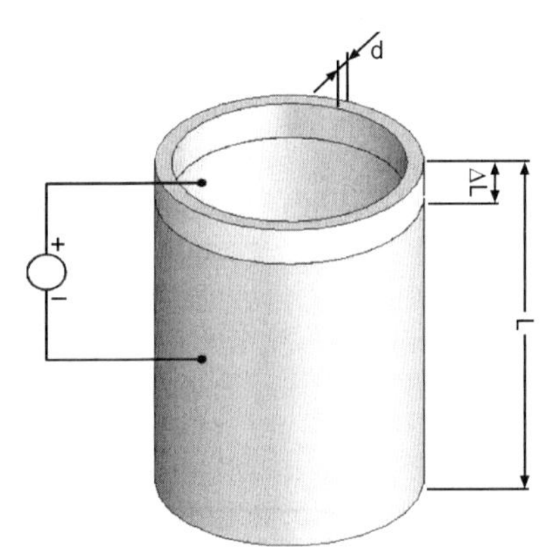

Fig. 2.3. Tube design

Partial-Rotation Design

The design employs a piezoelectric cylinder from LPZT (lithium-lead-zirconium-titanate) ceramic with radial polarization. The sketch of the design is presented in Figure 2.4, in which a piezoelectric cylinder (2) is mounted on a rigid base (1). The movement at the free end of the piezoelectric cylinder is transmitted to a friction pad (3). A spring (4) provides a normal force against the friction pad to generate friction force. A stopper (5) and a rod (6) serve to act as the axis of rotation.

The actuator is designed based on the indirect mode of actuation, therefore reducing the effect of hysteresis in the actuator output. The outer surface of the piezoelectric cylinder is divided into several sections. In the design, four sections are used, although the number of sections can be increased if a finer resolution is desired. The inner surface is a common electrical contact point. With this construction, the cylinder will expand unevenly when a voltage

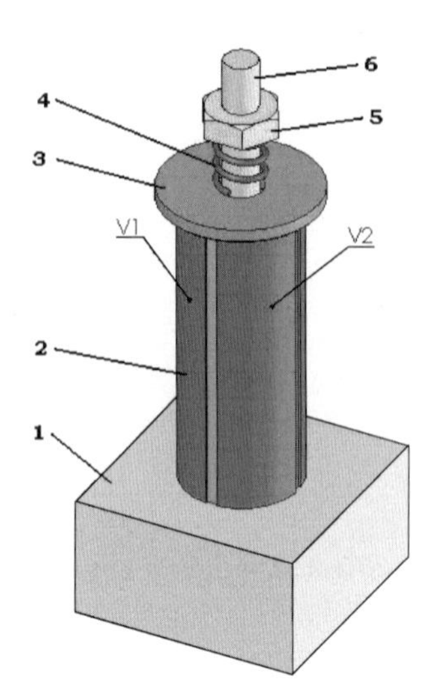

Fig. 2.4. Construction of partially-rotating actuator

is applied across a particular section. By applying a voltage orderly across adjacent sections in a controlled manner (*i.e.*, V_1, V_2, V_1, V_2, and so on), an incomplete rotating wave can be produced at the end of the piezoelectric cylinder, ready to be transmitted to the friction disc mounted on top of the cylinder. To reverse the direction of the rotation, a negative voltage can be applied across the sections. By controlling the duration of the positive and negative voltage applied, the sweeping angle of the motor can be controlled.

2.1.2 Mathematical Model

In this section, the dynamics of a linear and partially-rotating PA will be discussed. The piezo design presented can be used for linear or partial-rotation application. The dynamic model is derived from the physical characteristics of the systems, including its piezoelectric effect and the structure and construction of the piezoelectric element.

Linear PA System

The linear system considered here is based on a piezoelectric stack. The analysis may begin from the constitutive equations of piezoelectricity, followed by the generation of force, and then the movement of the object.

The effect of piezoelectricity is taken into account by its constitutive equations as follows:

$$\sigma_{ij} = c^E_{ijkl} s_{kl} - e_{kij} V_k, \tag{2.1}$$

$$D_i = e_{ikl} s_{kl} + \epsilon^S_{ij} V_k, \tag{2.2}$$

where σ is stress vector, s is strain vector, V is electric field vector, D is electric displacement vector, c is elastic stiffness constant matrix, e is piezoelectric constant matrix, and ϵ is permittivity constant matrix. Superscripts E and S denote constant electric field and constant strain, respectively.

The driving force of the linear PA is generated by the piezoelectric stack cylinder, where each layer exerts a force of F_i. Taking into account the hysteresis of the piezoelectric stack, the driving force can be expressed according to Goldfarb and Celanovic (1997) as follows:

$$F_i = \begin{cases} k_i(x - x_{b_i}) & if\, |k_i(x - x_{b_i})| < f_i \\ f_i sgn(\dot{x}) & else \end{cases}, \tag{2.3}$$

$$F_D = \sum_{i=1}^{n} F_i, \tag{2.4}$$

where k is the stiffness, x is the position, x_b is the position of the block, f is the breakaway force, and index $(.)_i$ denotes each layer of the stack.

Partially-Rotating PA System

In this section, the dynamics of a partially-rotating system is discussed. The analysis includes the piezoelectric effect and external effect from the spring. The piezoelectric effect is analyzed in a similar manner to that used for the linear system as explained in Equations (2.1) and (2.2).

Such a partial-rotation system has been used for piercing of eggs in intracytoplasmic sperm injection. Detailed dynamic of the system is presented as follows.

Because the motor is pressed by a spring force, an axial loading is present in the cylinder. The cylinder is regarded as a tube.

The following force equation applies:

$$F_s = k_s y_s, \tag{2.5}$$

where F_s is the spring force, k_s is the spring constant, and y_s is the spring displacement. The compression force is within control since the spring displacement, y_s, is adjustable. The spring constant depends on the type of the spring.

The spring force causes the cylinder to contract axially due to stress:

$$\sigma_{ax} = \frac{F_s}{A_a x} = \frac{k_s y_s}{\frac{1}{4}\pi\left(d_2^2 - d_1^2\right)}, \tag{2.6}$$

where σ_{ax} is the axial stress (stress due to axial force), A_{ax} is the area of the cylinder, and d_2 and d_1 are the outer and inner diameter of the cylinder, respectively; and the strain:

$$s_{ax} = \frac{\sigma_(ax)}{E} = \frac{k_s y_s}{\frac{1}{4}\pi E\left(d_2^2 - d_1^2\right)}, \tag{2.7}$$

where s_{ax} is the axial strain and E is the modulus of elasticity.

Assuming that the piezoelectric cylinder is only loaded within its elasticity limit, the axial displacement can be formulated to be

$$\begin{aligned} y_{ax} &= s_{ax} L = \frac{\sigma_{ax}}{E} L \\ &= \frac{L}{E} \frac{k_s y_s}{\frac{1}{4}\pi\left(d_2^2 - d_1^2\right)}. \end{aligned} \tag{2.8}$$

The infinitesimal element analysis can be used to analyse the displacement. The expression of position in cylindrical coordinates is as follows:

$$d\mathbf{r} = \hat{r}\mathbf{dr} + \hat{\phi} r\mathbf{d}\phi + \hat{z}\mathbf{dz}. \tag{2.9}$$

The stator of the motor can also be regarded as a cantilever undergoing a bending load. Due to the electrical voltage applied to the cylinder, the cylinder deforms along the $\hat{r}$, $\hat{\phi}$, and $\hat{z}$ direction. The deformation along $\hat{r}$ and $\hat{z}$ directions is neglected since they do not contribute to the rotation of the motor.

The electric field to be used in Equation (2.1) is as follows:

$$V = \frac{U}{r_2 - r_1}, \tag{2.10}$$

since the voltage is applied across its thickness, where U is the applied voltage, and r_2 and r_1 are the outer and inner radii, respectively.

The voltage to be applied is along $\hat{r}$ direction, or equivalently direction *1*, whereas the displacement to be observed is along $\hat{\phi}$ direction. Accordingly, e_{21} and c_{21} constants will be used.

From Equation (2.1), taking into account that no additional force/stress is applied to the cylinder

$$c_{21}s_2 - e_{21}V = 0. \tag{2.11}$$

Following Equation (2.11), the tangential strain is

$$s_\phi = \frac{e_{21}}{c_{21}} \frac{U}{r_2 - r_1}. \tag{2.12}$$

From the definition of strain in tangential direction

$$s_2 = s_\phi = \frac{r_1 \delta\phi}{r_1 d\phi}, \tag{2.13}$$

$$\delta x_\phi = r_1 \delta\phi = s_2 r_1 d\phi. \tag{2.14}$$

The tip of the element deflects as shown in Figure 2.5.

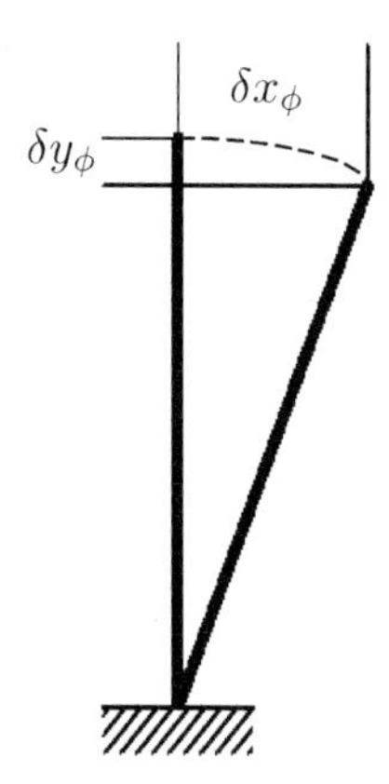

Fig. 2.5. Deflection of a piezoelectric element

As depicted above, an axial displacement accompanies the tangential displacement in flexural deflection. The amount of δy_ϕ corresponds directly to the amount of δx_ϕ. Since $\delta x_\phi << L$, a trajectory of a circle can be assumed with δx_ϕ and δy_ϕ as the orthogonal displacements and L as the radius, resulting in

$$\delta y_\phi = L - \sqrt{L^2 - (\delta x_\phi)^2}. \tag{2.15}$$

The result expressed in Equation (2.15) resembles the pressure distribution. The amount of δy_ϕ corresponds directly to the normal force to generate friction force between the stator and the rotor.

The amount of rotation and the dynamic equation of the motor depends on the normal force, which is in turn depends on δy_ϕ. The distribution of normal force, according to Equation (2.15) can be depicted in Figure 2.6.

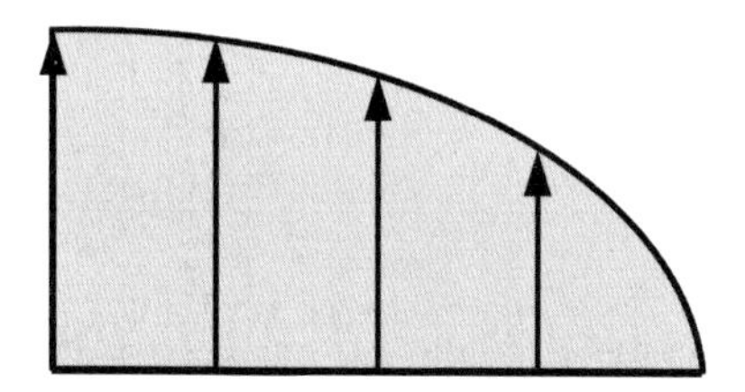

Fig. 2.6. Distribution of normal force

The total displacement along the axial direction is the summation of the axial and flexural deformation as follows:

$$\delta y = \delta y_\phi + y_{ax}. \tag{2.16}$$

It can be assumed that the rubber disc is hard enough so that it will not deform in the plastic region. Under this assumption, the strain of the rubber disc is as follows:

$$s_{rb} = \frac{\delta y}{l_{rb}}, \tag{2.17}$$

where s_{rb} is the strain of the rubber, δy is the total axial displacement, and l_{rb} is the thickness of the rubber. Thus, the pressure/stress imposed on the rubber is as follows:

$$\sigma_{rb} = E_{rb} s_{rb}, \tag{2.18}$$

where E_{rb} is the modulus of elasticity of the rubber and σrb is the pressure on the rubber.

The distribution of the pressure is obtained from Equation (2.18) by modifying Equation (2.15) as follows:

$$y_{\phi} = L - \sqrt{L^2 - r_1^2 \phi^2 s_2^2}, \tag{2.19}$$

resulting in pressure distribution of

$$\sigma_{rb} = \frac{E_{rb}}{l_{rb}} \left(y_{ax} + L - \sqrt{L^2 - r_1^2 \phi^2 s_2^2} \right). \tag{2.20}$$

The frictional force generated by this pressure is then

$$\begin{aligned} dF &= \mu \sigma_{rb} dA = \mu \sigma_{rb} dr r_1 d\phi \\ &= \mu r_1 \frac{E_{rb}}{l_{rb}} \left(y_{ax} + L - \sqrt{L^2 - r_1^2 \phi^2 s_2^2} \right) dr d\phi, \end{aligned} \tag{2.21}$$

$$\begin{aligned} F &= \mu r_1 \frac{E_{rb}}{l_{rb}} \times \\ &\left\{ \int_{r_1}^{r_2} \int_0^{\frac{1}{2}\pi} \left(y_{ax} + L - \sqrt{L^2 - r_1^2 \phi^2 s_2^2} \right) dr d\phi \right\}. \end{aligned} \tag{2.22}$$

The equation of motion is developed from fundamental equation of rotation of rigid body as follows:

$$I_R \ddot{\phi} = F r_1, \tag{2.23}$$

where I_R is the moment of inertia of the rotor and F is the friction force of the stator. The moment of inertia of the rotor is calculated by only considering the metallic part of the rotor, therefore neglecting the friction pad due to its much smaller mass.

Integrating Equation (2.23) over a short period of time (treating F as an impulse force), the angular velocity immediately after the application of force is as follows:

$$I_R \dot{\phi}_0 = F r_1. \tag{2.24}$$

The opposing force to this motion is the friction force between the shaft and the rotor, where the normal force is given by the spring force, added to the friction force between the friction pad and the stator, which initially causes the motion of the rotor. This can be formulated as follows:

$$I_R\ddot{\phi} = \mu_R F_s r_s + \mu F_s r_1, \tag{2.25}$$

where μ_R is the friction coefficient of the rotor and r_s is the mean radius of the contacting shaft.

Having obtained $\omega_0 = \dot{\phi}_0$ from Equation (2.24) and $\alpha = \ddot{\phi}$ from Equation (2.25), the equations of motion can be constructed as follows:

$$\alpha(t) = \ddot{\phi}, \tag{2.26}$$

$$\omega(t) = \omega_0 - \alpha t, \tag{2.27}$$

$$\theta(t) = \omega_0 t - \frac{1}{2}\alpha t^2. \tag{2.28}$$

Uniform Modeling of the Piezoelectric Actuator

The dynamical models described in the ensuing sections are useful during simulation, design and pre-design stages of the PAs to allow rapid performance verification and prediction. However, for control design purposes, a general model will be more amenable. In this section, a general model of the linear and partial-rotation PA is given. Some properties of the model are discussed.

The mathematical model for a voltage controllable PA system can be approximately described by the differential equation

$$m\ddot{x} = -K_f\dot{x} - K_g x + K_e(u(t) - F), \tag{2.29}$$

where $u(t)$ is the time-varying motor terminal voltage; $x(t)$ is the piezo position; K_f is the damping coefficient produced by the motor; K_g is the mechanical stiffness; K_e is the input control coefficient; m is the effective mass; F is the system nonlinear disturbance.

It is well-known that the dominant disturbance of the PA is the hysteresis phenomenon. Figure 2.7 shows a typical hysteresis phenomenon present in linear piezoelectric actuator. The magnitude of this hysteresis can constitute about 10-15% of the motion range. Hysteresis generally impedes high precision motion and hysteresis analysis is thus a key step towards the realization of high performance PA systems. As elaborated in Gilles *et al.* (2001), the hysteresis is a friction-like phenomenon. In Canudas-de-Wit *et al.* (1995), a complex dynamical friction model from a frictional force was presented. The resulting model shows most of the known friction behavior like hysteresis, friction lag, varying break-away force and stick-slip motion. It is comprehensive enough to capture dynamical hysteresis effects. This dynamical friction model is given by

$$F = \sigma_0 z + \sigma_1 \dot{z} + \sigma_2 \dot{x}, \tag{2.30}$$

with

$$\dot{z} = \dot{x} - \frac{|\dot{x}|}{h(\dot{x})} z,$$

$$h(\dot{x}) = \frac{F_c + (F_s - F_c)e^{-(\dot{x}/\dot{x}_s)^2}}{\sigma_0},$$

where $F_c, F_s, \dot{x}_s, \sigma_0, \sigma_1, \sigma_2$ are positive constants which are typically unknown.

The nonlinear function F can also be written as

$$F = (\sigma_1 + \sigma_2)\dot{x} + \sigma_0 z - \sigma_1 \frac{|\dot{x}|}{h(\dot{x})} z = (\sigma_1 + \sigma_2)\dot{x} + F_d(z, \dot{x}). \tag{2.31}$$

The first part $(\sigma_1 + \sigma_2)\dot{x}$ is a simple function of the velocity. The second part $(\sigma_0 - \frac{\sigma_1 |\dot{x}|}{h(\dot{x})})z$ is scaled by the z due to the dynamical perturbations in hysteresis. Since z and $h(\dot{x})$ are bounded,

$$|F_d(z, \dot{x})| = |(\sigma_0 - \sigma_1 \frac{|\dot{x}|}{h(\dot{x})})z| \le k_1 + k_2 |\dot{x}|, \tag{2.32}$$

where k_1 and k_2 are constants.

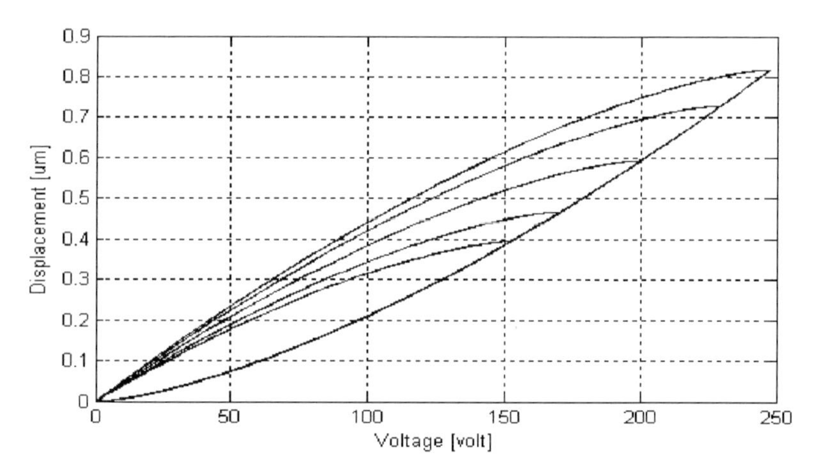

Fig. 2.7. Hysteresis phenomenon

2.1.3 Adaptive Control

In this section, an adaptive controller is designed based on the general model in Equation (2.29). The controller is robust against the modeling uncertainties.

Consider the system at Equation (2.29). The tracking control objective is stated as follows: to find a control mechanism for every bounded smooth output reference $x_d(t)$ with bounded time derivatives so that the controlled output $x(t)$ converges to the reference $x_d(t)$ as closely as possible, where the desired trajectories x_d as well as $\dot{x}_d, \ddot{x}_d$ are continuous and available, and they are bounded.

Define the position tracking error $e(t)$ and the corresponding filtered tracking error $s(t)$ as follows

$$e(t) = x_d(t) - x(t), \tag{2.33}$$

and error

$$s = K_I \int_0^t e(\tau)d\tau + K_p e + \dot{e}, \tag{2.34}$$

where $K_I, K_p > 0$ are chosen such that the polynomial $s^2 + K_p s + K_I$ is Hurwitz. Differentiating $s(t)$ and using Equation (2.29), one finds the dynamics in terms of $s(t)$ as

$$\begin{aligned}\frac{m}{K_e}\dot{s} &= \frac{m}{K_e}(K_I e + K_p \dot{e} + \ddot{x}_d) + \frac{K_f}{K_e}\dot{x} + \frac{K_g}{K_e}x - (u - F) \\ &= \frac{m}{K_e}(K_I e + K_p \dot{e} + \ddot{x}_d) + (\frac{K_f}{K_e} + \sigma_1 + \sigma_2)\dot{x} \\ &\quad + \frac{K_g}{K_e}x - u + F_d(z, \dot{x}).\end{aligned} \tag{2.35}$$

By using a straightforward exact model knowledge, one defines a control input as

$$u = K_v s + a_m(K_I e + K_p \dot{e} + \ddot{x}_d) + a_{k\sigma}\dot{x} + a_{ge}x + F_d, \tag{2.36}$$

where $K_v > 0$ is a constant, $a_m = \frac{m}{K_e}, a_{k\sigma} = \frac{K_f}{K_e} + \sigma_1 + \sigma_2$, and $a_{ge} = \frac{K_g}{K_e}$. Substituting the control input given by Equation (2.36) into the open-loop expression of Equation (2.29), the closed-loop filtered tracking error system is obtained, *i.e.*, $\frac{m}{K_e}\dot{s} = -K_v s$. Since $K_v > 0$, the resulting system is asymptotically stable. Unfortunately, the hysteresis is unknown *a priori* in practice. In addition, it is also difficult to obtain the precise values of m, K_e, K_f, K_g. Motivated by this observation, an adaptive control technique, by replacing $\frac{m}{K_e}, \frac{K_f}{K_e} + \sigma_1 + \sigma_2, \frac{K_g}{K_e}$ and F_d with the estimates $\hat{a}_m, \hat{a}_{k\sigma}, \hat{a}_{ge}$ and $\hat{k}_1 sgn(s) + \hat{k}_2|\dot{x}|sgn(r)$, respectively, is designed as follows:

$$\begin{aligned}u = K_v s &+ \hat{a}_m(K_I e + K_p \dot{e} + \ddot{x}_d) + \hat{a}_{k\sigma}\dot{x} \\ &+ \hat{a}_{ge}x + \hat{k}_1 sgn(s) + \hat{k}_2|\dot{x}|sgn(s),\end{aligned} \tag{2.37}$$

Substituting the control at Equation (2.37) into Equation (2.35), one has

$$\frac{m}{K_e}\dot{s} = -K_v r + \tilde{a}_m(K_I e + K_p\dot{e} + \ddot{x}_d) + \tilde{a}_{k\sigma}\dot{x} + \tilde{a}_{ge}x$$
$$-\hat{k}_1 sgn(s) - \hat{k}_2|\dot{x}|sgn(s) + F_d \quad (2.38)$$

where $\tilde{a}_m = \frac{m}{K_e} - \hat{a}_m, \tilde{a}_{k\sigma} = \frac{K_f}{K_e} + \sigma_1 + \sigma_2 - \hat{a}_{k\sigma}, \tilde{a}_{ge} = \frac{K_g}{K_e} - \hat{a}_{ge}$.

The following adaptive laws are chosen:

$$\dot{\hat{a}}_m = \gamma_1[(K_I e + K_p\dot{e} + \ddot{x}_d)s - \gamma_{11}\hat{a}_m], \quad (2.39)$$
$$\dot{\hat{a}}_{k\sigma} = \gamma_2[\dot{x}s - \gamma_{21}\hat{a}_{k\sigma}], \quad (2.40)$$
$$\dot{\hat{a}}_{ge} = \gamma_3[xs - \gamma_{31}\hat{a}_{ge}], \quad (2.41)$$
$$\dot{\hat{k}}_1 = \gamma_4[|s| - \gamma_{41}\hat{k}_1], \quad (2.42)$$
$$\dot{\hat{k}}_2 = \gamma_5[|\dot{x}||s| - \gamma_{51}\hat{k}_2]. \quad (2.43)$$

where $\gamma_1, \gamma_{11}, \gamma_2, \gamma_{21}, \gamma_3, \gamma_{31}, \gamma_4, \gamma_{41}, \gamma_5, \gamma_{51} > 0$,

Stability Analysis

The following stability result is established.

Theorem 2.1.

Consider the plant at Equation (2.29) and the control objective of tracking the desired trajectories, $x_d, \dot{x}_d, \ddot{x}_d$. The control law given by Equation (2.37) with Equations (2.39)-(2.43) ensures that the system states and parameters are uniformly bounded.

Proof.

Taking a positive definite function

$$V(t) = \frac{1}{2}\frac{m}{K_e}s^2 + \frac{1}{2\gamma_1}\tilde{a}_m^2 + \frac{1}{2\gamma_2}\tilde{a}_{k\sigma}^2 + \frac{1}{2\gamma_3}\tilde{a}_{ge}^2 + \frac{1}{2\gamma_4}\tilde{k}_1^2 + \frac{1}{2\gamma_5}\tilde{k}_2^2, \quad (2.44)$$

where $\tilde{k}_1 = k_1 - \hat{k}_1, \tilde{k}_2 = k_2 - \hat{k}_2$, its time derivative becomes

$$\dot{V} = -K_v s^2 + [\tilde{a}_m(K_I e + K_p\dot{e} + \ddot{x}_d) + \tilde{a}_{k\sigma}\dot{x} + \tilde{a}_{ge}x]s$$
$$+[-\hat{k}_1 sgn(r) - \hat{k}_2|\dot{x}|sgn(r) + F_d]s$$
$$+\frac{1}{\gamma_1}\tilde{a}_m\dot{\tilde{a}}_m + \frac{1}{\gamma_2}\tilde{a}_{k\sigma}\dot{\tilde{a}}_{k\sigma} + \frac{1}{\gamma_3}\tilde{a}_{ge}\dot{\tilde{a}}_{ge} + \frac{1}{\gamma_4}\tilde{k}_1\dot{\tilde{k}}_1 + \frac{1}{\gamma_5}\tilde{k}_2\dot{\tilde{k}}_2. \quad (2.45)$$

By using the inequality at Equation (2.32), it is shown that

$$\dot{V} \leq -K_v s^2 + [\tilde{a}_m(K_I e + K_p\dot{e} + \ddot{x}_d) + \tilde{a}_{k\sigma}\dot{x} + \tilde{a}_{ge}x]s + \tilde{k}_1|r| + \tilde{k}_2|\dot{x}||s|$$
$$+\frac{1}{\gamma_1}\tilde{a}_m\dot{\tilde{a}}_m + \frac{1}{\gamma_2}\tilde{a}_{k\sigma}\dot{\tilde{a}}_{k\sigma} + \frac{1}{\gamma_3}\tilde{a}_{ge}\dot{\tilde{a}}_{ge} + \frac{1}{\gamma_4}\tilde{k}_1\dot{\tilde{k}}_1 + \frac{1}{\gamma_5}\tilde{k}_2\dot{\tilde{k}}_2 \quad (2.46)$$

Substituting the adaptive laws into the above equation, it follows that

$$\begin{aligned}
\dot{V} &\leq -K_v s^2 + \gamma_{11}\tilde{a}_m\hat{a}_m + \gamma_{21}\tilde{a}_{k\sigma}\hat{a}_{k\sigma} + \gamma_{31}\tilde{a}_{ge}\hat{a}_{ge} + \gamma_{41}\tilde{k}_1\hat{k}_1 + \gamma_{51}\tilde{k}_2\hat{k}_2 \\
&= -K_v\{s^2 + \frac{\gamma_{11}}{K_v}(\tilde{a}_m - \frac{1}{2}a_m)^2 + \frac{\gamma_{21}}{K_v}(\tilde{a}_{k\sigma} - \frac{1}{2}a_{k\sigma})^2 \\
&\quad + \frac{\gamma_{31}}{K_v}(\tilde{a}_{ge} - \frac{1}{2}a_{ge})^2 \\
&\quad + \frac{\gamma_{41}}{K_v}(\tilde{k}_1 - \frac{1}{2}k_1)^2 + \frac{\gamma_{51}}{K_v}(\tilde{k}_2 - \frac{1}{2}k_2)^2 - \frac{\gamma_{11}}{4K_v}a_m^2 - \frac{\gamma_{21}}{4K_v}a_{k\sigma}^2 \\
&\quad - \frac{\gamma_{31}}{4K_v}a_{ge}^2 - \frac{\gamma_{41}}{4K_v}k_1^2 - \frac{\gamma_{51}}{4K_v}k_2^2\}
\end{aligned} \tag{2.47}$$

which is guaranteed negative as long as either

$$|s| > \sqrt{\frac{\gamma_{11}}{4K_v}a_m^2 - \frac{\gamma_{21}}{4K_v}a_{k\sigma}^2 - \frac{\gamma_{31}}{4K_v}a_{ge}^2 - \frac{\gamma_{41}}{4K_v}k_1^2 - \frac{\gamma_{51}}{4K_v}k_2^2} \tag{2.48}$$

or

$$\begin{aligned}
|\tilde{a}_m| &> \sqrt{\frac{1}{4}a_m^2 - \frac{\gamma_{21}}{4\gamma_{11}}a_{k\sigma}^2 - \frac{\gamma_{31}}{4\gamma_{11}}a_{ge}^2 - \frac{\gamma_{41}}{4\gamma_{11}}k_1^2 - \frac{\gamma_{51}}{4\gamma_{11}}k_2^2} + \frac{1}{2}a_m \\
|\tilde{a}_{k\sigma}| &> \sqrt{\frac{\gamma_{11}}{4\gamma_{21}}a_m^2 - \frac{1}{4}a_{k\sigma}^2 - \frac{\gamma_{31}}{4\gamma_{21}}a_{ge}^2 - \frac{\gamma_{41}}{4\gamma_{21}}k_1^2 - \frac{\gamma_{51}}{4\gamma_{21}}k_2^2} + \frac{1}{2}a_{k\sigma} \\
|\tilde{a}_{ge}| &> \sqrt{\frac{\gamma_{11}}{4\gamma_{31}}a_m^2 - \frac{\gamma_{21}}{4\gamma_{31}}a_{k\sigma}^2 - \frac{1}{4}a_{ge}^2 - \frac{\gamma_{41}}{4\gamma_{31}}k_1^2 - \frac{\gamma_{51}}{4\gamma_{31}}k_2^2} + \frac{1}{2}a_{ge} \\
|\tilde{k}_1| &> \sqrt{\frac{\gamma_{11}}{4\gamma_{41}}a_m^2 - \frac{\gamma_{21}}{4\gamma_{41}}a_{k\sigma}^2 - \frac{\gamma_{31}}{4\gamma_{41}}a_{ge}^2 - \frac{1}{4}k_1^2 - \frac{\gamma_{51}}{4\gamma_{41}}k_2^2} + \frac{1}{2}k_1 \\
|\tilde{k}_2| &> \sqrt{\frac{\gamma_{11}}{4\gamma_{51}}a_m^2 - \frac{\gamma_{21}}{4\gamma_{51}}a_{k\sigma}^2 - \frac{\gamma_{31}}{4\gamma_{51}}a_{ge}^2 - \frac{\gamma_{41}}{4\gamma_{51}}k_1^2 - \frac{1}{4}k_2^2} + \frac{1}{2}k_2
\end{aligned}$$

Thus,$s, \tilde{a}_m, \tilde{a}_{k\sigma}, \tilde{a}_{ge}, \tilde{k}_1, \tilde{k}_2$ are uniformly ultimately bounded. The value of $|s|$ can be made small by increasing K_v. Therefore, the conclusions are reached.

Experimental Results

For the linear PA, a direct servo motor manufactured by Physik Instrumente (PI) is used, which has a travel length of 80 μm and it is equipped with a linear variable displacement transformer (LVDT) sensor with an effective resolution of 5 nm. The dSPACE control development and rapid prototyping platform is used. MATLAB®/Simulink® can be used from within a dSPACE environment. The position and switch signals are feedback to the control system by the analog input and digital input of the dSPACE card, respectively. The trajectory mode is forwarded to the adaptive controller as a reference signal and then the controller output is sent out to the piezo actuator. The control system is implemented by running Real-Time Windows of MATLAB®.

One of the key roles in the control system design is to establish a model. To build the model, the first thing is to collect the experimental data from the piezo actuator system. A chirp signal is used for the model identification. By choosing analog output channel 1, the signal is sent to piezo motor. The sample rate is chosen as 2000 Hz and the frequency of the chirp signal is selected as from 0 to 200 Hz. The dominant linear model is

$$\ddot{x} = -1081.6\dot{x} - 5.9785 \times 10^5 x + 4.2931 \times 10^6 u.$$

This model gives the general characteristic of the linear PA, but it does not include uncertainties such as hysteresis.

The adaptive controller is applied to the PA. The parameters of the controller are selected as

$$K_v = 0.00001, K_I = 400000, K_p = 100. \tag{2.49}$$

The initial values for $\hat{a}_m, \hat{a}_{k\sigma}, \hat{a}_{ge}$ can be chosen based on the identified model. They are $\hat{a}_m(0) = 2.3293 \times 10^{-7}$, $\hat{a}_{k\sigma}(0) = 2.5194 \times 10^{-4}$, and $\hat{a}_{ge}(0) = 0.1184$. The initial values for $\hat{k}_1, \hat{k}_2$ are chosen as $10^{-7}, 10^{-8}$, respectively. Since the mechanical structure and other components in the system have inherent unmodeled high-frequency dynamics which should not be excited, small adaptation factors are used, where we choose $\gamma_1 = \gamma_2 = \gamma_3 = 10^{-22}, \gamma_4 = \gamma_5 = 10^{-20}, \gamma_{11} = \gamma_{21} = \gamma_{31} = \gamma_{41} = \gamma_{51} = 0.0001$.

The reference signal for tracking is the sinusoidal trajectory $A\sin(wt)$ where $A = 3\mu$m, $w = 6$ rad/sec. The actual response for the controller is shown in Figure 2.8. It can be observed that, under the proposed control, the actual response to the sinusoidal trajectories is good. The tracking error is about 0.3 μm. It is observed that a high micro level accuracy can be achieved using the proposed adaptive control. If only the PID control component in the proposed controller is used, the result is shown in Figure 2.9 and the tracking error is about 1.0 μm. This shows that the adaptive controller can achieve better tracking performance than that of PID control.

2.2 Permanent Magnet Linear Motors (PMLM)

Among the electric motor drives, permanent magnet linear motors (PMLM) are probably the most naturally akin to applications involving high speed and high precision motion control. The increasingly widespread industrial applications of PMLM in various semiconductor processes, precision metrology and miniature system assembly are self-evident testimonies of the effectiveness of PMLM in addressing the high requirements associated with these application areas. The main benefits of a PMLM include the high force density achievable, low thermal losses and, most importantly, the high precision and accuracy associated with the simplicity in mechanical structure. PMLM is designed by

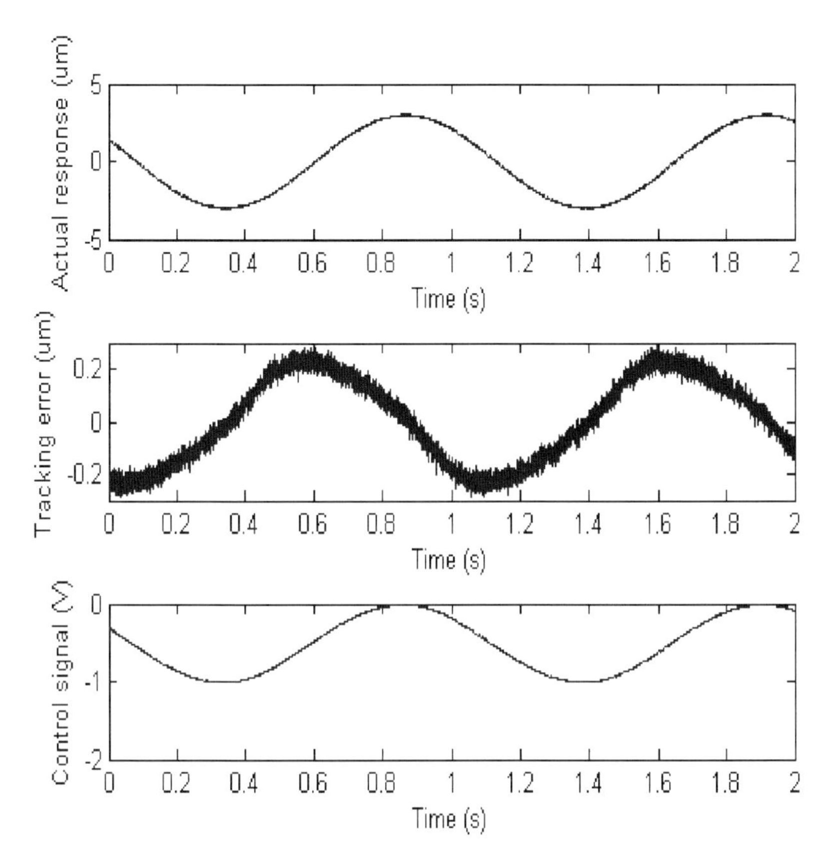

Fig. 2.8. Sine wave responses with the control scheme

cutting and unrolling their rotary counterparts, literally similar to the imaginary process of cutting a conventional motor rotary armature and rotary stator along a radial plane and unroll to lay it out flat, as shown in Figure 2.10. The result is a flat linear motor that produces linear force, as opposed to torque, because the axis of rotation no longer exists. The same forces of electromagnetism that produce torque in a rotary motor are used to produce direct linear force in linear motors. Compared to asynchronous linear induction motors, PMLM incorporates rare earth permanent magnets with very high flux density and are able to develop much higher flux without heating Unlike rotary machines, linear motors require no indirect coupling mechanisms as in gear boxes, chains and screws coupling. This greatly reduces the effects of contact-type nonlinearities and disturbances such as backlash and frictional forces, especially when they are used with aerostatic or magnetic bearings. However, the advantages of using mechanical transmission are also consequently lost, such as the inherent ability to reduce the effects of model

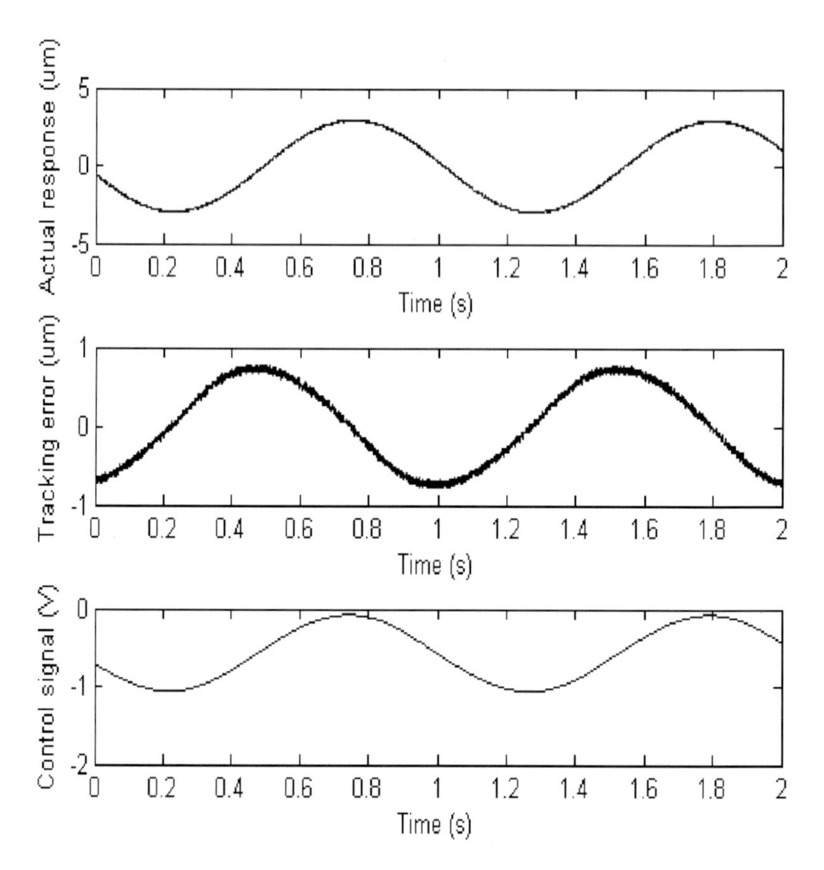

Fig. 2.9. Sine wave responses with the control scheme (PID)

Fig. 2.10. Unrolling a rotary motor

uncertainties and external disturbances. Therefore, a reduction of these effects, either through proper physical design or *via* the control system, is of paramount importance if high-speed and high-precision motion control is to be achieved.

This chapter presents various control schemes to enable precision motion tracking for PMLM.

2.2.1 Types of PMLM

The first few patents on linear motors dated back to the mid-twentieth century, but more recent innovations in materials and architecture have yielded performance and cost improvements to widen the application domains for these devices. The main PMLM types available commercially today are forcer-platen, U-shaped and the the tubular design. The following subsections will briefly review each of these designs.

Forcer-platen Linear Motors

Forcer-platen linear motors, as shown in Figure 2.11, are common brushless DC linear servo-motors which have been around for over 25 years. Forcer-platen motors are popular in automotive and machine tool applications where high continuous and peak forces are required. These linear motors consist of two main elements: the moving forcer and the stationary platen. The forcer-platen motor incorporates permanent magnets in the stator oriented at right angles to the thrust axis (like ties on a railed rack), but slightly skewed in the vertical plane, which has the effect of reducing the thrust ripple. The forcer-platen linear motors feature a low height profile and a wide range of available sizess.

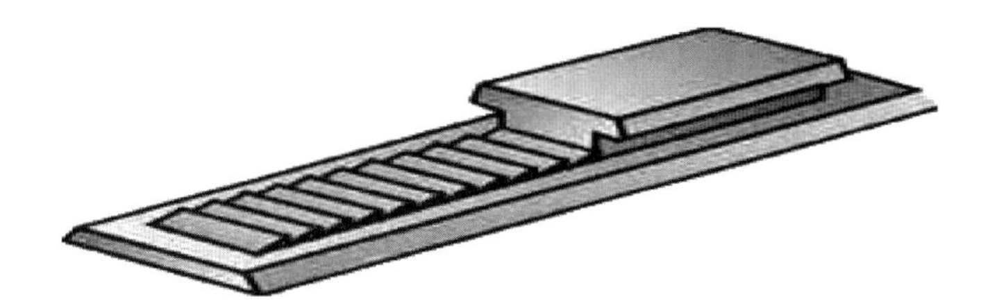

Fig. 2.11. Forcer-platen linear motor

Typically the coils in the forcer contain an iron core to increase the electromagnetic flux density, and hence the resultant motor force output. However, an iron core results in "coggy" movement due to the presence of significant detent (or cogging) force. The iron core also causes eddy current losses which are a function of motor velocity. This thermal energy must be dispersed effectively into the ambient environment to prevent the motor from overheating, and to avoid magnetic saturation. To accomplish this effectively, forced cooling (by air or water) is required in stringent applications. In addition, the magnetic flux utilisation is sub-optimal in forcer-platen designs. Segments of the coils which are not perpendicular to the magnetic field B generate only a fraction of the maximum force; whereas segments parallel to the motion axis do not contribute at all to the thrust output. To compensate for poor flux utilisation, these motors draw more current while entailing significant heat loss to achieve a given force level in comparison to other architectures. Consistency of the force output is dependent upon maintaining a close consistent air gap (≤ 0.5 mm); fluctuations in the air gap over the length of travel cause flux variations, and hence force output variations. These variations in force output must be compensated for in order to maintain good trajectory tracking performance. The high attractive forces between the forcer and platen, coupled with the precise air gap requirements, also lead to a relatively complex installation process.

U-shaped Linear Motors

Figure 2.12 shows another popular linear motor design used today— the U-shaped linear motor. U-shaped linear motors are widely used in high precision operations requiring smoothness of motion. The U-shaped motor armature consists a planar winding epoxy bonded to a plastic "blade" which projects between a double row of magnets. The permanent magnetic fields generated by the track works in conjunction with the electromagnetic fields in the blade to produce linear motion. This design is advantageous for its zero detent force and resultant smoothness as well as the absence of attractive forces between armature and stator. Besides the excellent smoothness, U-shaped linear motors also offer the general cost effectiveness and a wide range of travel length capabilities for motion control. Very long travel lengths are possible with U-shaped linear motors since there is no precision air gap requirements between the blade and the track.

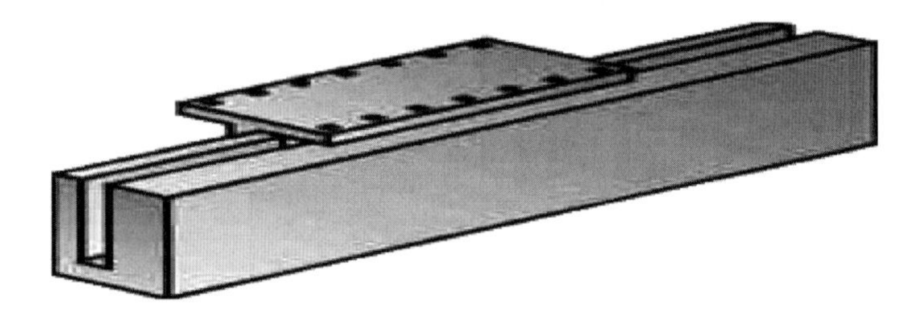

Fig. 2.12. U-shaped linear motor

One of the drawbacks of this architecture is the low mechanical stiffness of the epoxy-filled armature blade which might lead to resonance under servo control in high acceleration applications. The U-shaped geometry captures and traps the hot air next to the coils, and therefore U-shaped linear motors can only be efficiently cooled by mounting a heat sink on the motor blade or via forced cooling. In addition, the U-shaped motor is characterized by magnetic flux utilisation inefficiencies similar to those of the forcer-platen type.

Tubular Linear Motors

Tubular motors, as shown in Figure 2.13, consist of two main elements: the thrust rod containing the permanent magnets (typically stationary) and the thrust block containing the motor coils (typically the moving element). From a force generation and energy efficiency perspective, these motors have significant design advantages over other linear motor architectures.

The device consists of a single conductive wire cylindrically wound and encapsulated comprising the motor armature (thrust block) and a cylindrical assembly of sintered NdFeB high performance permanent magncts arrayed in an on-axis N-S stack contained within an encasing tube which comprises the stator. The thrust block does not ride on the stator; these two components are typically separated by a relatively large air gap ($\simeq 1$ mm) with an independent bearing system to support the moving thrust block. As with forcer-platen motors, multiple thrust blocks may be independently controlled on a single stator assembly.

Radial symmetry of the tubular geometry confers several advantages when compared to other linear motor types: all of the magnetic flux intersecting the coils generates thrust. The circular windings in the thrust block and the magnetic flux pattern are inherently perpendicular which maximizes the linear force attainable. For a given current rating and magnetic strength, the tubular configuration produces higher force than other less efficient designs. Symmetrical geometry inherently balances the magnetic fields to minimize the attractive forces between the translator and stator. Typical attractive forces for tubular motors are in the range of several pounds in comparison to the several hundred pounds characteristic of conventional forcer-platen types. Absence of high attractive forces in these motors provide for simplified installation and reduce the loading requirements for the thrust block support bearings. Eddy current losses are insignificant in tubular motors due to their slot-less design. Furthermore, the thrust block is designed to serve as a radiator and requires no forced air or circulating liquid cooling; it facilitates sufficient passive cooling such that the continuous force capacity increases in relation to the root mean square (RMS) velocity of the application. The relatively large allowance in the air gap reduces stringency of the alignment tolerances when installing the tubular motor. By omitting iron core elements (iron-less design), tubular motors are optimized for smoothness at the expense of a 30% reduction of force output capacity.

Despite some unique application advantages, tubular linear motors also have certain limitations when compared to other linear motor technologies, *e.g.*, limited travel lengths, tall overall height, limited size and force range. The performance advantages of the cylindrical configuration entail some compromises in size. Since the stator magnet assembly can only be supported at its extreme ends, sagging of the assembly under its own weight limits the stroke to approximately 2500 mm with a 38 mm diameter stator. The fully enveloping motor thrust block also results in a profile height greater than that afforded by the forcer-platen or side mounted U-shaped linear motors. The relative newness of the tubular topology results in a limited range of available sizes and forces when compared to other linear motor technologies. However, new design refinements and a downward trend in manufacturing costs would eventually promote the availability of various motor sizes and force ratings.

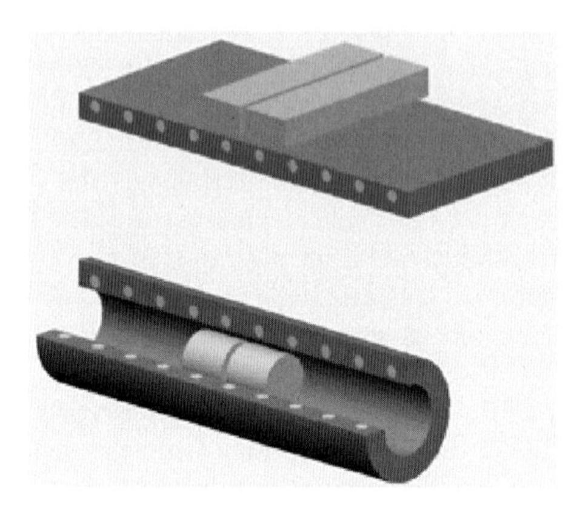

Fig. 2.13. Tubular linear motor

2.2.2 Mathematical Model

The dynamics of the PMLM can be viewed as comprising of two components: a dominantly linear model, and an uncertain and non-linear remnant which nonetheless must be considered in the design of the controller if high precision motion control is to be efficiently realised.

In the dominant linear model, the mechanical and electrical dynamics of a PMLM can be expressed as follows:

$$M\ddot{x} + D\dot{x} + F_{load} = F_m, \tag{2.50}$$

$$K_e\dot{x} + L_a\frac{dI_a}{dt} + R_aI_a = u, \tag{2.51}$$

$$F_m = K_fI_a, \tag{2.52}$$

where x denotes position; M, D, F_m, F_{load} denote the mechanical parameters: inertia, viscosity constant, generated force and load force respectively; u, I_a, R_a, L_a denote the electrical parameters: input DC voltage, armature current, armature resistance and armature inductance respectively; K_f denotes an electrical-mechanical energy conversion constant. K_e is the back EMF constant of the motor.

Since the electrical time constant is typically much smaller than the mechanical one, the delay due to electrical transient response may be ignored, giving the following simplified model:

$$\ddot{x} = -\frac{K_1}{M}\dot{x} + \frac{K_2}{M}u - \frac{1}{M}F_{load}, \tag{2.53}$$

where

$$K_1 = \frac{K_e K_f + R_a D}{R_a}, K_2 = \frac{K_f}{R_a}. \tag{2.54}$$

Clearly, this is a second-order linear dynamical model.

The dominant linear model has not included extraneous non-linear effects which may be present in the physical structure. Among them, the two prominent non-linear effects associated with PMLM are due to ripple and frictional forces, arising from the magnetic structure of PMLM and other physical imperfections. Figure 2.14 depicts a block diagram model of the motor, including explicitly the various exogenous disturbance signals present.

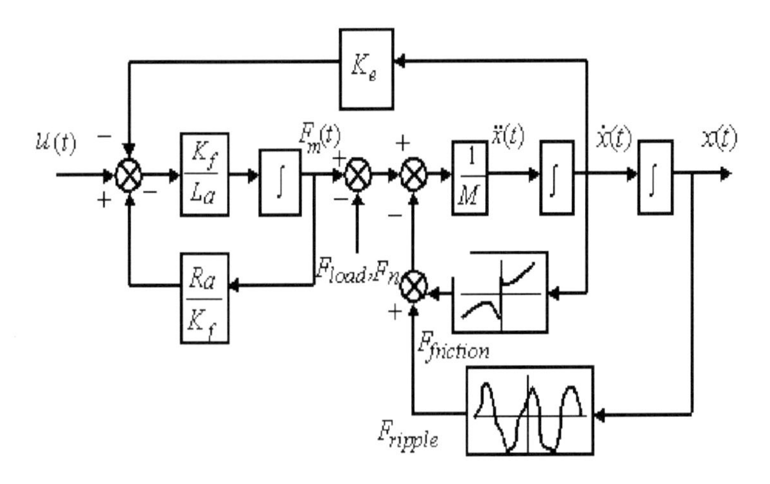

Fig. 2.14. Model of PMLM

2.2.3 Force Ripples

The thrust force transmitted to the translator of a PMLM is generated by a sequence of attracting and repelling forces between the poles and the permanent magnets when a current is applied to the coils of the translator. In addition to the thrust force, parasitic ripple forces are also generated in a PMLM due to the magnetic structure of PMLM. This ripple force exists in almost all variations of PMLM (flat, tubular, moving-magnet *etc.*), as long as a ferromagnetic core is used for the windings.

The two primary components of the force ripple are the cogging (or detent) force and the reluctance force. The cogging force arises as a result of the mutual attraction between the magnets and iron cores of the translator. This force exists even in the absence of any winding current and it exhibits a periodic relationship with respect to the position of the translator relative to the magnets. Cogging manifests itself by the tendency of the translator to align in a number of preferred positions regardless of excitation states. There are two potential causes of the periodic cogging force in PMLM, resulting from the slotting and the finite length of iron-core translator. The reluctance force is due to the variation of the self-inductance of the windings with respect to the relative position between the translator and the magnets. Thus, the reluctance force also has a periodic relationship with the translator-magnet position.

Collectively, the cogging and reluctant force constitute the overall force ripple phenomenon. Even when the PMLM is not powered, force ripples are clearly existent when the translator is moved along the guideway. There are discrete points where minimum/maximum resistance is experienced. At a lower velocity, the rippling effects are more fully evident due to the lower momentum available to overcome the magnetic resistance.

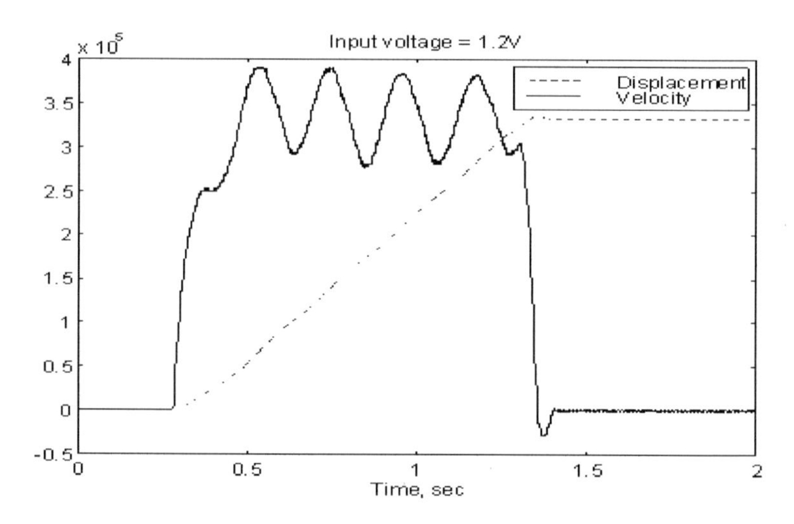

Fig. 2.15. Open-loop step response of a PMLM

Due to the direct-drive principle behind the operation of a linear motor, the force ripple has significant effects on the position accuracy achievable and it may also cause oscillations and yield stability problems, particularly at low velocities or with a light load (low momentum). Figure 2.15 shows the real-time open-loop step response of a tubular type PMLM manufactured by Linear Drive, UK. Figure 2.16 shows the velocity-position characteristics of the PMLM with different step sizes (*i.e.*, different steady-state velocities).

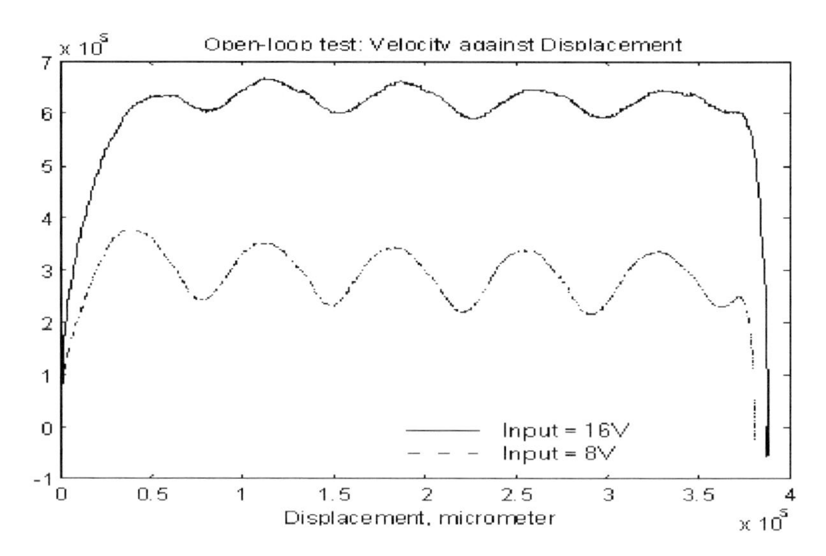

Fig. 2.16. Velocity ($\mu m/s$) against position (μm) for different step voltages

Interesting observations may be inferred from these responses. First, the ripple periodicity is independent of the step size (*i.e.*, independent of the velocity), but it exhibits a fixed relationship with respect to position. Second, the ripple amplitude is dependent on both position and velocity. At a higher velocity, the ripple amplitude decreases compared to when the PMLM is run at a lower velocity, when the higher dosage of ripple effects is experienced.

A first-order model for the force ripple can be described as a periodic sinusoidal type signal:

$$F_{ripple}(x) = A(x)\sin(\omega x + \phi). \tag{2.55}$$

Higher harmonics of the ripple may be included in higher order models.

2.2.4 Friction

Friction is inevitably present in nearly all moving mechanisms, and it is one major obstacle to achieving precise motion control. Several characteristic properties of friction have been observed, which can be broken down into two categories: static and dynamic. The static characteristics of friction, including the stiction friction, the kinetic force, the viscous force, and the Stribeck effect, are functions of steady state velocity. The dynamic phenomena include pre-sliding displacement, varying breakaway force, and frictional lag. Many empirical friction models have been developed which attempt to capture specific components of observed friction behaviour, but generally it is acknowledged that a precise and accurate friction model is difficult to be obtained in an explicit form, especially for the dynamical component. For many purposes, however, the Tustin model has proven to be useful and it has been validated adequately in many successful applications. The Tustin model may be written as

$$F_{friction} = [F_c + (F_s - F_c)e^{-(|\dot{x}/\dot{x}_s|)^{\delta}} + F_v|\dot{x}|]sgn(\dot{x}), \tag{2.56}$$

where F_s denotes static friction, F_c denotes the minimum value of Coulomb friction, $\dot{x}_s$ and F_v are lubricant and load parameters, and δ is an additional empirical parameter. Figure 2.17 graphically illustrates this friction model.

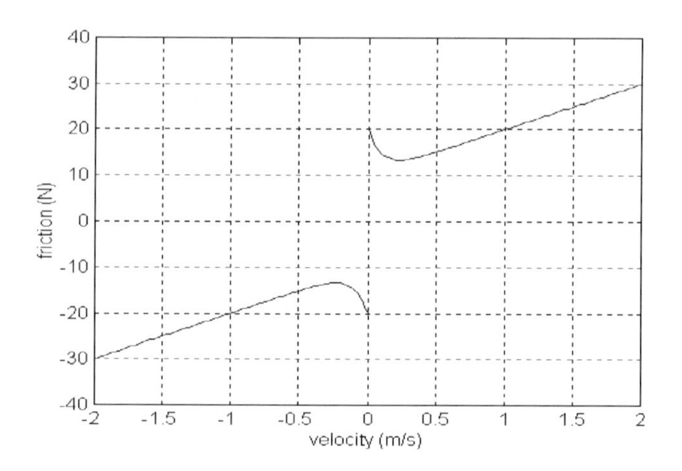

Fig. 2.17. The Tustin friction model

Considering these non-linear effects, the PMLM dynamics may be described by

$$\ddot{x} = -\frac{K_1}{M}\dot{x} + \frac{K_2}{M}u - \frac{1}{M}\left(F_{load} + F_{ripple} + F_{friction}\right). \tag{2.57}$$

The effects of friction can be greatly reduced using high quality bearings such as aerostatic or magnetic bearings.

2.2.5 Composite Control

In this scheme, a three tier composite control structure is adopted, as shown in Figure 2.18, comprising three control components: feedforward control, feedback control and a non-linear Radial Basis Function (RBF) based compensator.

To facilitate the formulation of this approach, a common non-linear function $F_1^*(x, \dot{x})$ may be used to represent the non-linear dynamical effects due to force ripple, friction and other unaccounted dynamics collectively. The servo system at Equation (2.57) can thus be alternatively described by

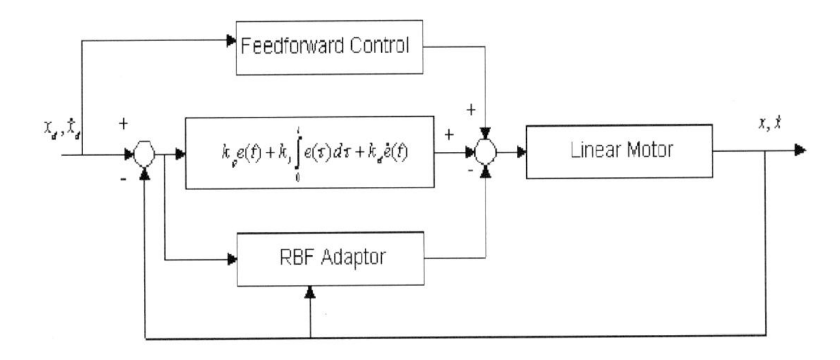

Fig. 2.18. Composite control scheme

$$\ddot{x} = -\frac{K_1}{M}\dot{x} + \frac{K_2}{M}u - \frac{1}{M}F_{load} + F_1^*(x,\dot{x}). \tag{2.58}$$

Let

$$\frac{K_2}{M}f(x,\dot{x}) = -\frac{1}{M}F_{load} + F_1^*(x,\dot{x}). \tag{2.59}$$

It follows that

$$\ddot{x} = -\frac{K_1}{M}\dot{x} + \frac{K_2}{M}u + \frac{K_2}{M}f(x,\dot{x}). \tag{2.60}$$

$f(x,\dot{x})$ is assumed to be a smooth non-linear function which may be unknown. To this end, however, it may be pointed out that many discontinuous non-linear functions may be adequately approximated by a continuous one.

With the tracking error e defined as

$$e = x_d - x,$$

Equation (2.60) may be expressed as

$$\ddot{e} = -\frac{K_1}{M}\dot{e} - \frac{K_2}{M}u - \frac{K_2}{M}f(x,\dot{x}) + \frac{K_2}{M}\left(\frac{M}{K_2}\ddot{x}_d + \frac{K_1}{K_2}\dot{x}_d\right). \tag{2.61}$$

Since

$$\frac{d}{dt}\int_0^t e(t)dt = e, \tag{2.62}$$

the system state variables are assigned as $x_1 = \int_0^t e(t)dt, x_2 = e$ and $x_3 = \dot{e}$. Denoting $x = [x_1\ x_2\ x_3]^T$, Equation (2.61) can then be put into the equivalent state space form:

$$\dot{x} = Ax + Bu + Bf(x,\dot{x}) + B\left(-\frac{M}{K_2}\ddot{x}_d - \frac{K_1}{K_2}\dot{x}_d\right), \tag{2.63}$$

$$A = \begin{bmatrix} 0 & 1 & 0 \\ 0 & 0 & 1 \\ 0 & 0 & -K_1/M \end{bmatrix}, B = \begin{bmatrix} 0 \\ 0 \\ -K_2/M \end{bmatrix}. \tag{2.64}$$

Feedforward Control

The design of the feedforward control law is straightforward. From Equation (2.63), the term $B(-\frac{M}{K_2}\ddot{x}_d - \frac{K_1}{K_2}\dot{x}_d)$ may be neutralised using a feedforward control term in the control signal. The feedforward control is thus designed as

$$u_{FF}(t) = \frac{M}{K_2}\ddot{x}_d + \frac{K_1}{K_2}\dot{x}_d. \tag{2.65}$$

Clearly, the reference position trajectory must be continuous and twice differentiable, otherwise a pre-compensator to filter the reference signal will be necessary. The only parameters required for the design of the feedforward control are the parameters of the second-order linear model. Additional feedforward terms may be included for the non-linear effects if the appropriate models are available. For example, if a good signal model of the ripple force is available Equation (2.55), then an additional static term in the feedforward control signal $u_{FFx} = \frac{1}{K_2}F_{ripple}(x_d)$ can effectively compensate for the ripple force. In the same way, a static friction feedforward pre-compensator can be installed if a friction model is available. Characteristic of all feedforward control schemes, the performance is critically dependent on the accuracy of the model parameters. Therefore, feedforward is usually augmented with suitable feedback control schemes.

PID Feedback Control

In spite of the advances in mathematical control theory over the last 50 years, industrial servo control loops are still essentially based on the three-term PID controller. The main reason is due to the widespread field acceptance of this simple controller which has been effective and reliable in most situations if adequately tuned. More complex advanced controllers have fared less favourably under practical conditions, despite the higher costs associated with implementation and the higher demands in control tuning. It is very difficult for operators unfamiliar with advanced control to adjust the control parameters. Given these uncertainties, there is little surprise that PID controllers continue to be manufactured by the hundred thousands yearly and still increasing. In the composite control system, PID is used as the feedback control term. While the simplicity in a PID structure is appealing, it is also often proclaimed as the reason for poor control performance whenever it occurs. In this design, advanced optimum control theory is applied to tune PID control gains. The PID feedback controller is designed using the Linear Quadratic Regulator (LQR) technique for optimal and robust performance of the nominal system. The feedforward plus feedback configuration is often also referred to as a two-degree-of-freedom (2-DOF) control.

The nominal portion of the system (without uncertainty) is given by

$$\dot{x}(t) = Ax(t) + Bu(t), \tag{2.66}$$

where

$$u_{PID} = Kx = kx_1 + k_{d1}x_2 + k_{d2}x_3. \tag{2.67}$$

This is a PID control structure which utilises a full-state feedback.

The optimal PID control parameters are obtained using the LQR technique that is well known in modern optimal control theory and has been widely used in many applications. It has a very nice robustness property, *i.e.*, if the process is of single-input and single-output, then the control system has at least a phase margin of 60^0 and a gain margin of infinity. This attractive property appeals to the practitioners. Thus, the LQR theory has received considerable attention since the 1950s.

The PID control is given by

$$u_{PID} = -(r_0 + 1)B^T Px(t), \tag{2.68}$$

where P is the positive definite solution of the Riccati equation

$$A^T P + PA - PBB^T P + Q = 0, \tag{2.69}$$

and $Q = H^T H$ where H relates to the states weighting parameters in the usual manner. Note that r_0 is independent of P and it is introduced to weigh the relative importance between control effort and control errors. Note for this feedback control, the only parameters required are the parameters of the second-order model and a user-specified error weight r_0.

One practically useful feature associated with LQR design is that under mild assumptions, the resultant closed-loop system is always stable. This feature is summarised in the following lemma.

Lemma 2.1.

For the system at Equation (2.66), if (A, B) is controllable and (H, A) is observable, then the control law given in Equation (2.68) stabilises the system at Equation (2.66).

RBF Compensation

The 2-DOF control may suffice for many practical control requirements. However, if further performance enhancement is necessary, a third control component may be enabled. A Radial Basis Function (RBF) is applied to model the non-linear remnant, and this is subsequently used to linearise the closed-loop system by neutralising the non-linear portion of the system. RBF, the basis of many neural networks, is often utilised for modelling non-linear functions that are not explicitly defined. For the RBF, the hidden units in within the neural network provide a set of basis functions as these units are expanded into the higher dimensional hidden-unit space.

Since $f(\dot{x}, x)$ is a non-linear smooth function (unknown), it can be represented as

$$f(\dot{x}, x) = \sum_{i=0}^{m} w_i \phi_i(\dot{x}, x) + \epsilon = \sum_{i=0}^{m} \phi_i(\dot{x}, x) w_i + \epsilon, \tag{2.70}$$

with $|\epsilon| \leq \epsilon_M$, where $\phi_i(\dot{x}, x)$ is the RBF given by

$$\phi_i(\dot{x}, x) = exp(-\frac{||x_{vect} - c_i||^2}{2\sigma_i^2}) / \sum_{j=0}^{m} exp(-\frac{||x_{vect} - c_j||^2}{2\sigma_j^2}), \tag{2.71}$$

where $x_{vect} = [\dot{x}\ x]^T$.

The following assumptions are made.

Assumption 2.1.

The ideal weights are bounded by known positive values so that $|w_i| \leq w_M, i = 1, 2, ...m$.

Assumption 2.2.

There exists an $\epsilon_M > 0$ such that $|\epsilon(x, \dot{x})| \leq \epsilon_M, \forall x_{vect} \in \Im$ on a compact region $\Im \subset R^2$.

Let the RBF functional estimates for $f(\dot{x}, x)$ be given by

$$\hat{f}(\dot{x}, x) = \sum_{i=0}^{m} \phi_i(\dot{x}, x) \hat{w}_i, \tag{2.72}$$

where $\hat{w}_i$ are estimates of the ideal RBF weights which are provided by the following weight-tuning algorithm:

$$\dot{\hat{w}}_i = r_1 x^T P B \phi_i - r_2 \hat{w}_i, \tag{2.73}$$

where $r_1, r_2 > 0$, and P is the solution of Equation (2.69). Therefore, the RBF adaptive control component is given by

$$u_{RBF} = -\hat{f}(\dot{x}, x). \tag{2.74}$$

The overall control signal is $u = u_{FF} + u_{PID} + u_{RBF}$.

Parameter Estimation of Nominal System

Under the control structure, a second-order model is necessary for computing the control parameters. Many approaches are available to yield the required linear model (Ljung 1997). The parameter estimation approach is one popular approach. Consider an ARX model given by

$$y(t) + a_1 y(t-1) + ... + a_{n_a} y(t-n_a) = b_1 u(t-1) + ... \\ + b_{n_b} u(t-n_b) + e(t), \tag{2.75}$$

where n_a is the number of poles, $n_b + 1$ is the number of zeros, and $e(t)$ represents the error term of the system. Based on this model, the linear predictor is given by

$$\hat{y}(t/\psi)) = \Psi^T(t)\psi + e(t), \tag{2.76}$$

where

$$\Psi(t) = [-y(t-1) \ -y(t-2)... -y(t-n_a)u(t-1)...u(t-n_b)]^T;$$
$$\psi = [a_1...a_{n_a} \ b_1...b_{n_b}]^T.$$

With Equation (2.76), the prediction error is given by

$$\varepsilon(t,\psi) = y(t) - \Psi^T(t)\psi. \tag{2.77}$$

Define the model fitting criteria function as

$$V_N(\psi) = \frac{1}{N}\sum_{t=1}^{N}\frac{1}{2}[y(t) - \Psi^T(t)\psi]^2, \tag{2.78}$$

which is the least-square criterion. The criterion can be minimised analytically, giving the least squares estimates of the model parameters as

$$\hat{\psi}_N^{LS} = \left[\frac{1}{N}\sum_{t=1}^{N}\Psi(t)\Psi^T(t)\right]^{-1}\frac{1}{N}\sum_{t=1}^{N}\Psi(t)y(t). \tag{2.79}$$

Persistently exciting input signals should be used. This may come in the form of explicit input sequences (*e.g.*, Pseudo-random Binary Sequences) or it may arise naturally from control signals generated in the closed-loop, in which case no explicit experiment needs be conducted.

The parameter estimation described above, using explicit user-defined input signals, will yield an initial set of parameters for the linear model. Thereafter, the model may be refined using incremental measurements of the input and output signals of the system under closed-loop control with a recursive version of the least square estimation algorithm (Ljung 1997), *i.e.*, the refinement may occur online with the linear actuator under normal motion operations.

Stability Analysis

It is required to demonstrate that the state x and weights will remain bounded under the composite control action. The following theorem will be useful to illustrate the stability.

Theorem 2.2.

Assume that the desired references x_d, and $\dot{x}_d$ are bounded. Consider the case where the controller given by Equations (2.65), (2.67) and (2.74) is applied to the system at Equation (2.63). Then the state and the RBF estimation errors are uniformly ultimately bounded.

Proof.

The differential equation at Equation (2.63) can be written as (upon applying the controls at Equations (2.65), (2.67) and (2.74)):

$$\dot{x} = Ax + B(u_{FF} + u_{PID} + u_{RBF}) + BF(\ddot{x}, \dot{x}, x) + B(-\frac{M}{K_2}\ddot{x}_d - \frac{K_1}{K_2}\dot{x}_d) \tag{2.80}$$

$$= (A - (r_0 + 1)BB^T P)x - B\sum_{i=0}^{m} \phi_i \hat{w}_i + B[\sum_{i=0}^{m} \phi_i w_i + \epsilon] \tag{2.81}$$

$$= [A - BB^T P - r_0 BB^T P]x + B[\sum_{i=0}^{m} \phi_i \tilde{w}_i + \epsilon]. \tag{2.82}$$

Now consider the following Lyapunov function candidate:

$$V(x, \tilde{w}) = x^T P x + \frac{1}{r_1}\sum_{i=0}^{m} \tilde{w}^2. \tag{2.83}$$

Taking the time derivative of v along the solution of Equation (2.82), it can be shown:

$$\begin{aligned}
\dot{V} &= x^T(A^T P + PA - PBB^T P)x + x^T(-2r_0 - PBB^T P)x \\
&\quad +2x^T PB\sum_{i=0}^{m} \phi_i \tilde{w}_i + 2x^T PB\epsilon + \frac{2}{r_1}\sum_{i=0}^{m} \tilde{w}_i \dot{\tilde{w}}_i, \\
&= -\lambda_{min}[Q + (2r_0 + 1)PBB^T P]||x||^2 + 2x^T PB\sum_{i=0}^{m} \phi_i \tilde{w}_i \\
&\quad +2x^T PB\epsilon + 2\sum_{i=0}^{m}(-x^T PB\phi_i + \frac{r_2}{r_1}\hat{w}_i)\tilde{w}_i, \\
&= -\lambda_{min}[Q + (2r_0 + 1)PBB^T P]||x||^2 + 2x^T PB\epsilon \\
&\quad +2\frac{r_2}{r_1}\sum_{i=0}^{m} \hat{w}_i \tilde{w}_i, \\
&= -\lambda_{min}[Q + (2r_0 + 1)PBB^T P]||x||^2 + 2x^T PB\epsilon
\end{aligned}$$

$$-2\frac{r_2}{r_1}\sum_{i=0}^{m}\tilde{w}_i^2 + 2\frac{r_2}{r_1}\sum_{i=0}^{m} w_i\tilde{w}_i. \tag{2.84}$$

From

$$\begin{aligned} 2x^T PB\epsilon &\leq \eta x^T PBB^T Px + \frac{1}{\eta}\epsilon^2 \\ &\leq \eta x^T PBB^T Px + \frac{1}{\eta}\epsilon_M^2, \end{aligned} \tag{2.85}$$

$$2\frac{r_2}{r_1} w_i\tilde{w}_i \leq \frac{r_2}{\beta r_1} w_i^2 + \frac{\beta r_2}{r_1}\tilde{w}_i^2, \tag{2.86}$$

it follows

$$\begin{aligned} \dot{V} \leq -\{\lambda_{min}[Q + (2r_0 + 1)PBB^T P] + \eta\lambda_{max}(PBB^T P)\}||x||^2 \\ -2\frac{r_2}{r_1}\sum_{i=0}^{m}(1 - \frac{1}{2}\beta)\tilde{w}_i^2 + \frac{1}{\eta}\epsilon_M^2 + \frac{r_2}{\beta r_1} w_i^2. \end{aligned} \tag{2.87}$$

Define $\theta = [x^T \tilde{w}_0 \tilde{w}_1 ... \tilde{w}_m]^T$ and it follows:

$$\dot{V} \leq -2\gamma||\theta||^2 + \lambda_1, \tag{2.88}$$

where

$$\begin{aligned} \gamma = &\frac{1}{2} min\{\lambda_{min}[Q + (2r_0 + 1)PBB^T P)] - \eta\lambda_{max}(PBB^T P), \\ &2\frac{r_2}{r_1}(1 - \frac{1}{2}\beta)\}, \end{aligned} \tag{2.89}$$

$$\lambda_1 = \frac{1}{\eta}\epsilon_M^2 + \frac{r_2}{\beta r_1} w_i^2. \tag{2.90}$$

Clearly, $\gamma > 0$ for sufficiently small η, β. The following condition must hold for $\dot{V}$ to be negative:

$$||\theta|| > (\frac{\lambda_1}{2\gamma})^{1/2}. \tag{2.91}$$

In order to show the boundedness of the states and weights, note from

$$\mu(||\theta||^2) \leq V \leq \nu(||\theta||^2), \tag{2.92}$$

that

$$||\theta|| \leq [\frac{\nu}{\theta}||\theta(0)||^2 e^{-2\gamma t/\nu} + \frac{\lambda_1\nu}{2\mu\gamma}(1 - e^{-2\gamma t/\nu})]^{-1/2}, \tag{2.93}$$

where $\mu = min\{\lambda_{min}(P), 1/r_1\}, \nu = max\{\lambda_{max}(P), 1/r_1\}$.

From Equation (2.93), it may be concluded that the states and weights are bounded. The proof is completed.

Remark 2.1.

For the adaptive scheme at Equation (2.73), the first term grows rapidly as $x^T PB\phi_i$ increases (which reflects the "poor" system performance). This will result in a high increase rate of $\hat{w}_i$ and, therefore, strong feedback. The parameters r_1, r_2 offer a design trade-off between the relative eventual magnitudes of $||\theta||$ and $|\tilde{w}_i|$.

Experiments

Experimental results are provided to illustrate the effectiveness of the control scheme. The linear motor used is a direct thrust tubular servo motor manufactured by Linear Drives Ltd (LDL)(LD 3810), which has a travel length of 500 mm and it is equipped with a Renishaw optical encoder with an effective resolution of 1 μm. The dSPACE control development and rapid prototyping system based on DS1102 board is used. dSPACE integrated the entire development cycle seamlessly into a single environment, so that individual development stages between simulation and test can be run and rerun, without any frequent re-adjustment. Figure 2.19 shows the experimental set-up. The functional block diagram is given in Figure 2.20. Many of the experimental results subsequently discussed are based on this testbed.

Fig. 2.19. Experimental set-up

The tracking performance achieved from the use of composite control is given in Figure 2.21. A sinusoidal reference trajectory is chosen. A maximum error of less than 4 μm is achieved.

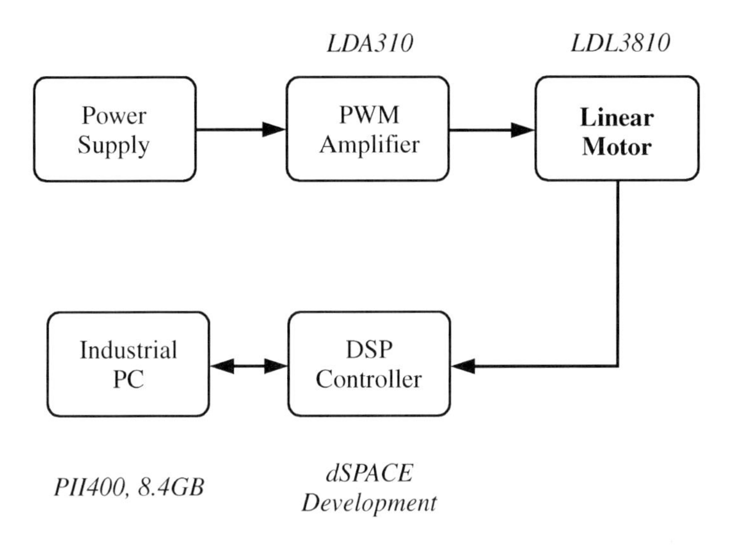

Fig. 2.20. Block diagram of the experimental set-up

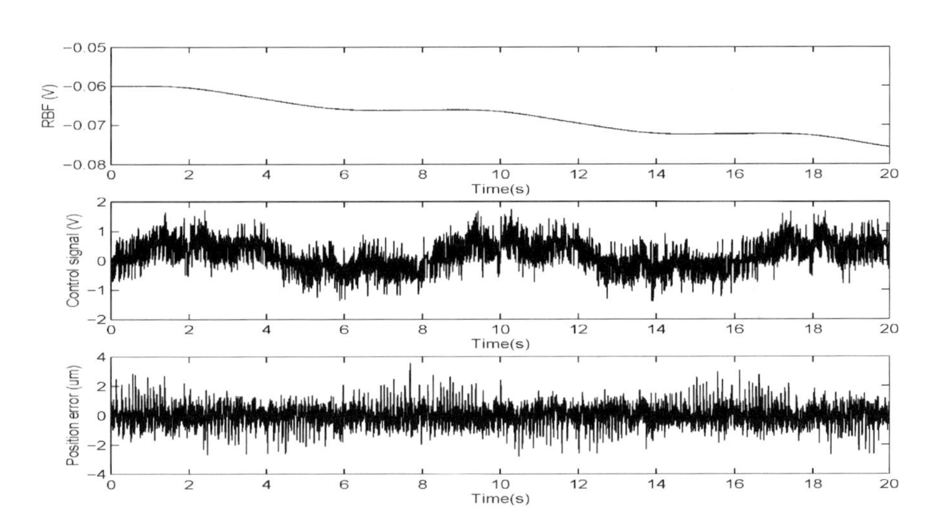

Fig. 2.21. Tracking performance under composite control

2.2.6 Control Enhancement with Accelerometers

Acceleration signals are often assumed to be available, either in a measured form or a derived one, in many literature pertaining to advanced motion control. However, in practice, one can rarely be encountered which effectively uses acceleration measurements of the object concerned. There are several reasons for this phenomenon. First, until recently, quality accelerometers were very expensive and they were also unavailable as standard off-the-shelves components. However, the trend is different today. The cost of accelerometers has reduced dramatically, and they are now widely available, due to advances in the mass production processes of these miniature devices. Second, despite the high associated cost, the signals from accelerometers are very noisy, thus limiting their applications to vibration, shock and force measurements where the envelope of the signal is more important than the actual signal contents. Other possible uses of acceleration measurements can be found documented in McInroy and Saridis (1990), de Jager (1994) and White and Tomizuka (1997). As will be shown, with adequate digital filtering techniques the noise-related problems can be reduced to a level where the signal may be made available for direct analysis by the controller. Third, as discussed in de Jager (1994), acceleration feedback may introduce an algebraic loop due to the direct feedthrough of the input to the acceleration measurement. This concern is, however, of little significance in practice for the following reasons:

- A filter is normally necessary for filtering the acceleration signal and this filter will introduce some time delay which will break the algebraic loop,
- Even in the absence of a filter, the actuator and system dynamics are of inherently low-pass characteristics which act as a natural filter, thus eliminating the algebraic loop.

Acceleration measurements may be used in the controller either in the feedforward or feedback mode. In the feedforward mode (White and Tomizuka 1997), a feedforward control signal is constructed from the acceleration measurements which provides a direct and immediate compensation to the system to be controlled. This method, however, relies heavily on the accuracy of the system model. The other method uses the acceleration measurement in the feedback control law. As mentioned in de Jager (1994), apart from the *direct* use of acceleration measurement discussed above, it is possible to make use of acceleration measurement in an observer *indirectly* to improve the estimate of velocity signal, otherwise derived from position measurements.

Acceleration Sensor

Accelerometers typically employ a fully active or a half-active Wheatstone Bridge, consisting of either semiconductor strain gauges, diffused, implanted (ion), thin film or epitaxial piezo-resistive strain gauges, or metal foil strain gauges. The strain gauges are either bonded to an acceleration sensing member which is clamped at one or more surfaces or atomically diffused, implanted or grown as part of the sensing member. Other techniques, using a force sensitive medium such as a micro-machined diaphragm or a force collector, can also be used. The acceleration member supports a mass at some point along the free length and when the mass experiences acceleration, it produces bending or flexing which in turn creates strain in the strain gauge. The strain is proportional to the applied acceleration. Because the strain gauges are placed in areas which are both tensile and compressive, the developed strain produces a bridge imbalance with a bridge excitation voltage applied. This imbalance, in turn, produces a voltage change at the bridge output proportional to the acceleration acting on the mass.

The EGCS-A series accelerometer from Entran Sensors & Electronics is chosen for this work due to its high sensitivity and heavy duty. With a built-in instrumentation amplifier, this series of accelerometers provide a conditioned robust signal from within an impressive compact package (16 mm×16 mm×16 mm for EGCS-A2).

The basic specifications are listed in Table 2.1. The sensitivity increases as the g range decreases with a decreased frequency bandwidth and a smaller g over-range limit. For the application of concern, a frequency bandwidth of the acceleration sensor of around 200 Hz is desired, since the bandwidth of the PMLM control system is expected to be around 20–80 Hz. With this bandwidth, the same accelerometer can also be used, in parallel, for vibration analysis and monitoring of the linear actuator (see Chapter 7). For these reasons, EGCS-A2-10 is chosen. Other specifications of EGCS-A2-10 can be found in Table 2.2. It should also be noted that EGCS-A2-10 has a standard compensated (internally) temperature range of 20–80°C.

Table 2.1. Basic specifications of EGCS series accelerometers

g ranges "FS"	g overrange limit	Frequency response $\pm\frac{1}{2}$dB, nom./min. Hz	Natural frequency nom. (Hz)	Sensitivity EGCS-A mV/g nom.	Output "FSO" EGCS-A V
±2	±200	0–90/50	170	2500	±5
±5	±500	0–150/80	300	1000	±5
±10	±1000	0–200/120	400	500	±5
±25	±2000	0–400/240	800	200	±5
±50	±5000	0–600/350	1200	100	±5
±100	±10000	0–900/500	1800	50 ±5	
±500	±10000	0–1700/1000	3500	10	±5
±1000	±10000	0–2500/1500	5000	5	±5

The EGCS-A2-10 accelerometer is first calibrated in the frequency domain to verify its specifications. This is necessary and important as the sub-

Table 2.2. Specifications of EGCS-A2-10 accelerometer

Contents	Value	Unit	Remark
g range	±10	g	9.81 m/s^2
Excitation	15	V	DC
Impedance in	15	mA	max.
Impedance out	1	KΩ	nom.
Zero offset	±250	mV	typ. (at 20°C)
Thermal zero shift	±50	mV	per 50°C
Thermal sensitivity shift (TSS)	±2.5%		per 50°C
Nonlinearity and hysteresis	±1%	FSO	
Transverse sensitivity	2%		max.
Damping ratio	0.7 (0.5-0.9)		nom. at 20°C
Overrange stops			integral
Operation temp.	-40 to 120	°C	
Compensated temp.	20 to 80	°C	

sequent design of digital signal conditioners hinges critically on the frequency response of the accelerometer. The experiment is performed on a dedicated vibrational table and the power spectrum analysis is done by an HP FFT Analyzer. The output is shown in Figure 2.22 for an illustrated frequency of 150 Hz. From Figure 2.22, the magnitude response at this frequency is $20log(0.2856/0.3035) = -0.03$ dB. This is verified positively as lying between the nominal and minimal frequency bandwidth of the accelerometer.

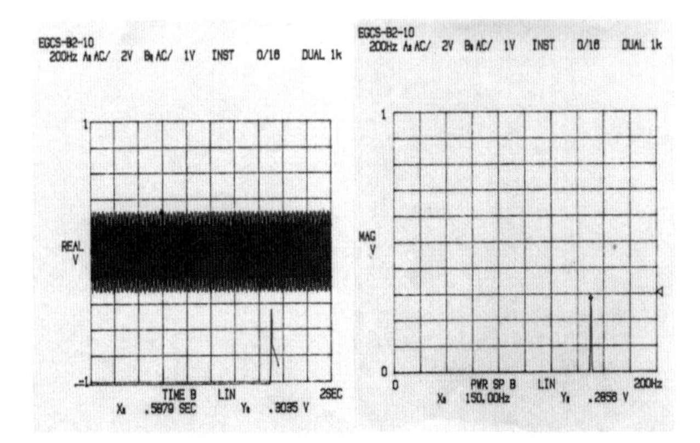

Fig. 2.22. Calibration result: vibration excitation versus power spectrum of response at 150 Hz

In order to achieve accurate measurements with the accelerometer, it is important that the accelerometer be firmly mounted to the test object. Epoxy potting or external clamping is the typical mounting method for accelerometers without mounting holes or studs. In the set-up, the adhesive alternative is used. It should be pointed out that should the mounting becomes loose, the frequency characteristics of the accelerometer may be degraded or distorted, so great care must be exercised over the mounting process. A zoomed-in view of the mounting is shown in Figure 2.23. The testbed used here is a brushed DC PMLM produced by Anorad Corp. The g direction arrow mark on the accelerometer body must be strictly aligned to the measurement axis.

Fig. 2.23. Accelerometer mounting on the testbed: top view (*left*) and close-up (*right*)

It is possible to consider a permanent mounting for EGCS-A2-10 as it has provisions for mounting screws or studs. In this case, the entire surface must be firmly in contact with the test object, and no dirt or other particles should be clamped between the accelerometer and its mounting surface. Otherwise, the accelerometer may vibrate on the test surface, giving erroneous results on one hand, and damaging the unit on the other.

After mounting, the EGCS-A2-10 sensor is powered from a 24 VDC power source. When connected to the A/D channel of the control system, zeroing tests are carried out *via* the A/D calibration. Meanwhile, the scaling factor is also determined to be 43 (decimal) per meter per square second.

A typical record of the raw acceleration signal is shown in Figure 2.24. Clearly, it is too noisy to be practically useful for the purpose of providing crisp motion information for control purposes. However, the high frequency fluctuation is unbiased so that the acceleration profile may be smoothened out by the naked eye. This implies that proper filtering of the noisy signal is essential. Acceleration may also be derived from the position data via a double differentiation. The derived signal is also shown in Figure 2.24 with an overall higher SNR (signal to noise ratio).

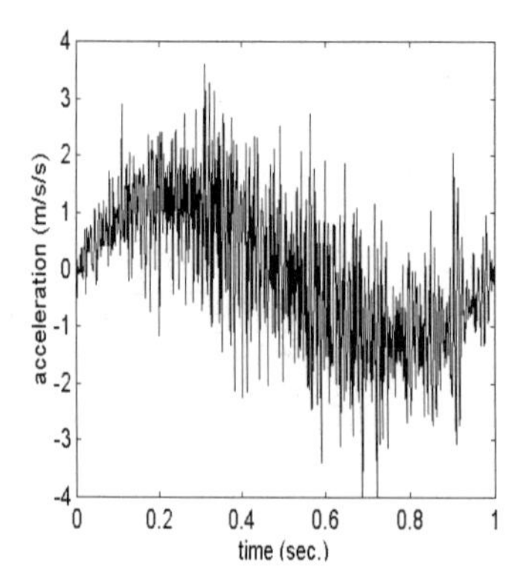

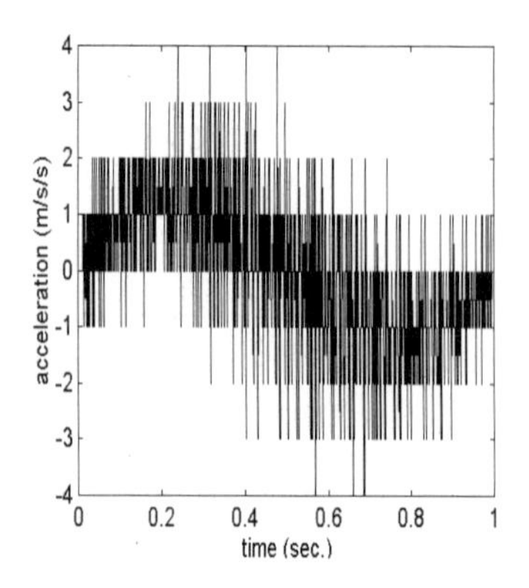

Fig. 2.24. Sample acceleration signals: from accelerometer (*left*) and from double numerical differentiator of position measurement (*right*)

A Butterworth low pass digital filter is used because Butterworth filters are characterised by a magnitude response that is maximally flat in the passband and strictly monotonic over the entire frequency range. A cut-off frequency ω_n, which is normalised to between 0.0 and 1.0, with 1.0 corresponding to half the sample rate, has to be specified. In the system under study, the sampling rate is set to 0.001 s. Considering the frequency characteristics of the accelerometer and closed-loop control bandwidth, the cut-off frequency is set to 80 Hz and the filter order is set to three. The squared magnitude response function of the Butterworth analog low pass filter prototype is given by

$$\mid H_B(\omega) \mid^2 = \frac{1}{1+\omega^6}. \tag{2.94}$$

The equivalent digital filter is

$$H_B(z^{-1}) = \frac{0.0102 + 0.0305z^{-1} + 0.0305z^{-2} + 0.0102z^{-3}}{1 - 2.0038z^{-1} + 1.4471z^{-2} - 0.3618z^{-3}}. \tag{2.95}$$

The filtered acceleration signals are shown in Figure 2.25, with a comparison of the signal quality from both the accelerometer and the double differentiator. Comparisons to the ideal acceleration signal are also given in Figure 2.25. The ideal acceleration is obtained from the desired position tracking trajectory which is pre-planned as a polynomial function of time t satisfying the rest-to-rest movement with a bell-type velocity profile.

There exists a phase shift between the original and the filtered signals. A zero-phase post-filter can be used to illustrate this phase shift. In Figure 2.26, the solid line represents the filter output which clearly has some phase delay compared to the output of the zero-phase filter $H_B(z^{-1})H_B(z)$, the dotted line. The real-time implementation of the zero-phase filter is not possible due to its non-causality. In the real-time implementation of $H_B(z^{-1})$, which is a recursive IIR (infinite impulse response) filter, the delay is inevitable. However, the influence of the delay in the filtering should not be too large so as not to dilute the benefits associated with acceleration feedback.

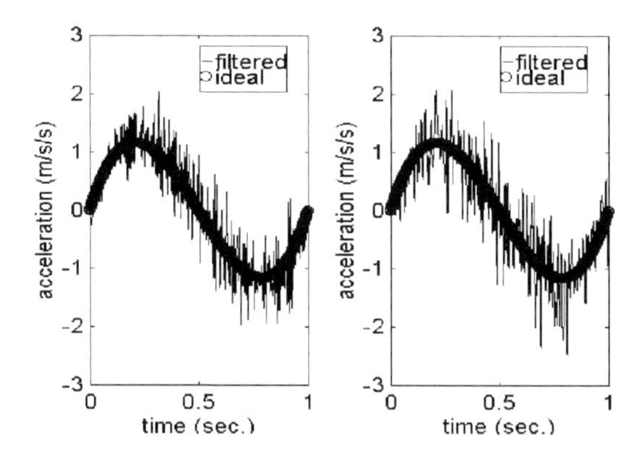

Fig. 2.25. Filtered acceleration signals: from accelerometer (*left*) and from double numerical differentiator of position measurements (*right*)

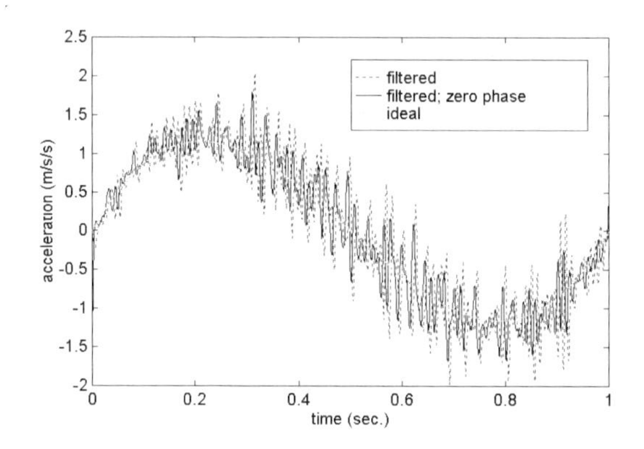

Fig. 2.26. Phase delay in the filtered acceleration signals

Full-state Feedback Control

With the additional acceleration signal now available, a full-state feedback controller can be designed and used in place of the PID controller. Some improvement in the results can be achieved with this additional state feedback. The pole-placement design method, based on Ackermann's formula, is adopted as it is able to provide a natural association of specifications to practical requirements in terms of desired natural frequency and damping of the closed loop. A second-order dominant desired response with a damping ratio of $\zeta = 0.707$ and natural frequency of 25 Hz serves as the reference model.

To illustrate clearly the enhancement to control performance, the full-state feedback control is compared with the PID control using only a feedforward-feedback configuration. The reference signal is a sinusoid given by

$$x_d(t) = 0.1 \sin(\pi t).$$

Figure 2.27 shows the tracking error comparison. More than 50% improvement is observed in the closed-up plot shown in Figure 2.28.

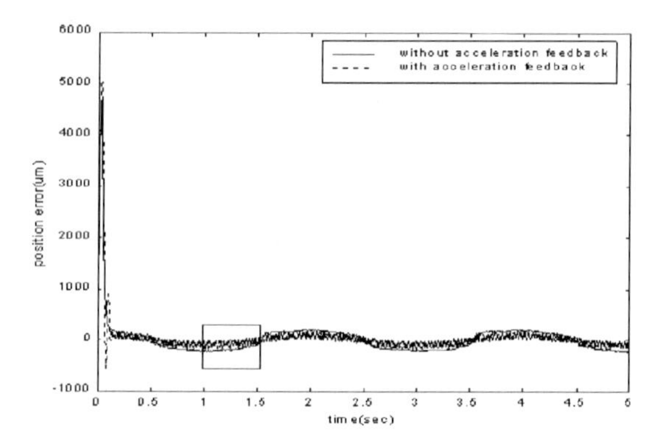

Fig. 2.27. Tracking error comparison: full-state feedback (with accelerometer) vs conventionally tuned PID controllers (without accelerometer)

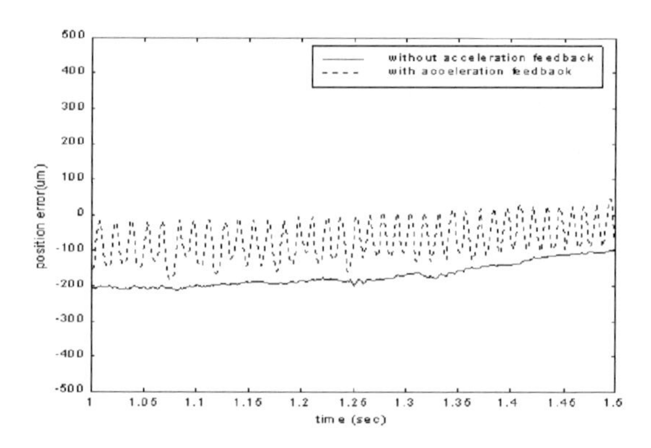

Fig. 2.28. Tracking error comparison: a closed-up view

2.2.7 Ripple Compensation

From motion control viewpoints, force ripples are highly undesirable, and yet they are predominantly present in PMLM. They can be minimised or even eliminated by an alternative design of the motor structure or spatial layout of the magnetic materials such as skewing the magnet, optimising the disposition and width of the magnets *etc*. These mechanisms often increase the complexity of the motor structure. PMLM, with a slotless configuration, as shown schematically in Figure 2.29 is a popular alternative since the cogging force component due to the presence of slots is totally eliminated. Nevertheless, the motor may still exhibit significant cogging force owing to the finite length of the iron-core translator. Finite element analysis confirms that the force produced on either end of the translator is sinusoidal and unidirectional. Since the translator has two edges (leading and trailing edges), it is possible to optimise the magnet length so that the two sinusoidal force waveform of each edge cancel out each other. However, this would again contribute some degree of complexity to the mechanical structure. A more practical approach to eliminate cogging force would be to adopt a sleeve-less or an iron-less design in the core of the windings. Linear Drives, UK, for example, offers two versions of PMLM: the sleeve type and the sleeveless type of PMLM. However, this approach results in a highly inefficient energy conversion process with a high leakage of magnetic flux due to the absence of material reduction in the core. As a result, the thrust force generated is largely reduced (typically by 30% or more). This solution is not acceptable for applications where high acceleration is necessary. In addition, iron-core motors, which produce high thrust forces, are ideal for accelerating and moving large masses while maintaining stiffness during the machining and processing operations.

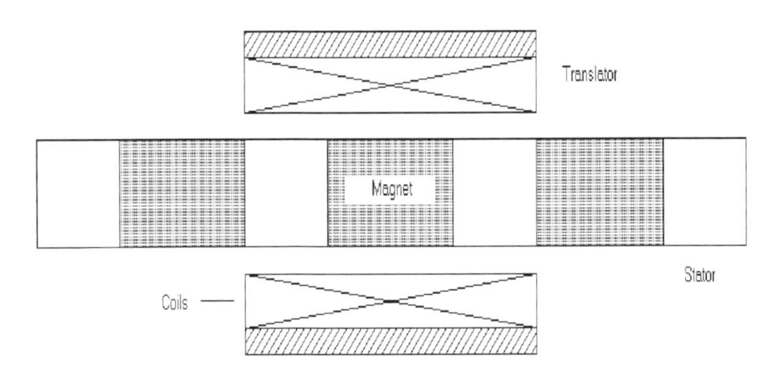

Fig. 2.29. Tubular slotless PMLM

In this section, a simple approach will be developed based on the use of a dither signal as a "Trojan Horse" to cancel the effects of force ripples. The construction of dither signal requires knowledge of the characteristics of force ripples which can be obtained from simple step experiments. For greater robustness, real-time feedback of motion variables can be used to adaptively refine the dither signal characteristics. The approach will be described in greater details in the next section.

A three-tier composite control structure is adopted with the configuration as shown in Figure 2.30. It is similar to the configuration of Figure 2.18, except for the presence of an additional adaptive feedforward control component and the exclusion of the RBF compensator. The composite controller comprises a feedback component (PID), a feedforward component (FFC) and an adaptive feedforward compensator (AFC). The feedforward controller (FFC) is designed as the inverse of the *a priori* dominant linear model for fast response; PID feedback control enhances the stability and robustness of the system; AFC is the adaptive dither signal generator which aims to eliminate or suppress the effects of the force ripple.

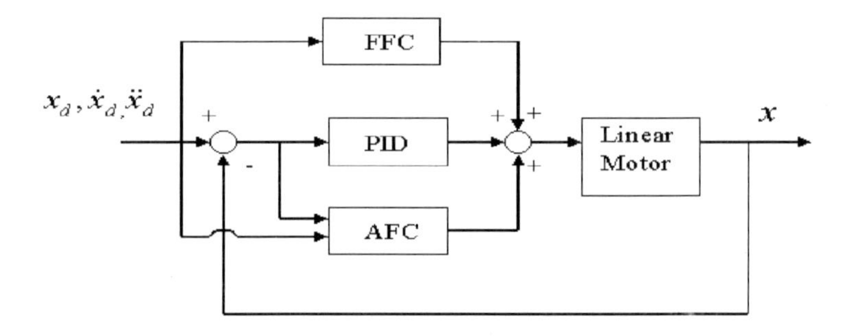

Fig. 2.30. Block diagram of the control scheme

Adaptive Feedforward Component

Following the characteristics highlighted in Figure 2.16, it is assumed that the force ripple can be equivalently viewed as a response to a virtual input described in the form of a periodic sinusoidal signal:

$$u_{ripple} = A(x)\sin(\omega x + \phi) = A_1(x)\sin(\omega x) + A_2(x)\cos(\omega x). \qquad (2.96)$$

The dither signal is thus designed correspondingly to eradicate this virtual force as

$$u_{AFC} = a_1(x(t))\sin(\omega x) + a_2(x(t))\cos(\omega x). \qquad (2.97)$$

Perfect cancellation will be achieved when

$$a_1{}^*(x) = -A_1(x), a_2{}^*(x) = -A_2(x). \qquad (2.98)$$

Feedforward compensation schemes are well known to be sensitive to modelling errors which inevitably result in significant remnant ripples. An adaptive approach is thus adopted so that a_1 and a_2 will be continuously adapted based on desired trajectories and prevailing tracking errors.

Let

$$a = \begin{bmatrix} a_1(x) \\ a_2(x) \end{bmatrix}, \theta = \begin{bmatrix} \sin(\omega x) \\ \cos(\omega x) \end{bmatrix}, a^* = \begin{bmatrix} -A_1(x) \\ -A_2(x) \end{bmatrix}. \tag{2.99}$$

The system output due to AFC is then given by

$$x_a = P[a - a^*]^T \theta, \tag{2.100}$$

where P denotes the system.

Equation (2.100) falls within the standard framework of adaptive control theory. Possible update laws for the adaptive parameters will therefore be

$$\dot{a_1}(x(t)) = -ge\sin(\omega x), \tag{2.101}$$

$$\dot{a_2}(x(t)) = -ge\cos(\omega x), \tag{2.102}$$

where $g > 0$ is an arbitrary adaptation gain.

Differentiating Equations (2.101) and (2.102) with respect to time, it follows that

$$\dot{a_1}(t) = -ge\dot{x}_d\sin(\omega x), \tag{2.103}$$

$$\dot{a_2}(t) = -ge\dot{x}_d\cos(\omega x). \tag{2.104}$$

In other words, the adaptive update laws at Equations (2.103) and (2.104) can be applied as an adjustment mechanism such that $a_1(t)$ and $a_2(t)$ in Equation (2.97) converge to their true values.

Simulation Results

The results of a quick simulation study will be presented in this section. The trajectory profile is prescribed as

$$x_d(\tau) = 10^6[x_0 + (x_0 - x_f)(15\tau^4 - 6\tau^5 - 10\tau^3)], \tag{2.105}$$

$$\dot{x}_d(\tau) = 10^6(x_0 - x_f)(60\tau^3 - 30\tau^4 - 30\tau^2), \tag{2.106}$$

where 10^6 is used to normalise the system units to μm. x_d and $\dot{x}_d$ denote the desired position and velocity trajectories, x_0 and x_f denote the initial and final positions, respectively. $\tau = t/(t_f - t_0)$, where t_0 and t_f are the initial time and final time of the trajectory.

The linear motor model considered for simulation is a slotless tubular linear permanent magnet motor manufactured by Linear Drive, UK. The simulation results of the control scheme are shown in Figure 2.31. For comparison, the simulations results for the composite control minus the adaptive dither are shown in Figure 2.32. It is clearly evident that the adaptive dither has contributed to a much superior control performance.

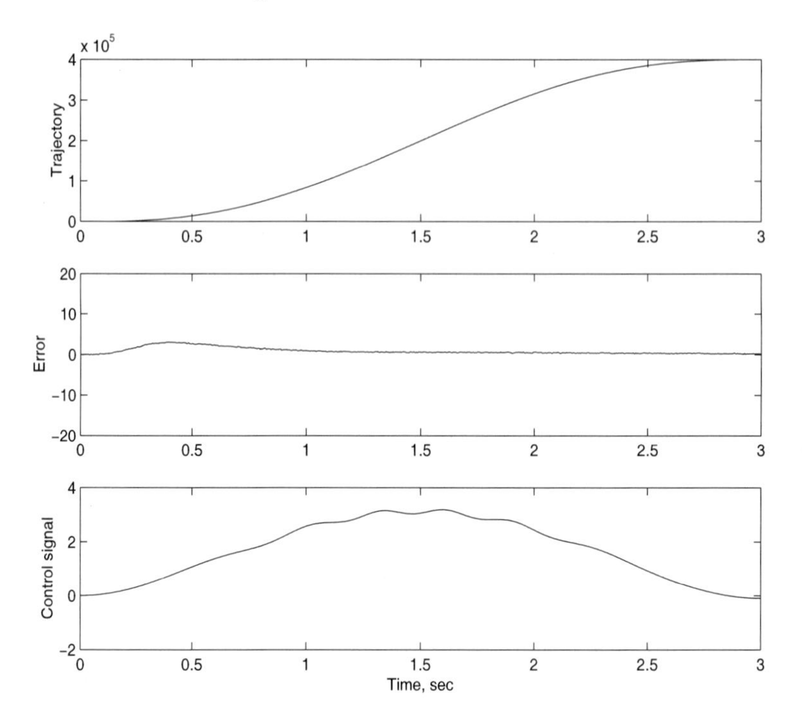

Fig. 2.31. Simulation results with the control scheme: trajectory (μm)(*top*), error (μm)(*middle*), and control signal (V)(*bottom*)

Experiments

The experimental results, from the use of an adaptive ripple compensator (ARC), are shown in Figure 2.33 and Figure 2.34, showing a maximum tracking error of around 5 μm. It should be noted that this is achieved without any RBF. To illustrate further the effectiveness of the adaptive dither, the control results without the dither signal are shown in Figure 2.35. Apart from reduced tracking errors, note the w-shaped characteristics in the feedback signal of Figure 2.35 have been removed in Figure 2.34 as a result of the adaptive dither.

Fast Fourier Transform (FFT) shown in Figure 2.36 is performed on the error signals to examine its frequency content. With the adaptive dither, the error spectrum is much reduced particularly at the lower frequency range, verifying the improved trajectory tracking performance.

2.2.8 Disturbance Observation and Cancellation

The achievable performance of PMLM is unavoidably limited by the amount of disturbances present. These disturbances may arise due to load changes, system parameter perturbation owing to prolonged usage, measurement noise

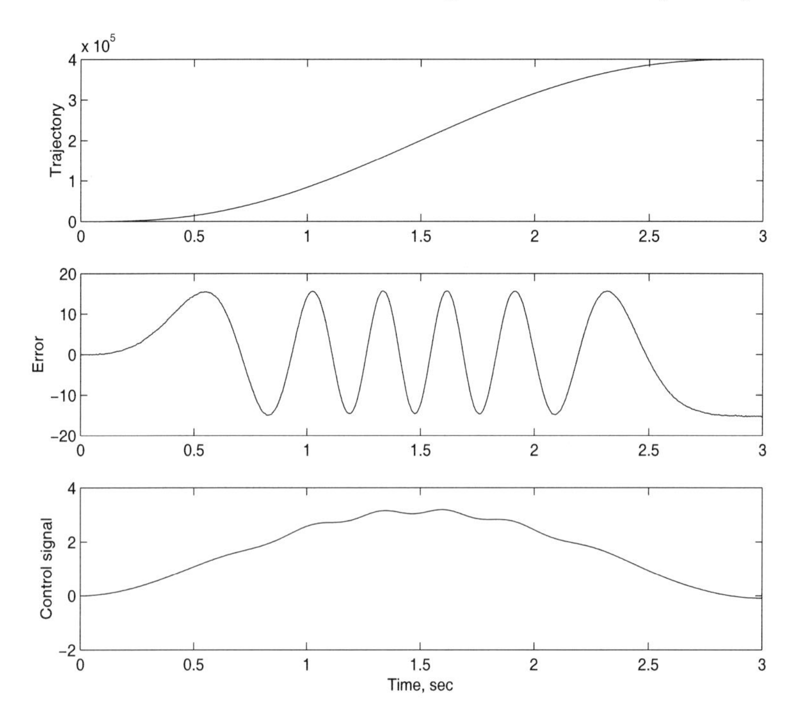

Fig. 2.32. Simulation results without adaptive dither: Trajectory (μm)(*top*), error (μm)(*middle*), and control signal (V)(*bottom*)

and high frequencies generated from the amplifiers (especially when a *Pulse Width Modulated* (PWM) amplifier is used), or inherent non-linear dynamics such as the force ripples and frictional forces. Incorporating a higher resolution in the measurement system *via* the use of high interpolation electronics on the encoder signals can only achieve improvement in positioning accuracy to a limited extent. Thereafter, the amount of disturbances present will ultimately determine the achievable performance. In this section, this important issue of disturbance compensation for precision motion control systems will be addressed, focusing particularly on systems using PWM amplifiers.

Based on a describing function approach, an overall model for a DC PMLM driven by a sinusoidal PWM amplifier will be established. For practical control design purposes, the desirable system model to be applied is usually a linear one. The amplifier model used is a first order quasi-linear approximation, considering only the fundamental frequency of the multitude of frequencies generated by the amplifier. The PMLM model is a second-order transfer function.

Therefore, the overall mathematical model will inevitably inherit modelling uncertainties and a robust control scheme is thus necessary. In this section, one such scheme is proposed which augments a disturbance observa-

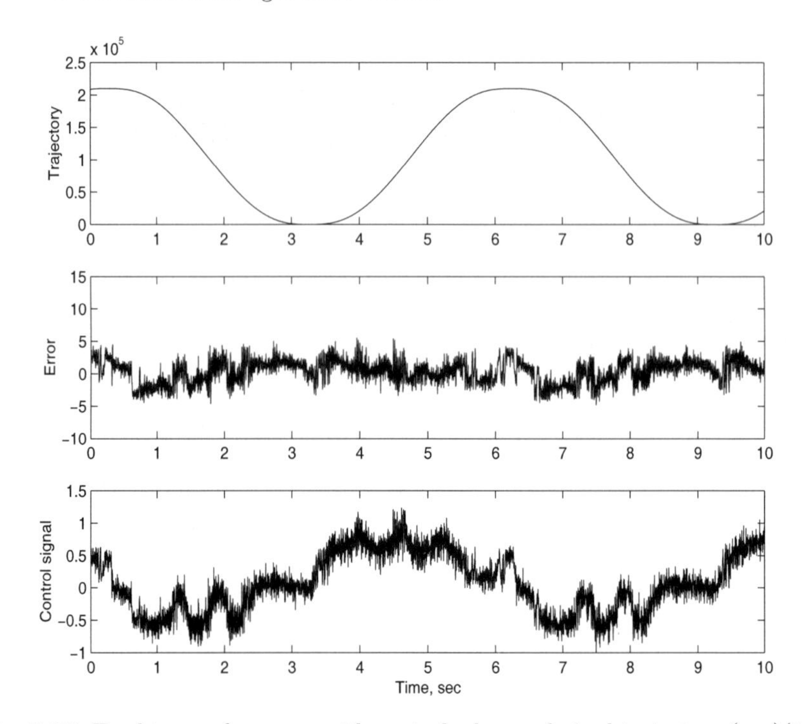

Fig. 2.33. Tracking performance with control scheme: desired trajectory (μm)(*top*), error (μm)(*middle*), and control signal (V)(*bottom*)

tion/cancellation scheme to the motion controller to reduce the sensitivity of the control performance to disturbances arising. The control and disturbance compensator design is relatively simple and directly amenable to practical applications. Full experimental results will be provided to illustrate the effectiveness of the proposed robust control system for PMLM precision motion control.

Overall System Model (PWM+PMLM)

In transfer function form, the dynamics of a PMLM according to Equation (2.53) can be described by

$$P_{PMLM}(s) = \frac{K_p}{s(T_p s + 1)}, \tag{2.107}$$

where $K_p = K_2/K_1$ and $T_p = M/K_1$ are the parameters of the model. This choice of model also greatly facilitates the use of automatic control tuning approaches (Chapter 3).

The recent availability of high-voltage and high-current PWM amplifiers in hybrid packages has attracted the interest of many servo drive designers who

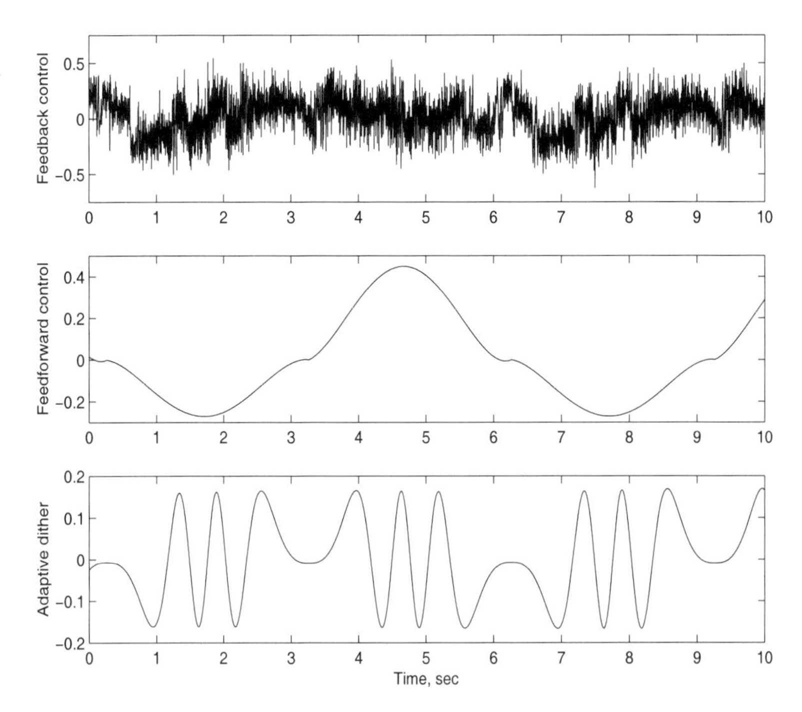

Fig. 2.34. Control signal components with control scheme: feedback control (V)(*top*), feedforward control (V)(*middle*), and adaptive Dither (V)(*bottom*)

traditionally use linear amplifiers. The main advantage of PWM amplifiers over linear amplifier is clearly in the power transfer efficiency. An efficiency of 70–90% can readily be achieved with PWM amplifiers. High efficiency also translates into lower internal power loss, smaller heat sinks, and therefore a reduced overall physical size. One can also use PWM amplifiers to emulate linear constant-voltage amplifiers or linear constant-current amplifiers, both at much higher levels of efficiency compared to linear amplifiers.

The PWM amplifier of concern here is a sinusoidal PWM with bipolar switching, and it uses a triangular carrier. This PWM amplifier converts a sinusoidal signal (which may be from the output of a DC-AC converter) into a pulse train of variable duty cycle. The PWM input controls the duty cycle of the output pulse train, which switches on and off once during each cycle. When a high output is required, the pulse train will have a duty cycle which approaches 100%. Figure 2.37 shows the basic structure of a PWM amplifier. u_i is the analogue input to be modulated, whereas u_{oa} is a pulse train and u_{ob} is its inverse. The PWM oscillator provides the switching frequency (or carrier frequency). The bridge drive circuitry thus consists of a comparator which compares the input signal u_i and the triangular carrier u_c to generate a sequence of pulses to trigger the H-bridge as depicted in Figure 2.38.

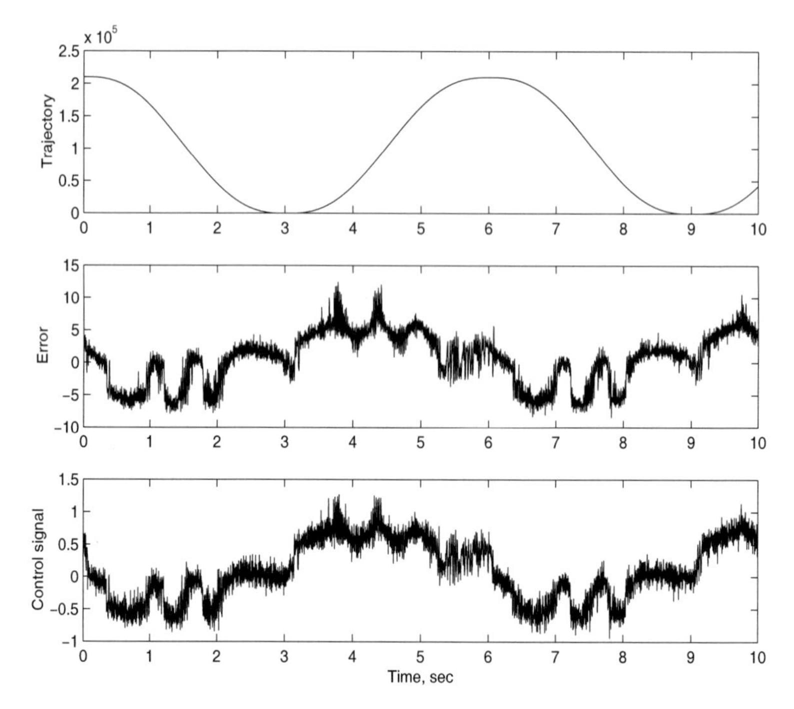

Fig. 2.35. Experimental results without adaptive dither: desired trajectory (μm)(*top*), and error (μm)(*middle*), and control signal (V)(*bottom*)

The switching moments are determined by the crossover points of the two waveforms.

The MOSFETs of the H-bridge (S_1, S_2, S_3 and S_4) simply act as switches. The gating signal (TTL), u_g turns the MOSFETs on and off in diagonal sets, *i.e.*, when S_1 and S_4 are on, S_2 and S_3 are off, and $vice-versa$. As u_i changes from its minimum to its maximum, the duty cycle of u_{oa} changes from 0 to 100% and from its inverse u_{ob} changes from 100% to 0, and $vice - versa$. The differential voltage $u_{oa} - u_{ob}$ has the same waveform as the pulse train u_{oa}, but it has the amplitude of the motor rated voltage. This simple voltage amplification feature is another advantage which PWM amplifiers offer for high voltage applications over the linear amplifiers.

The pulse trains at outports A and B can be connected directly to a motor because the motor, which is essentially a low-pass filter, would screen out the high harmonics of the pulsed voltage and hence produce an analogue signal. It is however advisable to connect LC filters next to the amplifier module or to have built-in LC filters. The filters are useful for EMI cancellation (Electromagnetic Interference) and EMC (Electromagnetic Compatibility) purposes. Without the filters, the long cables to the motor carry high-voltage switching pulses and acts as antennas. If an external filter is required, a rule of thumb

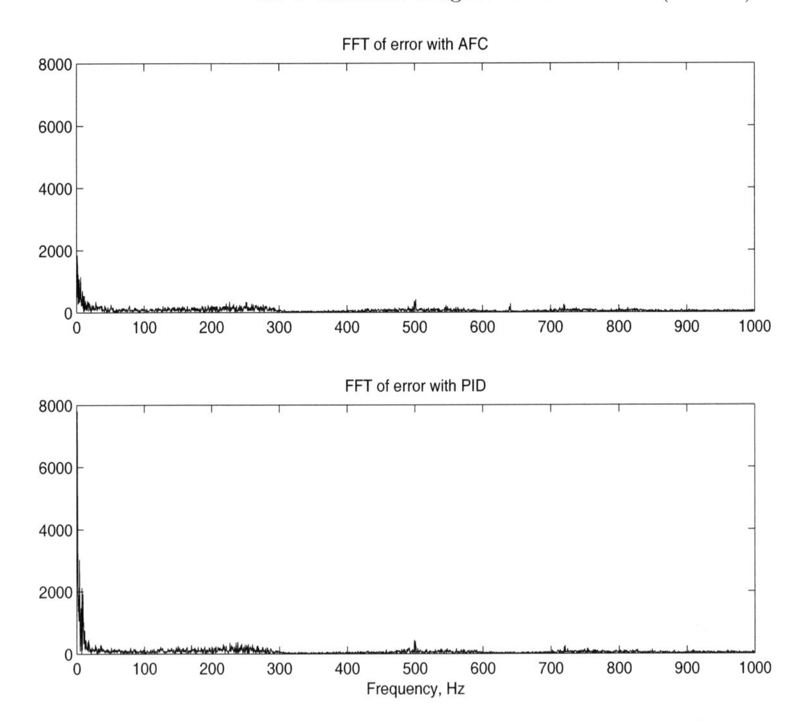

Fig. 2.36. Fast Fourier Transform of error signal: with dither(top)and without dither(bottom)

is to set the corner frequency (or cut-off frequency) of the LC filter to be one decade below the PWM switching frequency.

In the testbed system, the linear actuator is driven by a sinusoidal PWM amplifier (bipolar switching) with a triangular carrier of 20 kHz. It has been observed that the harmonic frequency contents in the PWM output are not negligible, and they can adversely affect positioning accuracy. In many motion control systems where the control performance requirements are modest, the dynamics of the PWM amplifier are simply ignored and only the PMLM is modelled. However, when the performance requirements become more stringent, it is necessary to account explicitly for the dynamics associated with the PWM amplifier. Thus, a model for the PWM amplifier will be derived based on a describing function approach. Based on the overall model, consisting of the PWM amplifier and the PMLM, the controller will be designed.

As in a typical describing function analysis, a sinusoidal input to the PWM amplifier is assumed, given by

$$u_i(t) = u_{im} \cos(\omega t). \tag{2.108}$$

This is also a natural input to the PWM amplifier, since the usual DC input from the controller will be converted into a sinusoid of an appropriate amplitude *via* a DC-AC converter. The PWM output will consist of a sequence of

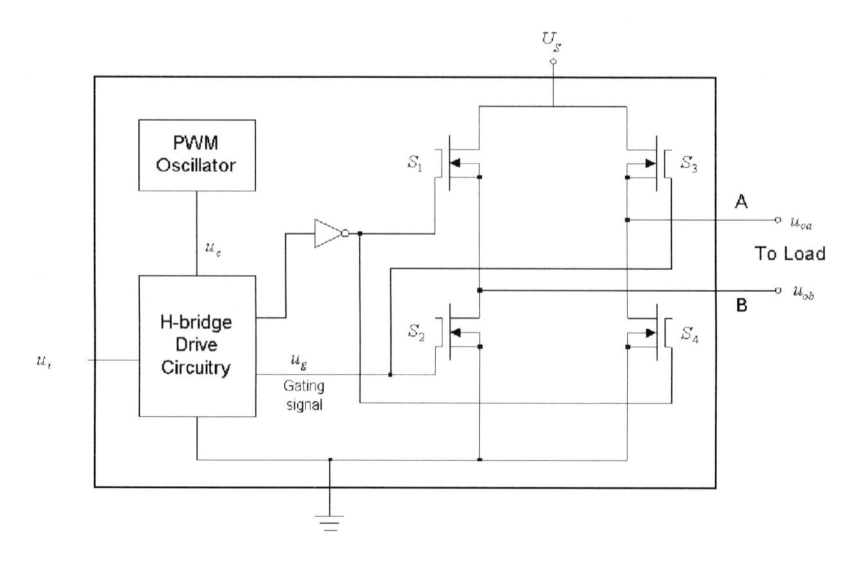

Fig. 2.37. Basic structure of a PWM amplifier

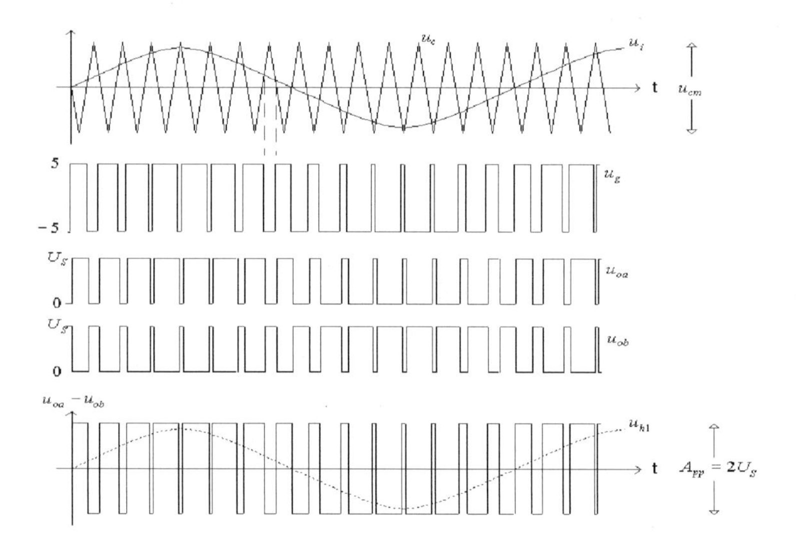

Fig. 2.38. Pulse width modulated signal

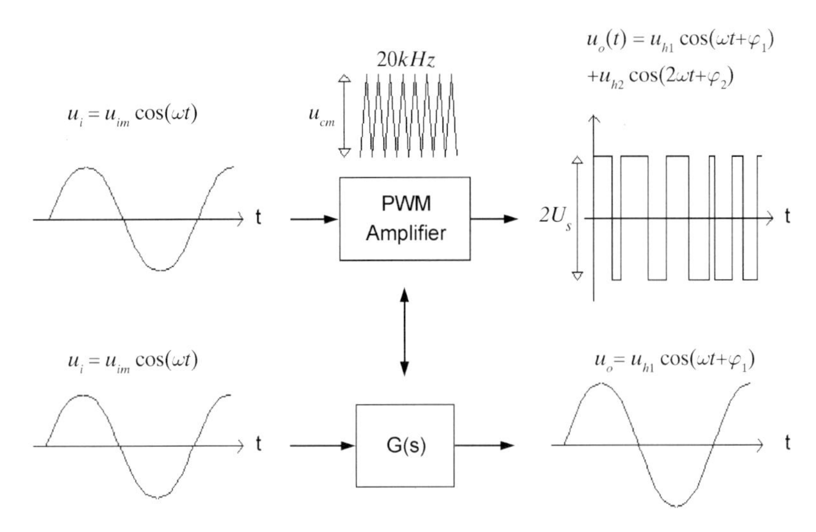

Fig. 2.39. Describing function analysis of the PWM amplifier

pulses at the fundamental frequency of ω (Figure 2.39). Using Fourier analysis, this periodic signal can be equivalently decomposed into a fundamental sinusoid and its high frequency harmonics, *i.e.*,

$$u_o(t) = \sum_{i=1}^{\infty} u_{hi}\cos(i\omega t + \varphi_i), \tag{2.109}$$

where u_{hi} is the Fourier coefficient corresponding to the ith harmonic.

If it is assumed that the higher harmonics will be naturally filtered *via* the low-pass characteristics of the PMLM (filtering hypothesis), only the fundamental frequency, *i.e.*, $u_o(t) = u_{h1}\cos(\omega t + \varphi_1)$ may be considered.

The approximate frequency response of the PWM amplifier is thus given by

$$|G(j\omega)| = \frac{u_{h1}}{u_{im}}, \tag{2.110}$$

$$\arg[G(j\omega)] = \varphi_1. \tag{2.111}$$

It may be assumed that the gain is linear so that

$$\frac{u_{h1}}{u_{im}} = \frac{A_{pp}}{u_{cm}}.$$

The frequency response can be converted to a parametric transfer function with a delay (φ_1/ω):

$$G(s) = \frac{A_{pp}}{u_{cm}} e^{-s(\varphi_1/\omega)}. \tag{2.112}$$

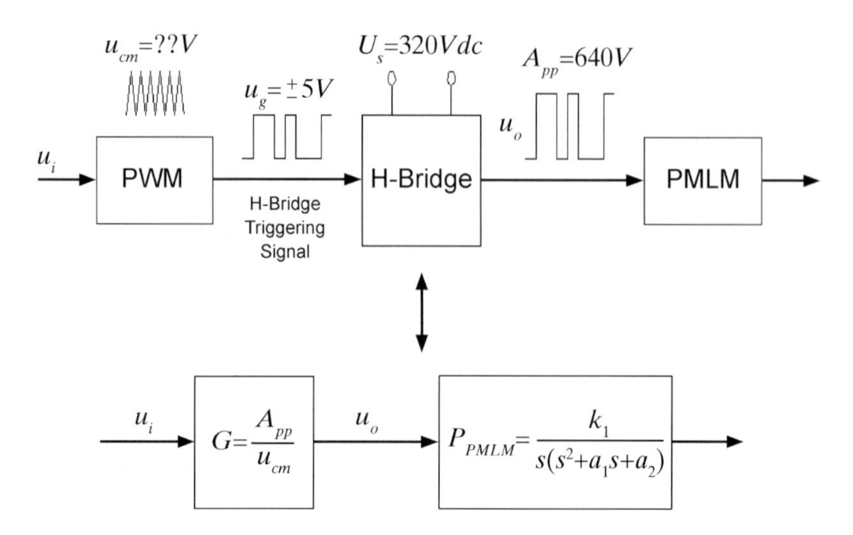

Fig. 2.40. The overall model

Usually, φ_1 is small so that the delay term may be ignored.

Figure 2.40 shows the schematic diagram of the overall model of a PMLM driven by a PWM amplifier. The PWM amplifier is described by a constant gain whereas the PMLM is represented by a third-order transfer function. The PWM amplifier acts as a pulse generator. The gating signal generated, u_g will trigger the H-bridge to give an amplified pulsating signal. The output of the power electronics circuit, u_o, is finally filtered to produce an analogue signal which is in turn used to drive the PMLM.

Design of the Disturbance Observer

In this section, a robust control scheme employing a disturbance observer is presented to reduce the sensitivity of the control performance to disturbances, more notably in the form of high harmonics from PWM amplifiers, ripple forces and load changes. Figure 2.41 shows the block diagram of the control system which uses an estimate of the actual disturbance, deduced from a disturbance observer, to compensate for the disturbances. r, u, ξ, x, d and $\hat{d}$ denote the reference signal, control signal, measurement noise, system output, actual and estimated disturbance respectively. The disturbance observer, shown demarcated within the dotted box in Figure 2.41, estimates the disturbance based on the output x and the control signal u. P denotes the actual system. P_n denotes the nominal system which can be generally described by

$$P_n = \frac{a_0}{s^l\left(s^{m-l} + a_1 s^{m-l-1} + \ldots + a_{m-l-1}s + a_{m-l}\right)}, \tag{2.113}$$

where P_n is a m-th order delay system and has l poles at the origin. Here, a third order model will be used, *i.e.*,$l = 1, m = 2$:

$$P_n = \frac{K_p}{s(T_p s + 1)}. \tag{2.114}$$

The disturbance observer incorporates the inverse of the nominal system, and thus a low pass filter F is required to make the disturbance observer proper and practically realisable. For the choice of a second order model P_n, a suitable filter is

$$F(s) = \frac{f_2}{s^2 + f_1 s + f_2}. \tag{2.115}$$

Higher-order observers may be used which can predict the occurence of the disturbances earlier. For illustration, only a feedback controller C_f is used which is usually designed with respect to the nominal system P_n. Additional control components, such as those described in preceding sections, can be used too. The estimated disturbance is added to the overall input to cancel the effects of the disturbances. Thus, this function is similar to a feedforward compensator, and it can improve the transient performance to the disturbance as well as the steady state operations.

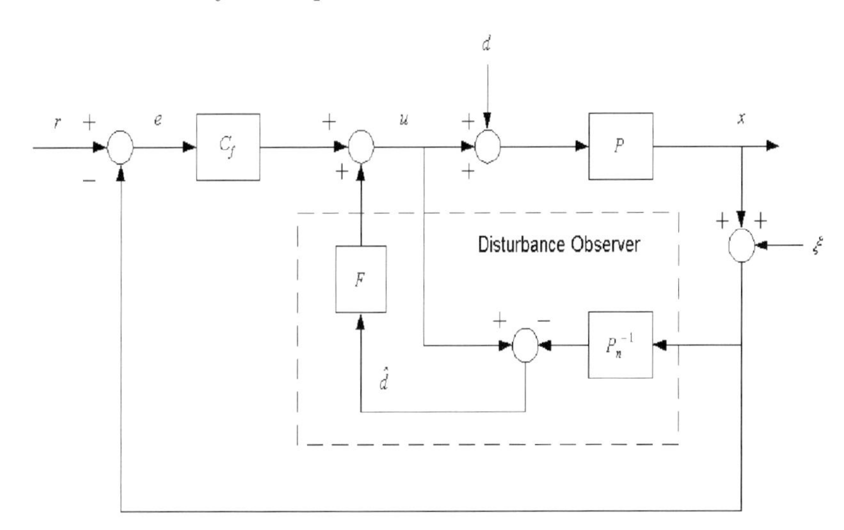

Fig. 2.41. Control system with disturbance observer

Within the bandwidth of the observer filter F, the control system with the disturbance observer as depicted in Figure 2.41 essentially approximates a nominal system without disturbances. This observation may be clearer by transforming Figure 2.41 to the equivalent configuration of a filter-type two-degree-of-freedom control system as depicted in Figure 2.42. Figure 2.42 shows that the disturbance observer is equivalent to an additional disturbance compensator C_{obsv}, which closes a fast inner loop. Consequently, it may be considered that the thus compensated inner loop constitutes essentially a nominal system without disturbances, since they have been compensated by C_{obsv}.

It can be shown that

$$C_{obsv} = \frac{F}{1-F}P_n^{-1}. \tag{2.116}$$

For the choice of P_n and F, it follows that

$$C_{obsv} = \frac{f_2(T_p s + 1)}{K_p(s + f_1)}. \tag{2.117}$$

Therefore, C_{obsv} can be considered as a lead/lag compensator by appropriately designing f_1 and f_2 relatively to K_p and T_p.

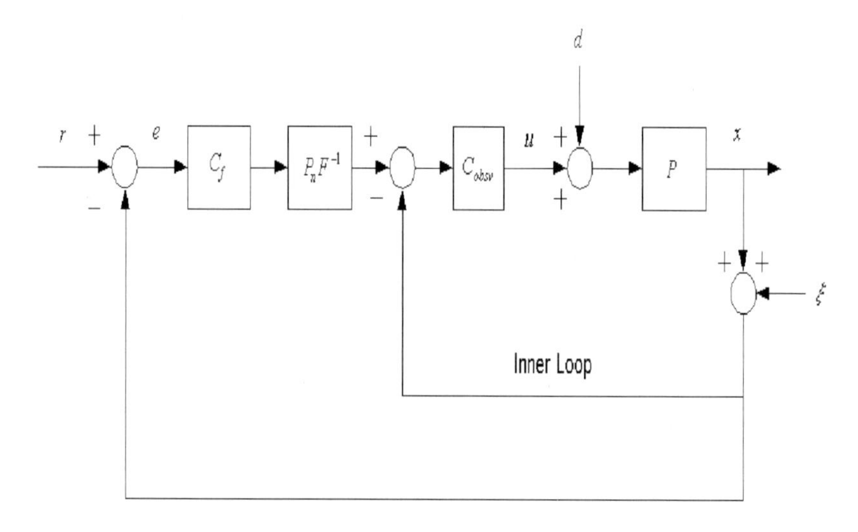

Fig. 2.42. Equivalent system to Figure 2.41

The disturbance observer can be designed in many ways. One possible approach is given as follows:

- Identify the nominal model (*i.e.*, K_p, T_p), based on which the outer loop controller C_f can be designed to achieve a desired command response. If C_f is a PID controller, many design methods are available.
- Adjust f_1 and f_2 of the disturbance compensator C_{obsv} to satisfy requirements for robustness and disturbance suppression characteristics. The system sensitivity function and the system transmission function can thus be

set independently. Such a feature is especially useful when there are strict requirements on both set-point tracking and disturbance suppression and an acceptable compromise between these sometimes conflicting requirements might not exist.

- Carry out simulation and fine tuning till the performance is acceptable.

Experiments

The PWM amplifier has a triangular carrier with peak-to-peak amplitude of 10V and the PWM output ranges from –5 V to +5 V. The pulses trigger the bridge circuit to give an amplified pulsating voltage ranging from –320 V to 320 V which is then used to drive the PMLM. The overall amplifier gain is thus given by

$$G = \frac{2 \times 320}{10} = 64. \tag{2.118}$$

Substituting the various constants associated with the manufacturer specifications into Equation (2.107) yields a overall dominant model (normalised to μm):

$$P_n = \frac{6.91 \times 10^7}{s(s + 136.5)}. \tag{2.119}$$

The real-time experimental results obtained as given in Figure 2.43 show that the controller with disturbance observer can achieve a tracking error of less than 7 μm. The controller performs satisfactorily even when a load disturbance is deliberately introduced into the system (*Box B* in Figure 2.43). For comparison, *Box A* highlights the performance of the system before the introduction of the load disturbance. The changes in control signal due to the introduction of disturbance are not reflected in the error signal. In other words, the control system is able to reject the external disturbance and the performance is not significantly affected.

To illustrate further the performance enhancement from the use of the disturbance observer, control results without the observer are shown in Figure 2.44. The deliberate load disturbance introduced into the system is clearly manifested in the error signal (*Box B* in Figure 2.44). A comparison between Figure 2.43 and Figure 2.44 shows that the use of the disturbance observer is not only effective in reducing the tracking error, but also useful in eliminating or reducing the inherent force ripple which is characterised by the w-shaped part of the control signal.

2.2.9 Robust Adaptive Control

Adaptive control systems include a controller, a performance index, and an automatic gains adjustment algorithm. The adjustment mechanism using the

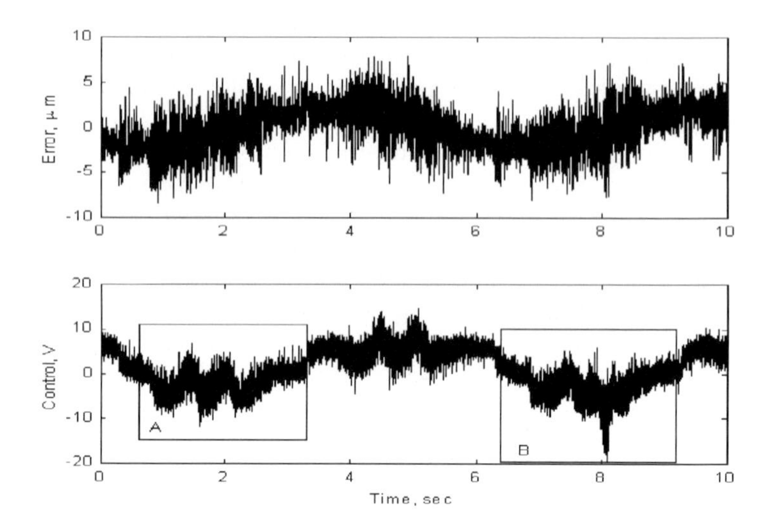

Fig. 2.43. Experimental results with disturbance observer

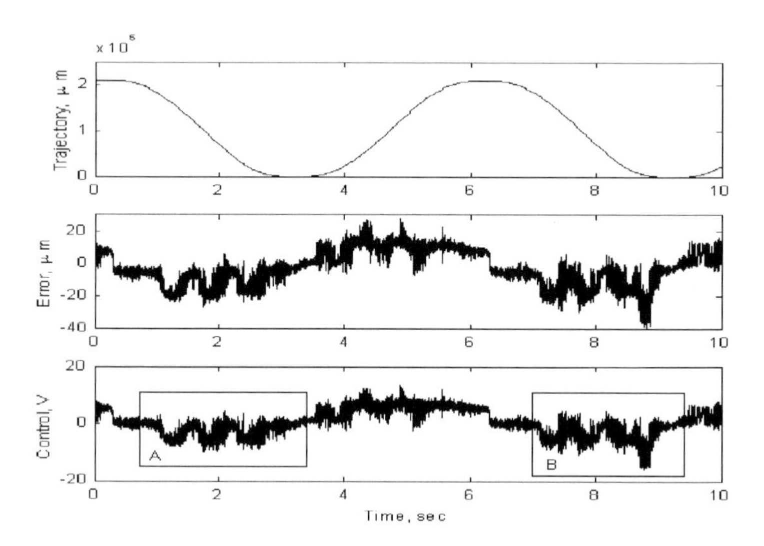

Fig. 2.44. Experimental results without disturbance observer: desired trajectory (μm)(*top*), error (μm)(*middle*), and control signal (V)(*bottom*)

control signal, system output, and the performance measure, continuously adjusts control parameters in order to improve system performance. The need for a system, which may be non-linear or time varying, to automatically optimise its performance over the entire operating range is the main motivation for considering adaptive control. This is unachievable using a fixed control law. The performance of pure adaptive control hinges on how efficiently accurate information of the system, implicit or explicit, can be inferred from the system signals and also whether the performance can be robust to acceptable modelling uncertainties, both parametric and non-parametric. Adaptive control may be augmented to sliding mode control input to achieve robust control with an uncertain model inheriting both parametric and non-parametric modelling error. Adaptive control provides for the adaptation of the model parameters to the actual system parameters, while sliding mode control can account effectively for residual non-parametric errors.

Control Design

In this scheme, a robust adaptive tracking control scheme is given for states trajectories tracking for PMLM. The control has an additional sliding mode control input to compensate for any remaining unmodelled residual dynamics The gain of the sliding control input is adjusted adaptively to estimates of the linear bound of the unmodelled dynamics. The block diagram of the control scheme is given in Figure 2.45.

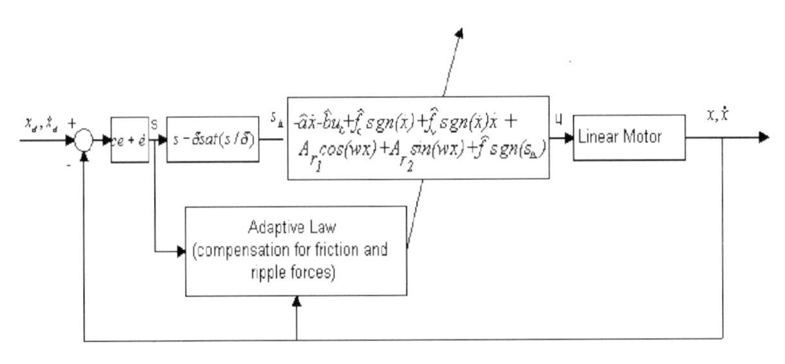

Fig. 2.45. Adaptive control scheme

Equation (2.57) can be simplified to a form more amenable for the development of the adaptive control scheme. Let

$$a = \frac{K_1}{K_2},$$

$$b = \frac{M}{K_2},$$
$$\bar{F}_{load} = \frac{F_{load}}{K_2},$$
$$\bar{F}_{friction} = \frac{F_{friction}}{K_2},$$
$$\bar{F}_{ripple} = \frac{F_{ripple}}{K_2},$$

and denote any remaining unmodelled dynamics as $\bar{F}_{res}$; it follows that

$$\ddot{x} = \frac{a\dot{x} + u - \bar{F}_{friction} - \bar{F}_{ripple} - \bar{F}_{load} - \bar{F}_{res}}{b} \tag{2.120}$$

Assume the load force is bounded as follows:

$$|\bar{F}_{load}(t)| < f_{lM} \; \forall \, t > 0 \tag{2.121}$$

Following Equation (2.57) with $\delta = 2$, $\bar{F}_{friction}$ can be written in terms of its equivalent components as

$$\begin{aligned} \bar{F}_{fric} = & [f_c + (f_s - f_c)e^{-(|\dot{x}/\dot{x}_s|)^2} \\ & + f_v|\dot{x}|]sgn(\dot{x}). \end{aligned}$$

f_c, f_s, and f_v are assumed to be constants. Since $0 < e^{-(\dot{x}/\dot{x}_s)^2} < 1$, the Stribeck effect is a bounded disturbance, *i.e.*, $|(f_s - f_c)e^{-(\dot{x}/\dot{x}_s)^2}| \leq f_{fM}$. The ripple force is described by a sinusoidal function of the load position with a period of w and an amplitude of A_r, *i.e.*,

$$f_{ripple} = A_r \sin(wx + \varphi) = A_{r1} \cos(wx) + A_{r2} \sin(wx), \tag{2.122}$$

where $A_r, w, \varphi, A_{r1}, A_{r2}$ are constants.

Define the tracking errors

$$e(t) = x_d(t) - x(t), \tag{2.123}$$
$$\dot{e}(t) = \dot{x}_d(t) - \dot{x}(t), \tag{2.124}$$

where x_d and $\dot{x}_d$ are the desired position and velocity, respectively. To achieve robust tracking control, a sliding surface is defined as

$$s = \Lambda_1 \int_0^t e(\tau)d\tau + \Lambda_2 e(t) + \dot{e}(t), \tag{2.125}$$

where Λ_1, Λ_2 are chosen such that the polynomial $\lambda^2 + \Lambda_2\lambda + \Lambda_1$ is Hurwitz.

Here, another error metric $s_\Delta(t)$ is defined:

$$s_\Delta(t) = s(t) - \delta sat(s(t)/\delta), \tag{2.126}$$

where $sat(.)$ is a saturation function defined as

$$sat(x) = \begin{cases} x & ,if\ |x| < 1 \\ sgn(x) & ,otherwise \end{cases} \tag{2.127}$$

The function s_Δ has the following useful properties:
(i) If $|s| < \delta$, then $\dot{s}_\Delta = s_\Delta = 0$,
(ii) If$|s| > \delta$, then $\dot{s}_\Delta = \dot{s}$ and $|s_\Delta| = |s| - \delta$,
(iii)$s_\Delta sat(s/\delta) = |s_\Delta|$.

Thus, the problem is to design a control law $u(t)$ which ensures that the tracking error metric $s(t)$ lies in the predetermined boundary δ for all time $t > 0$.

The following controller is constructed for the non-linear system at Equation (2.120):

$$u = -\hat{a}\dot{x} - \hat{b}u_c + \hat{f}_c sgn(\dot{x}) + \hat{f}_v sgn(\dot{x})\dot{x} + \hat{A}_{r1}\cos(wx) + \hat{A}_{r2}\sin(wx) + \hat{f} sgn(s_\Delta), \tag{2.128}$$

where $\hat{a}$ and $\hat{b}$ are the estimates of a and b respectively; $\hat{f}_c$ represents the estimate of f_c; $\hat{f}_v$ represents the estimate of f_v; $\hat{A}_{r1}$ and $\hat{A}_{r2}$ represents the estimates of A_{r1} and A_{r2}, respectively; $\hat{f}$ is the estimate of $f_{fM} + f_{lM}$. u_c is an additional control which is given by

$$u_c = -\Lambda_1 e - \Lambda_2 \dot{e} - \ddot{x}_d - K_v s_\Delta. \tag{2.129}$$

Differentiating $s(t)$ and applying the control law given by Equation (2.128), the system dynamics may be written in terms of the filtered tracking errors as

$$\begin{aligned} &\dot{s} + K_v s_\Delta \\ &= \Big(-\tilde{a}\dot{x} - \tilde{b}u_c + \tilde{f}_c sgn(\dot{x}) + \tilde{f}_v sgn(\dot{x})\dot{x} + \tilde{A}_{r1}\cos(wx) + \tilde{A}_{r2}\sin(wx) \\ &\quad -\hat{f} sgn(s_\Delta) + (f_s - f_c)e^{-(\dot{x}/\dot{x}_s)^2} + \bar{F}_{load}\Big)/b, \end{aligned} \tag{2.130}$$

where $\tilde{a} = a - \hat{a}$, $\tilde{b} = b - \hat{b}$, $\tilde{f}_c = f_c - \hat{f}_c$, $\tilde{f}_v = f_v - \hat{f}_v$, $\tilde{A}_{r1} = A_{r1} - \hat{A}_{r1}$, $\tilde{A}_{r2} = A_{r2} - \hat{A}_{r2}$.

The parameter update laws are now specified as

$$\dot{\hat{a}} = -k_a \dot{x} s_\Delta, \tag{2.131}$$

$$\dot{\hat{b}} = -k_b u_c s_\Delta, \tag{2.132}$$

$$\dot{\hat{f}}_c = k_{fc} sgn(\dot{x}) s_\Delta, \tag{2.133}$$

$$\dot{\hat{f}}_v = k_{fv} \dot{x} sgn(\dot{x}) s_\Delta, \tag{2.134}$$

$$\dot{\hat{A}}_{r1} = k_{r1}\cos(wx)s_{\Delta}, \tag{2.135}$$

$$\dot{\hat{A}}_{r2} = k_{r2}\sin(wx)s_{\Delta}, \tag{2.136}$$

$$\dot{\hat{f}} = k_f|s_{\Delta}|. \tag{2.137}$$

Convergence Analysis

The following theorem is given to establish the convergence of the tracking error under the controller.

Theorem 2.3.

Consider the non-linear system at Equation (2.120) and the control objective of tracking desired trajectories given by $x_d, \dot{x}_d$ and $\ddot{x}_d$. The control law given by Equation (2.128) with Equations (2.131)–(2.137) will ensure that the system states and parameters are uniformly bounded and that $s(t)$ asymptotically converge to the predetermined boundary δ.

Proof.

A Lyapunov function candidate $V(t)$ is first defined as

$$V(t) = \frac{1}{2}bs_{\Delta}^2 + \frac{1}{2k_a}\tilde{a}^2 + \frac{1}{2k_b}\tilde{b}^2 + \frac{1}{2k_{fc}}\tilde{f}_c^2 + \frac{1}{2k_{fv}}\tilde{f}_v^2 + \frac{1}{2k_{r1}}\tilde{A}_{r1}^2 + \frac{1}{2k_{r2}}\tilde{A}_{r2}^2 + \frac{1}{2k_f}\tilde{f}^2, \tag{2.138}$$

where $\tilde{f} = f - \hat{f}, f = f_{fM} + f_{lM}$. Noting that $\dot{s}_{\Delta} = \dot{s}$ outside the boundary layer, while $s_{\Delta} = 0$ inside the boundary layer, it follows that

$$\begin{aligned}
\dot{V} &= b\dot{s}s_{\Delta} + \frac{1}{k_a}\tilde{a}\dot{\tilde{a}} + \frac{1}{k_b}\tilde{b}\dot{\tilde{b}} + \frac{1}{k_{fc}}\tilde{f}_c\dot{\tilde{f}}_c + \frac{1}{k_{fv}}\tilde{f}_v\dot{\tilde{f}}_v + \frac{1}{k_{r1}}\tilde{A}_{r1}\dot{\tilde{A}}_{r1} \\
&\quad + \frac{1}{k_{r2}}\tilde{A}_{r2}\dot{\tilde{A}}_{r2} + \frac{1}{k_f}\tilde{f}\dot{\tilde{f}}, \\
&= -K_v b s_{\Delta}^2 - [\tilde{a}\dot{x} + \tilde{b}u_c - \tilde{f}_c sgn(\dot{x}) - \tilde{f}_v\dot{x}sgn(\dot{x}) - \tilde{A}_{r1}\cos(wx) \\
&\quad - \tilde{A}_{r2}\sin(wx)]s_{\Delta} + [(f_s - f_c)e^{-(\dot{x}/\dot{x}_s)^2}sgn(s_{\Delta}) + \bar{F}_{load} \\
&\quad + \frac{1}{k_a}\tilde{a}\dot{\tilde{a}} + \frac{1}{k_b}\tilde{b}\dot{\tilde{b}} + \frac{1}{k_{fc}}\tilde{f}_c\dot{\tilde{f}}_c + \frac{1}{k_{fv}}\tilde{f}_v\dot{\tilde{f}}_v + \frac{1}{k_{r1}}\tilde{A}_{r1}\dot{\tilde{A}}_{r1} \\
&\quad + \frac{1}{k_{r2}}\tilde{A}_{r2}\dot{\tilde{A}}_{r2} + \frac{1}{k_f}\tilde{f}\dot{\tilde{f}}, \\
&\leq -K_v b s_{\Delta}^2 - [\tilde{a}\dot{x} + \tilde{b}u_c - \tilde{f}_c sgn(\dot{x}) - \tilde{f}_v\dot{x}sgn(\dot{x}) - \tilde{A}_{r1}\cos(wx) \\
&\quad - \tilde{A}_{r2}\sin(wx)]s_{\Delta} + [f_{fM} + f_{lM} - \hat{f}]|s_{\Delta}| + \frac{1}{k_a}\tilde{a}\dot{\tilde{a}}
\end{aligned}$$

$$
\begin{aligned}
&+\frac{1}{k_b}\tilde{b}\dot{\tilde{b}}+\frac{1}{k_{fc}}\tilde{f}_c\dot{\tilde{f}}_c+\frac{1}{k_{fv}}\tilde{f}_v\dot{\tilde{f}}_v+\frac{1}{k_{r1}}\tilde{A}_{r1}\dot{\tilde{A}}_{r1}+\frac{1}{k_{r2}}\tilde{A}_{r2}\dot{\tilde{A}}_{r2}+\frac{1}{k_f}\tilde{f}\dot{\tilde{f}},\\
=&-K_v b s_\Delta^2-[\tilde{a}\dot{x}+\tilde{b}u_c-\tilde{f}_c sgn(\dot{x})-\tilde{f}_v\dot{x}sgn(\dot{x})-\tilde{A}_{r1}\cos(wx)\\
&-\tilde{A}_{r2}\sin(wx)]s_\Delta+\tilde{f}|s_\Delta|-\frac{1}{k_a}\tilde{a}\dot{\hat{a}}-\frac{1}{k_b}\tilde{b}\dot{\hat{b}}-\frac{1}{k_{fc}}\tilde{f}_c\dot{\hat{f}}_c-\frac{1}{k_{fv}}\tilde{f}_v\dot{\hat{f}}_v\\
&-\frac{1}{k_{r1}}\tilde{A}_{r1}\dot{\hat{A}}_{r1}-\frac{1}{k_{r2}}\tilde{A}_{r2}\dot{\hat{A}}_{r2}-\frac{1}{k_f}\tilde{f}\dot{\hat{f}}. \qquad (2.139)
\end{aligned}
$$

Substituting the expressions given by Equations (2.131)–(2.137) yields

$$
\dot{V} \leq -K_v b s_\Delta^2. \qquad (2.140)
$$

Since $b > 0$, it follows that $\dot{V} < 0$. This implies that s_Δ, $\hat{a}$, $\hat{b}$, $\hat{f}_c$, $\hat{f}_v$, $\hat{A}_{r1}$, $\hat{A}_{r2}$, and $\hat{f}$ are uniformly bounded with respect to t. To prove the boundedness of the tracking error, it is necessary to prove that x and $\dot{x}$ are bounded.

Define

$$
\sigma_0 = \int_0^t (x_d - x)d\tau. \qquad (2.141)
$$

From Equation (2.125), it follows that

$$
\begin{bmatrix} \dot{\sigma}_0 \\ \ddot{\sigma}_0 \end{bmatrix} = \begin{bmatrix} 0 & 1 \\ -\Lambda_1 & -\Lambda_2 \end{bmatrix} \begin{bmatrix} \sigma_0 \\ \dot{\sigma}_0 \end{bmatrix} + \begin{bmatrix} 0 \\ 1 \end{bmatrix} s. \qquad (2.142)
$$

Since Λ_1, Λ_2 are chosen such that the polynomial $\lambda^2 + \Lambda_2\lambda + \Lambda_1$ is Hurwitz, the free system of the above equation is asymptotically stable. This together with s_Δ bounded, implies that $x, \dot{x}$ are bounded.

By definition, $\dot{s}_\Delta(t)$ is either 0 or $\dot{s}(t)$, where $\dot{s}(t)$ is given in Equation (2.130). Since $(f_s - f_c)e^{-(\dot{x}/\dot{x}_s)^2}$, $\bar{F}_{load}$ and the system parameters are bounded, this implies that the right side of Equation (2.130) is bounded which in turn implies that $\dot{s}$ is bounded. Equation (2.140) and the positive definiteness of V further imply that

$$
\lim_{t\to\infty} \int_0^t -\dot{V}(\tau)d\tau = V(0) - \lim_{t\to\infty} V(t) < \infty. \qquad (2.143)
$$

By virtue of Barbalat's lemma,

$$
\lim_{t\to\infty} \dot{V}(t) = 0. \qquad (2.144)
$$

Applying Equation (2.140) further implies that

$$
\lim_{t\to\infty} s_\Delta = 0. \qquad (2.145)
$$

The proof is completed.

Remark 2.2.

Notice that $K_v s$ is actually standard PID control. In principle, any existing PID tuning methods can be employed to tune and determine $\Lambda_1, \Lambda_2, K_v$. This initial tuning can be a rough one, since the system performance can be further improved by the adaptive component.

Remark 2.3.

To achieve high-accuracy tracking, δ should be chosen to be small. However, a small δ may lead to control chattering. Therefore, there should be a trade-off between the desired tracking error and the discontinuity of the input which is tolerable.

Experiments

The tracking performance achieved from the use of adaptive control is given in Figure 2.46 where the desired trajectory is also shown. A maximum error of less than 8 μm is achieved compared to a maximum tracking error of 15 μm when pure PID control is used as shown in Figure 2.47.

2.2.10 Iterative Learning Control

When the motion to be executed by the PMLM is repetitive in nature, iterative learning control (ILC) can be used as a simple model-free learning enhancement to a PID feedback controller. The main objective of this feedforward term is to reject exogenous disturbances, and to compensate for the nonlinearities mentioned in Section 2.1 which would otherwise limit the accuracy achievable with simple feedback control systems. ILC exploits the repetitive nature of the tasks as experience gained to compensate for the poor or incomplete knowledge of the system model and the disturbances present. A recent comprehensive survey of ILC can be found in (Moore 1998). ILC is essentially a memory-based scheme which needs to store the tracking errors and control efforts of previous repetition in order to construct the control efforts of present cycle. Thus, a discrete-time implementation is necessary. There are two common updating laws for the ILC, a P-type updating law which only considers the tracking errors as input for learning and a D-type scheme which needs to differentiate the tracking errors (Longman 1998). For practical applications, the P-type updating law has proven to be more robust and effective in implementation.

Figure 2.48 shows a general block-diagram of such control scheme where the feedback controller is to stabilise the system while the ILC feedforward controller is to enhance the performance of the next cycle based on previous cycles.

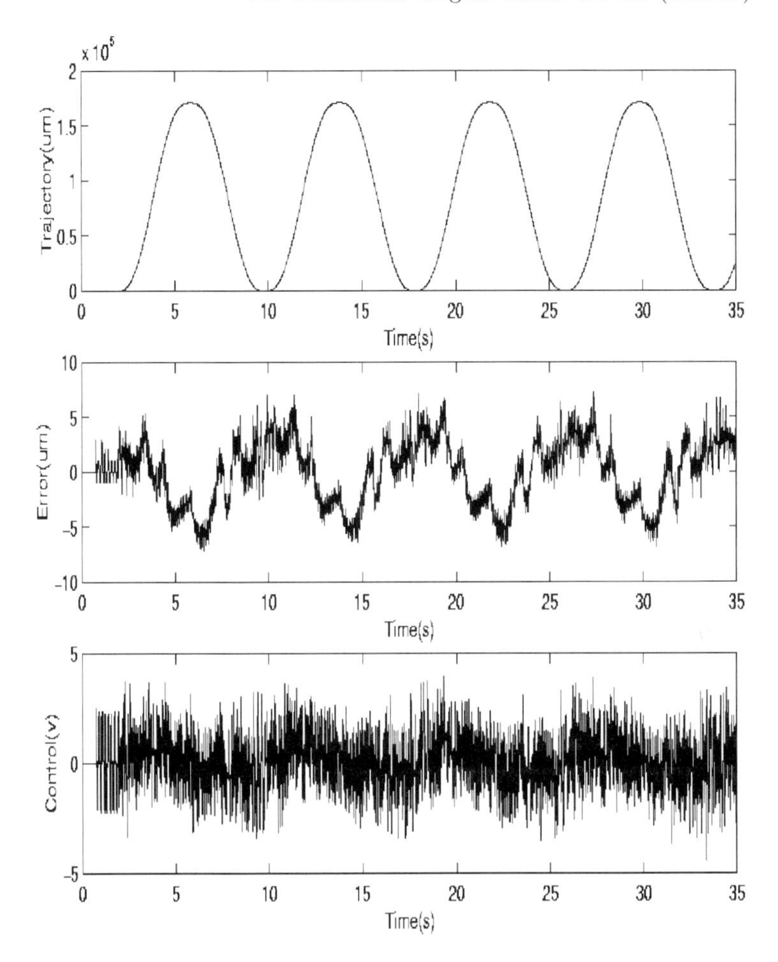

Fig. 2.46. Tracking performance based on adaptive control

The task is assumed to be executed repeatedly within a finite time duration denoted by T . At the i-th repetition, the tracking error is $e_i(t)$ for a given desired output trajectory $x_d(t)$ while the control effort is $u_i(t)$. The control input to the system is a cumulation of the feedforward input $u_i^f(t)$ and the feedback input $u_i^b(t)$. The feedforward input $u_i^f(t)$ is to be updated based on the control efforts and tracking errors of previous repetition(s), and in general, it can be described by

$$u_i^f(t) = \mathcal{L}(u_{i-1}^f(t), e_{i-1}(t)), \qquad (2.146)$$

where $\mathcal{L}$ is the learning operator to be determined.

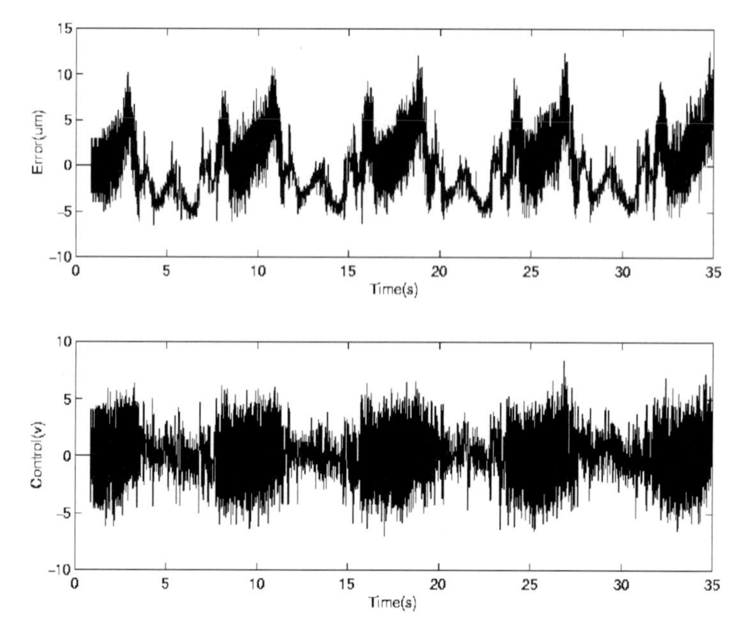

Fig. 2.47. Tracking performance based on PID controller

Preliminaries and Problem Formulation

Consider the following discrete-time uncertain non-linear time-varying system which has to perform a given task repeatedly:

$$\begin{cases} \chi_i(t+1) = f(\chi_i(t), t) + B(\chi_i(t), t)u_i(t) + w_i(t), \\ x_i(t) = C(t)\chi_i(t) + u_i(t), \end{cases} \tag{2.147}$$

where i denotes the i-th repetitive operation of the system; t is the discrete time index and $t \in [0, N]$ which means that $t \in \{0, 1, \cdots, N\}$; $\chi_i(t) \in R^n$, $u_i(t) \in R^m$, and $x_i(t) \in R^r$ are the state, control input, and output of the system, respectively; $C(t) \in R^{r\times n}$ is a time-varying matrix; the functions $f(\cdot,\cdot) : R^n \times [0, N] \mapsto R^n$ and $B(\cdot,\cdot) : R^n \times [0, N] \mapsto R^m$ are uniformly globally Lipschitz in χ, *i.e.*,$\forall t \in [0, N], \forall i, \exists$ constants k_f, k_B, such that

$$\|\Delta f_i(t)\| \le k_f \|\Delta\chi_i(t)\|, \quad \|\Delta B_i(t)\| \le k_B \|\Delta\chi_i(t)\|, \tag{2.148}$$

where $\Delta f_i(t) \triangleq f(\chi_i(t), t) - f(\chi_{i-1}(t), t)$, $\Delta B_i(t) \triangleq B(\chi_i(t), t) - B(\chi_{i-1}(t), t)$, $\Delta\chi_i(t) \triangleq \chi_i(t) - \chi_{i-1}(t)$; $w_i(t)$, $u_i(t)$ are uncertainties or disturbances to the system bounded with unknown bounds b_w, b_v defined as

$$b_w \triangleq \sup_{t\in[0,N]} \|w_i(t)\|, \quad b_v \triangleq \sup_{t\in[0,N]} \|u_i(t)\|, \ \forall i. \tag{2.149}$$

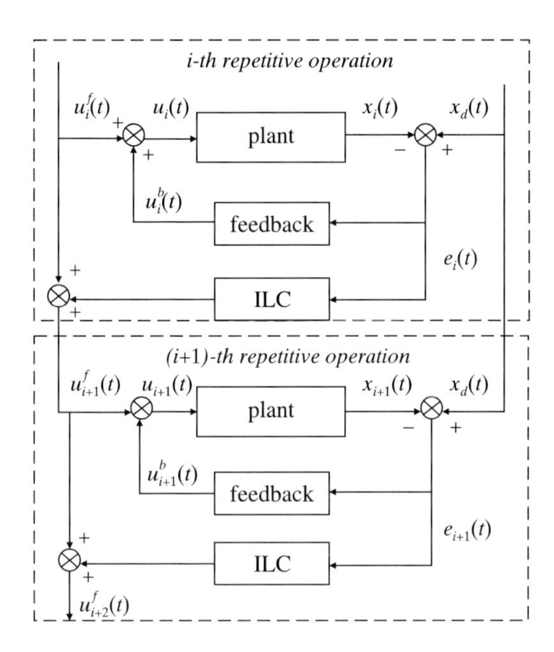

Fig. 2.48. Block diagram of learning enhanced control by ILC

The norms used in this paper are defined as

$$\|\bar{v}\| = \max_{1 \leq i \leq n} \mid \bar{v}_i \mid, \quad \|G\| = \max_{1 \leq i \leq m} \left(\sum_{j=1}^{n} \mid g_{i,j} \mid \right),$$

where $\bar{v} = [\bar{v}_1, \cdots, \bar{v}_n]^T$ is a vector, $G = [g_{i,j}]_{m \times n}$ is a matrix. Denote the output tracking error $e_i(t) \triangleq x_d(t) - x_i(t)$ where $x_d(t)$ is the given desired output trajectory, which is realisable, *i.e.*, given a bounded $x_d(t)$, there exists a unique bounded desired input $u_d(t),\ t \in [0, N]$ such that when $u(t) = u_d(t)$, the system has a unique bounded desired state $\chi_d(t)$ satisfying

$$\begin{cases} \chi_d(t+1) = f(\chi_d(t), t) + B(\chi_d(t), t)u_d(t) \triangleq f_d + B_d u_d, \\ x_d(t) = C(t)\chi_d(t) \triangleq C(t)\chi_d. \end{cases} \tag{2.150}$$

Denote the bound of the desired control u_d as $b_{u_d} \triangleq \sup_{t \in [0,N]} \|u_d(t)\|$. The control problem is formulated as follows. Starting from an arbitrary continuous initial control input $u_0(t)$, obtain the next control input $u_1(t)$ and the subsequent series $\{u_i(t) \mid i = 2, 3, \cdots\}$ for the system at Equation (2.147) by using a proper learning control updating law in such a way that when $i \to \infty, x_i(t) \to x_d(t) \pm \varepsilon^*$ in the presence of bounded uncertainty, disturbance and re-initialisation error.

At the i-th ILC iteration, the control input $u_i(t)$ to the system at Equation (2.147) is

$$u_i(t) = u_i^f(t) + u_i^b(t), \tag{2.151}$$

where $u_i^f(t)$ is from the feed*f*orward iterative learning controller and $u_i^b(t)$ is from the feed*b*ack stabilising controller. The feedback stabilising controller is assumed to be in the following general form.

$$z_i(t+1) = h_a(z_i(t)) + H_b(z_i(t))e_i(t), \tag{2.152}$$
$$u_i^b(t) = h_c(z_i(t)) + H_d(z_i(t))e_i(t), \tag{2.153}$$

where $z_i(t) \in R^{n_c}$ is the state of the feedback stabilising controller with $z_i(0) = 0, \ \forall i$. The vector-valued functions $h_a(\cdot) : R^{n_c} \mapsto R^{n_c}$ and $h_c(\cdot) : R^{n_c} \mapsto R^m$ are designed to be sector-bounded as

$$\|h_a(z_i(t))\| \leq b_{h_a}\|z_i(t)\|, \ \|h_c(z_i(t))\| \leq b_{h_c}\|z_i(t)\|.$$

The function matrices $H_b(\cdot) : R^{n_c} \mapsto R^{n_c \times r}$ and $H_d(\cdot) : R^{n_c} \mapsto R^{m \times r}$ are designed to be with uniform bounds, *i.e.*, $\forall t \in [0, N], \forall z_i(t) \in R^{n_c}$,

$$\|H_b(z_i(t))\| \leq b_{H_b}, \ \|H_d(z_i(t))\| \leq b_{H_d}.$$

The $b_{h_a}, b_{h_c}, b_{H_b}, b_{H_d}$ above are positive constants which are not necessarily known. A simple ILC updating law is used, *i.e.*,

$$u_{i+1}^f(t) = u_i(t) + Q(t)e_i(t+1), \tag{2.154}$$

where $Q(t) \in R^{m \times r}$ is the learning matrices which is to be determined to ensure the ILC convergence.

The following assumptions are made before further developments:

- The initialisation error is bounded as follows: $\forall t \in [0, N], \ \forall i, \ \|\chi_d(0) - \chi_i(0)\| \leq b_{\chi_0}, \ \|x_d(0) - x_i(0)\| \leq b_C b_{\chi_0} + b_v$, where $b_C \triangleq \sup_{t \in [0,N]} \|C(t)\|$.
- Matrix $C(\cdot)B(\cdot, \cdot)$ has a full column rank $\forall t \in [0, N], \ \chi(t) \in R^n$.
- Operator $B(\cdot, \cdot)$ is bounded, *i.e.*, $\exists$ a constant b_B such that for all i, $\sup_{t \in [0,N]} \|B(\chi_i(t), t)\| \triangleq \sup_{t \in [0,N]} \|B_i(t)\| \leq b_B$.
- The desired output $x_d(t), t \in [0, N]$ is achievable by the desired input $u_d(t), \forall t \in [0, N]$.

To analyse the robust convergence property of the proposed feedback-assisted P-type ILC algorithm, the following $\lambda-$norm equivalent to the infinity-norm is introduced for a discrete-time vector $h(t), t = 0, 1, \cdots, N$:

$$\|h(t)\|_\lambda \triangleq \sup_{t \in [0,N]} \hat{e}^{-\lambda t}\|h(t)\|, \tag{2.155}$$

where $\lambda > 0$ when $\hat{e} > 1$ or $\lambda < 0$ when $\hat{e} \in (0, 1)$.

The following recursive formula can be directly verified:

$$\bar{z}_{i+1} = a_1\bar{z}_i + a_2\bar{z}_i^* + a_3 = a_1^{i+1}\bar{z}_0 + \sum_{j=0}^{i} a_1^{i-j}(a_2\bar{z}_j^* + a_3), \tag{2.156}$$

where $\{\bar{z}_i, \bar{z}_i^* \mid i = 0, 1, \cdots\}$ are two series which are related to each other by coefficients a_1, a_2, a_3. Denote

$$\delta f_i(t) \triangleq f_d - f(\chi_i(t), t), \quad \delta B_i(t) \triangleq B_d - B_i(t).$$

Then, from Equations (2.147) and (2.150), it can be found that

$$\delta\chi_i(t+1) = \delta f_i(t) + \delta B_i(t)u_d + B_i(t)\delta u_i(t) - w_i(t). \tag{2.157}$$

Taking the norm for Equation (2.157) yields

$$\|\delta\chi_i(t+1)\| \le (k_f + b_{u_d}k_B)\|\delta\chi_i(t)\| + b_B\|\delta u_i(t)\| + b_w. \tag{2.158}$$

From Equation (2.153), it can be seen that

$$\|u_i^b(t)\| \le b_{h_c}\|z_i(t)\| + b_{H_d}b_C\|\delta\chi_i(t)\| + b_{H_d}b_v, \tag{2.159}$$

and

$$\begin{aligned} \|\delta u_i(t)\| &= \|u_d(t) - (u_i^f(t) + u_i^b(t))\| \\ &\le \|\delta u_i^f(t) - u_i^b(t)\| \le \|\delta u_i^f(t)\| + \|u_i^b(t)\|. \end{aligned} \tag{2.160}$$

Then Equation (2.158) becomes

$$\begin{aligned} \|\delta\chi_i(t+1)\| &\le (k_f + b_{u_d}k_B + b_Bb_{H_d}b_C)\|\delta\chi_i(t)\| \\ &+ b_Bb_{h_c}\|z_i(t)\| + b_B\|\delta u_i^f(t)\| + b_Bb_{H_d}b_v + b_w. \end{aligned} \tag{2.161}$$

On the other hand, it can be observed from Equation (2.152) that

$$\|z_i(t+1)\| \le b_{h_a}\|z_i(t)\| + b_{H_b}b_C\|\delta\chi_i(t)\| + b_{H_b}b_v. \tag{2.162}$$

Thus, adding Equations (2.162) and (2.161) yields

$$\begin{aligned} (\|\delta\chi_i(t+1)\| + \|z_i(t+1)\|) &\le \hat{e}(\|\delta\chi_i(t)\| + \|z_i(t)\|) \\ &+ b_B\|\delta u_i^f(t)\| + \hat{\varepsilon}, \end{aligned} \tag{2.163}$$

where

$$\begin{aligned} \hat{e} &\triangleq \max\{k_f + b_{u_d}k_B + b_Bb_{H_d}b_C + b_{H_b}b_C,\ b_{h_a} + b_Bb_{h_c}\} \ne 1; \\ \hat{\varepsilon} &\triangleq (b_{H_b} + b_Bb_{H_d})b_v + b_w. \end{aligned}$$

Applying Equation (2.156), it follows that

$$\begin{aligned} \|\delta\chi_i(t+1)\| + \|z_i(t+1)\| &\le \hat{e}^{t+1}b_{\chi_0} \\ &+ \sum_{j=0}^{t} \hat{e}^{t-j}(b_B\|\delta u_i^f(j)\| + \hat{\varepsilon}). \end{aligned} \tag{2.164}$$

A simpler relationship can be derived between $\|\delta\chi_i(t)\|_\lambda + \|z_i(t)\|_\lambda$ and $\|\delta u_i^f(t)\|_\lambda$, by noting the following relations:

- $\|c\|_\lambda \equiv | c |, \forall c \in R$;
- $\forall | \lambda |> 1$, $\sup_{t\in[0,N]} \hat{e}^{-(\lambda-1)t} = 1$;
- $\forall t_1 \in [0, N_1], t_2 \in [0, N_2]$, if $0 \leq N_1 \leq N_2 \leq N$, then $\|\delta h(t_1)\|_\lambda \leq \|\delta h(t_2)\|_\lambda$.

Taking the $\lambda-$norm ($| \lambda |> 1$) operation of Equation (2.164) gives

$$\|\delta\chi_i(t)\|_\lambda + \|z_i(t)\|_\lambda \leq b_{\chi_0} + b_B O(| \lambda |^{-1})\|\delta u_i^f(t)\|_\lambda + c_0\hat{\varepsilon}, \tag{2.165}$$

where

$$O(| \lambda |^{-1}) \triangleq \frac{1-\hat{e}^{-(\lambda-1)N}}{\hat{e}^\lambda - \hat{e}}, \quad c_0 \triangleq \sup_{t\in[0,N]} \frac{\hat{e}^{-(\lambda-1)t}(1-\hat{e}^{-t})}{\hat{e}-1}.$$

For brevity in the development in the sequel, the following notations are used:

$$b_Q \triangleq \sup_{t\in[0,N]} \|Q(t)\|,$$

$$\rho \triangleq \sup_{t\in[0,N]} \|I_m - Q(t)C(t+1)B_i(t)\|, \quad \forall i.$$

The Class K function is also defined.

Definition 2.1. (Class-K function (Inonnou and Sun, 1996):

A continuous function $\phi : [0, r] \mapsto R^+$ is said to be **class-K**, *i.e.*, $\phi \in K$ if

1) $\phi(0) = 0$,
2) ϕ is strictly increasing on $[0, r]$.

In the following subsection, it will be shown that the tracking error bound is a **class-K** function of the bounds of uncertainty, disturbance, and re-initialisation error. Moreover, under additional restrictions, the tracking error bound can also be shown to be a **class-K** function of the bounds of differences of uncertainties, disturbances, and re-initialisation errors between two successive ILC iterations.

Robust Convergence Analysis

A main result on error convergence is presented in the following theorem.

Theorem 2.4.

For the repetitive discrete-time uncertain time-varying non-linear system at Equation (2.147) under assumptions A1)–A3), given the realisable desired trajectory $x_d(t)$ over the fixed time interval $[0, NT_s]$, by using the ILC updating law at Equation (2.154) and the feedback controller at Equations (2.152)–(2.153), if the condition

$$\rho < 1, \tag{2.166}$$

is satisfied, then the λ-norm of the tracking errors $e_i(t)$, $\delta u_i(t)$, $\delta\chi_i(t)$ are bounded for all i. For a sufficiently large $|\lambda|$, $\forall t \in [0, N]$,

$$b_{u^f} \triangleq \lim_{i\to\infty} \|\delta u_i^f(t)\|_\lambda \le b_{u^f}(b_{\chi_0}, b_w, b_v), \tag{2.167}$$

$$b_u \triangleq \lim_{i\to\infty} \|\delta u_i(t)\|_\lambda \le b_u(b_{\chi_0}, b_w, b_v), \tag{2.168}$$

$$b_\chi \triangleq \lim_{i\to\infty} \|\delta\chi_i(t)\|_\lambda \le b_{\chi_0} + b_B O(|\lambda|^{-1}) b_{u^f} + c_0\hat{\varepsilon}, \tag{2.169}$$

$$b_e \triangleq \lim_{i\to\infty} \|e_i(t)\|_\lambda \le b_C b_\chi + b_v. \tag{2.170}$$

Moreover, b_u, b_χ, b_e are **class-K** functions of b_w, b_v, b_{χ_0}, *i.e.*, b_u, b_χ, b_e converge uniformly to zero as $i \to \infty$ in the absence of uncertainty, disturbance and initialisation error, *i.e.*, $b_w, b_v, b_{\chi_0} \to 0$.

Proof.

The tracking error at $(i+1)-$th repetition is

$$\begin{aligned} e_i(t) &= x_d(t) - x_i(t), \\ &= C(t)\delta\chi_i(t) - u_i(t). \end{aligned} \tag{2.171}$$

The learning control deviation at the $(i+1)$-th repetition, $\delta u_{i+1}^f(t)$, is given by

$$\begin{aligned} \delta u_{i+1}^f(t) &= \delta u_i(t) - Q(t)e_i(t+1), \\ &= \delta u_i(t) - Q(t)C(t+1)\delta\chi_i(t+1), \\ &+Q(t)u_i(t+1). \end{aligned} \tag{2.172}$$

Referring to Equation (2.157), Equation (2.172) can be written as

$$\begin{aligned} \delta u_{i+1}^f(t) &= \delta u_i(t) - Q(t)C(t+1) \\ &[\delta f_i(t) + \delta B_i(t)u_d + B_i(t)\delta u_i(t) \\ &-w_i(t)] + Q(t)u_i(t+1). \end{aligned} \tag{2.173}$$

Collecting terms and then performing the norm operation for Equation (2.173) yields

$$\begin{aligned} \|\delta u_{i+1}^f(t)\| &\le \rho\|\delta u_i(t)\| \\ &+b_Q b_C(k_f + b_{u_d}k_B)\|\delta\chi_i(t)\| \\ &+b_Q(b_C b_w + b_v). \end{aligned} \tag{2.174}$$

Based on Equations (2.160) and (2.159), Equation (2.174) becomes

$$\|\delta u_{i+1}^f(t)\| \leq \rho\|\delta u_i^f(t)\| \\ +\alpha(\|\delta\chi_i(t)\| + \|z_i(t)\|) + \varepsilon, \tag{2.175}$$

where $\alpha \triangleq \max\{b_Q b_C(k_f + b_{u_d} k_B) + b_{H_d} b_C \rho, b_{h_c}\rho\}$ and $\varepsilon \triangleq [b_Q(b_C b_w + b_v) + b_{H_d} b_v \rho]$.

By utilising the relationship in Equation (2.165), taking the λ–norm for Equation (2.174) gives

$$\|\delta u_{i+1}^f(t)\|_\lambda \leq \rho\|\delta u_i^f(t)\|_\lambda \\ +\alpha b_B O(|\ \lambda\ |^{-1})\|\delta u_i^f(t)\|_\lambda \\ +\alpha(b_{\chi_0} + c_0\hat{\varepsilon}) + \varepsilon. \tag{2.176}$$

Referring to Equation (2.166), it is clear that a sufficiently large $|\ \lambda\ |$ can be used to ensure that

$$\rho + \alpha b_B O(|\ \lambda\ |^{-1}) \triangleq \hat{\rho} < 1. \tag{2.177}$$

Therefore, it follows that

$$b_{u^f} = \lim_{i\to\infty} \|\delta u_i^f(t)\|_\lambda = \frac{\varepsilon_0}{1-\hat{\rho}} \triangleq b_{u^f}(b_{\chi_0}, b_w, b_v) \tag{2.178}$$

where $\varepsilon_0 \triangleq \varepsilon + \alpha(b_{\chi_0} + c_0\hat{\varepsilon})$. From Equations (2.165) and (2.171), Equations (2.169) and (2.170) can be verified. It can be observed from Equation (2.165) that

$$b_{\chi z} \triangleq \lim_{i\to\infty}(\|\delta\chi_i(t)\|_\lambda + \|z_i(t)\|_\lambda) \\ \leq b_{\chi_0} + b_B O(|\ \lambda\ |^{-1}) b_{u^f} + c_0\hat{\varepsilon}. \tag{2.179}$$

Therefore, from Equations (2.160) and (2.159), and by referring to Equation (2.179), it follows that

$$b_u \triangleq \lim_{i\to\infty} \|\delta u_i(t)\|_\lambda \leq b_{u^f} + \max\{b_{h_c}, b_{H_d} b_C\} b_{\chi z} + b_{H_d} b_v, \\ \triangleq b_u(b_{\chi_0}, b_w, b_v), \tag{2.180}$$

which verifies Eqution (2.168). Moreover, it is easy to observe that b_{u^f}, b_u, b_χ, and b_e will all tend to zero uniformly for $t = 0, 1, \cdots, N$ as $i \to \infty$ in the absence of uncertainty, disturbance and initialisation error, *i.e.*, when $b_w, b_v, b_{\chi_0} \to 0$.

Additional Issues

For an effective implementation of ILC, the following additional issues should be considered.

- *Initialisation.* As one of the postulates in ILC formulation, the system is required to be reset at the identical initial state after each repetition of the motion trajectory, since the error in initialisation will affect the final tracking performance directly and adversely. This is especially so for the PMLM system, where the ripple force varies with the position of the translator. It is thus critical to correct the value of set-point during the initialisation *via* adequate PID control. The terminal tracking error of the last repetition can be used for this correction.
- *Filtering.* As only the displacement measurement is available in the PMLM system, the velocity is obtained using a numerical difference method which is naturally subject to noise amplification. To reduce the noise effects, the position measurement used is an averaged value of three consecutive A/D samples. The velocity is obtained by using a simple finite difference formula
$$\tilde{\dot{x}}(t) = (x(t) - x(t-1))/t_s, \tag{2.181}$$
where t_s is the sampling period. To reduce further the noise effect, an averaging finite difference formula may be used instead:
$$\tilde{\dot{x}}(t) = (x(t) - x(t-N))/(Nt_s), \tag{2.182}$$
where a practical choice of N is $N = 3$. It has been validated in many robotic applications that the simple scheme at Equation (2.182) is equivalent to many advanced and complex schemes. Moreover, the pre-stored tracking error in the memory pool can be manipulated arbitrarily. Therefore, non-causal filtering is possible.

Experiments

The tracking performance achieved from the use of ILC is given in Figure 2.49 at the 50th iteration. A maximum error of 15 μm is achieved through learning. The advantage of the scheme is that the PID feedback controller need not be tightly tuned. However, the motion must be repetitive in nature for the ILC scheme to be applicable. The learning convergence curve relating tracking error (RMS) to number of iterations is given in Figure 2.50.

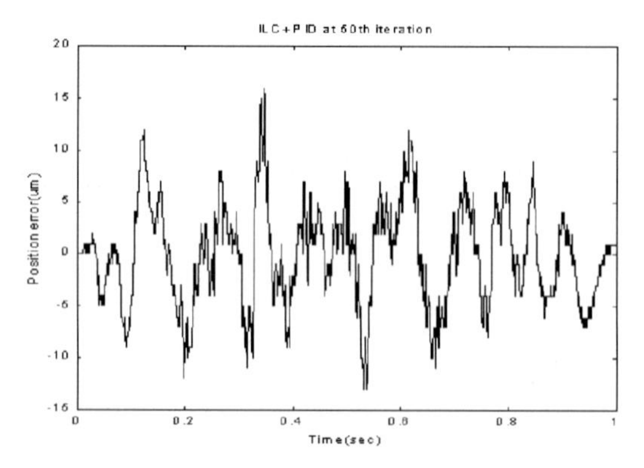

Fig. 2.49. Tracking performance under ILC (50th iteration)

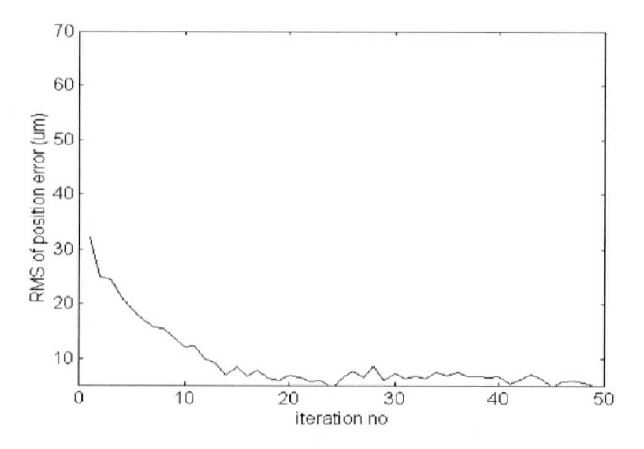

Fig. 2.50. Learning convergence of the tracking error (RMS)

3

Automatic Tuning of Control Parameters

Riding on the advances in adaptive control and techniques, modern industrial controllers are becoming increasingly intelligent. Many high-end controllers appearing in the market now come equipped with auto-tuning and self-tuning features. No longer is tedious manual tuning an inevitable part of control systems. The role of operators in PID tuning has been reduced to simple specifications and decision.

Different systematic methods for tuning controllers are available, but regardless of the design method, the following three phases are usually applicable:

1. The system is disturbed with specific control inputs or control inputs automatically generated in the closed loop.
2. The response to the disturbance is analysed, yielding a model of the system which may be non-parametric or parametric.
3. Based on this model and certain operation specifications, the control parameters are determined.

Automatic tuning of controllers means quite simply that the above procedures are automated so that the disturbances, model calculation and choice of controller parameters all occur within the same controller. In this way, the work of the operator is made simpler, so that instead of having to derive or calculate suitable controller parameters himself, he only needs to initiate the tuning process. He may have to give the controller some information about the system before the tuning is done, but this information will be considerably simpler to specify than the controller parameters.

In this chapter, relay tuning approaches towards tuning of control systems for servo-mechanisms are presented. The approaches are directly amenable to be used in conjunction with the various control scheme presented in Chapter 2.

3.1 Relay Auto-tuning

In order to commission the control schemes, a nominal system model is necessary. In this section, the development of an automatic tuning method for the parts of the control schemes needing the nominal model is considered. Among the various automatic tuning methods proposed in recent time, the work due to Aström and Hägglund (1995) is arguably the most attractive from a practical viewpoint. They use an on-off relay as a controller inserted in the control loop as shown in Figure 3.1. With this arrangement, it is conjectured that sustained oscillation will be generated in many systems. This conjecture has also been field-proven in many applications involving process control systems.

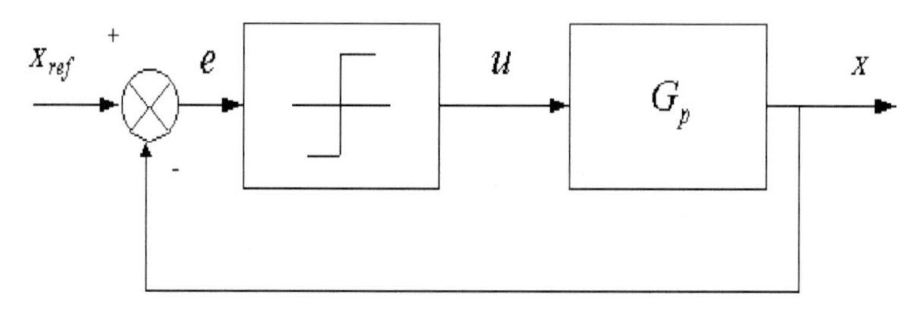

Fig. 3.1. Relay feedback system

Since most systems will exhibit low-pass characteristics, the output oscillation will approximate a sine wave with a period of t_u and an oscillation amplitude of a. Expressing the relay control signal, u by the first harmonic of the Fourier series expansion of the square wave, the frequency response of the system is given by

$$G_p\left(j\frac{2\pi}{t_u}\right) = -\frac{\pi a}{4h}. \tag{3.1}$$

where h represents the relay amplitude. Equation (3.1) is a complex equation. The system parameters may be inferred from this relay feedback arrangement by solving Equation (3.1), provided the number of parameters to be determined is less than or equal to two. This simple operation may be viewed in the frequency domain using the describing function analysis method. The necessary condition for oscillation is that the feedforward transmission must be equal to −1, equivalently described by

$$G_p\left(j\frac{2\pi}{t_u}\right) N(a) = -1. \tag{3.2}$$

Suppose the transfer function $G_p(s)$ is known. Equation (3.2) may be solved graphically by plotting the negative inverse of the describing function, $-\frac{1}{N(a)}$, with the Nyquist curve of $G_p(s)$. An intersection point will typically suggest the existence of a limit cycle oscillation. The period and amplitude of the oscillation are given by the frequency response parameters of that point. One problem with conventional relay tuning is that certain systems do not exhibit stable limit cycle oscillations. Typically, these systems have only low order dynamics and no transportation lag. This is especially true for servo-mechanical systems which rarely exhibits a phase lag of more than $-\pi$, and which probably explain why relay feedback methods have been mainly applied to process control systems thus far. This observation is visually clear from a frequency domain analysis. The describing function of a standard relay is given by Equation (3.3).

$$N(a) = \frac{4h}{\pi a}. \tag{3.3}$$

The negative inverse of the describing function is shown in Figure 3.2 as DF1. For the Nyquist curve of the system also shown in Figure 3.2, typical of a servo system, it is clear there is no intersection between the Nyquist curve and DF1 in the finite frequency range.

Another shortcoming associated with the standard autotuning method is that the experiment identifies the only point on the Nyquist curve that intersects the negative real axis. This point may not, however, provide adequate information on the system for control design.

To overcome the two shortcomings, some modifications to the conventional relay feedback arrangement are needed. From Figure 3.2, for the limit cycle oscillation to occur, it may be necessary to introduce a phase angle to the negative inverse describing function. The modified negative inverse describing function is shown pictorially in Figure 3.2 as DF2. Two possible ways of introducing this phase lag will be described in the following subsections.

3.1.1 Relay with Delay

A phase lag in the relay negative inverse describing function (frequency domain) may be associated with a pure delay in the time domain. If L is the additional time delay introduced, the resultant phase angle shift of the negative inverse describing function can be shown to be $\omega^* L$, where ω^* is the frequency of the point of intersection between the inverse describing function of the relay-delay element and the Nyquist curve of the system. The set-up is illustrated in Figure 3.3.

If a stable limit cycle oscillation exists, the period and the amplitude of the oscillation can be measured. The model parameters may be obtained by solving Equation (3.2) algebraically. For a second-order model given by

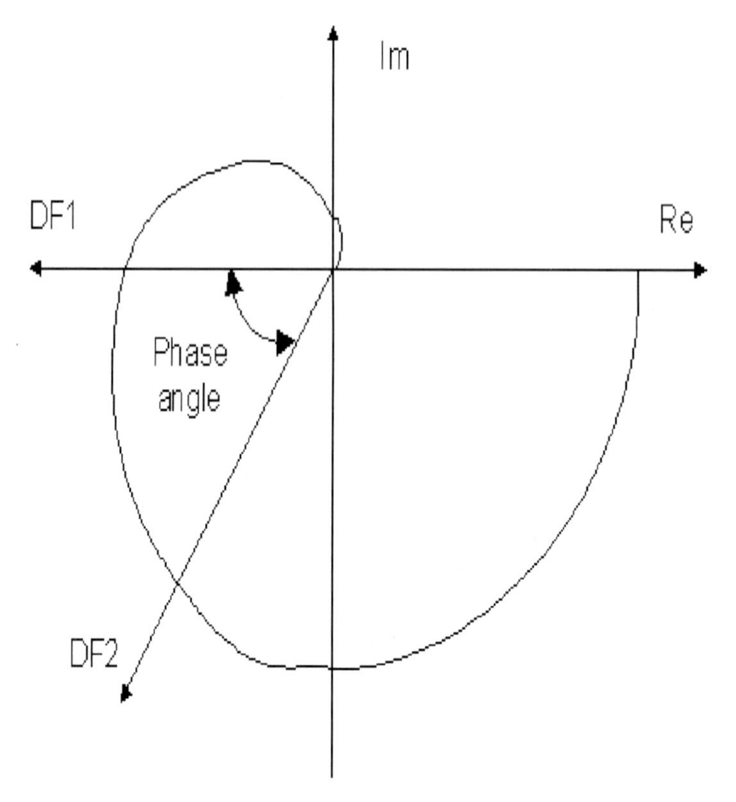

Fig. 3.2. Nyquist Plot where limit cycle does not exist with standard relay auto-tuning

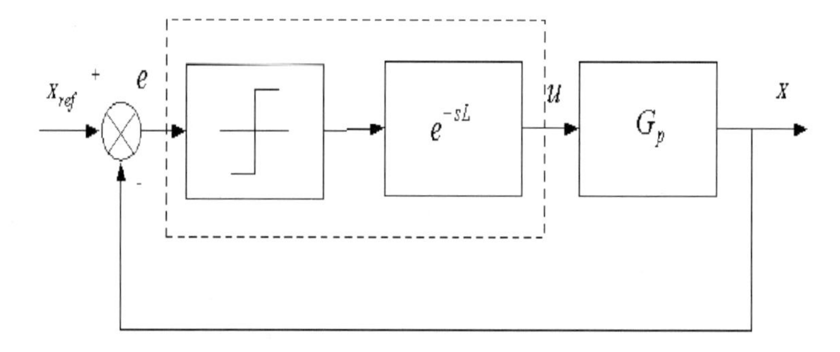

Fig. 3.3. Relay with a pure delay

$$G_p(s) = \frac{K_p}{s(T_p s + 1)},$$

with the relay experiment conducted on the position loop, the model parameters are given by Equations (3.4) and (3.5).

$$K_p = \frac{\omega^* \sqrt{1 + T_p^2 \omega^{*2}}}{K^*}, \tag{3.4}$$

$$T_p = -\frac{cot(\omega^* L)}{\omega^*}. \tag{3.5}$$

It is straightforward to show that $K_p = K_2/K_1$ and $T_p = M/K_1$, where K_1, K_2, M are defined as in Equation (2.53).

3.1.2 Two-channel Relay Tuning

A two-channel relay tuning method was first proposed by Friman and Waller (1997). A describing function with a phase lag may be broken down into two orthogonal components. These components may be conveniently chosen to be along the real and imaginary axes. In this method, an additional relay that operates on the integral of the error is added in parallel to the conventional relay loop. With this method, the phase lag can be specified by selecting proper design parameters h_1 and h_2. The basic construction is shown in Figure 3.4. A similar set of equations for the system parameters may be obtained as in the case of relay with a delay:

$$\begin{aligned} K_p &= \frac{\pi a}{4\sqrt{h_1^2 + h_2^2}} \omega^* \sqrt{1 + \omega^{*2} T_p^2}, \\ T_p &= \frac{h_1}{h_2 \omega^*}. \end{aligned} \tag{3.6}$$

3.2 Friction Modelling Using Relay Feedback

It has been noted in Equation (2.120) that, considering the frictional and load forces present, the dynamic model of a PMLM can be described by

$$\ddot{x} = \frac{a\dot{x} + u - \bar{F}_{fric} - \bar{F}_{load}}{b}. \tag{3.7}$$

Neglecting the Stribeck effect, the frictional force affecting the movement of the translator can be modelled as a combination of Coulomb and viscous friction. The mathematical model may be written as

$$\bar{F}_{fric} = [f_c + f_v |\dot{x}|] sgn(\dot{x}), \tag{3.8}$$

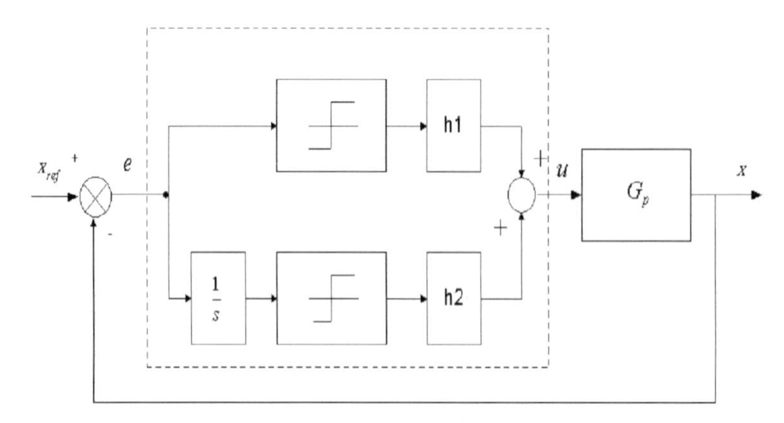

Fig. 3.4. Set-up of the two-channel relay tuning

where parameters f_c and f_v relate to the coefficients of Coulomb and viscous friction respectively.

For loading effects which are independent of the direction of motion, $\bar{F}_{load}$ can be described as

$$\bar{F}_{load} = f_l sgn(\dot{x}). \tag{3.9}$$

Cumulatively, the frictional and load forces can be described as one external disturbance F given by

$$F = [f_1 + f_2|\dot{x}|]sgn(\dot{x}), \tag{3.10}$$

where $f_1 = f_l + f_c$ and $f_2 = f_v$. Figure 3.5 graphically illustrates the characteristics of F. Figure 3.6 is a block diagram depicting the overall model of the servo-mechanical system. It is an objective in this section to estimate the key characteristics of F using a relay feedback experiment.

3.2.1 Friction Identification Method

Under the double channel relay feedback for servo-mechanical systems, the closed-loop arrangement depicted in Figure 3.7 may be posed equivalently in the configuration of Figure 3.8, consisting of a parallel relay construct acting on the linear portion of the servo-mechanical system. The second feedback relay ($FR2$ which is cascaded to an integrator) is necessary to excite oscillation at a finite frequency since the phase response of servo-mechanical systems rarely exceeds $-\pi$.

The parallel relay construct (henceforth called the equivalent relay ER) consists of feedback relays $FR1$ and $FR2$, as well as the inherent system relay SR due to frictional and load forces. The describing function (DF) approximation is thus directly applicable towards the analysis of the feedback system.

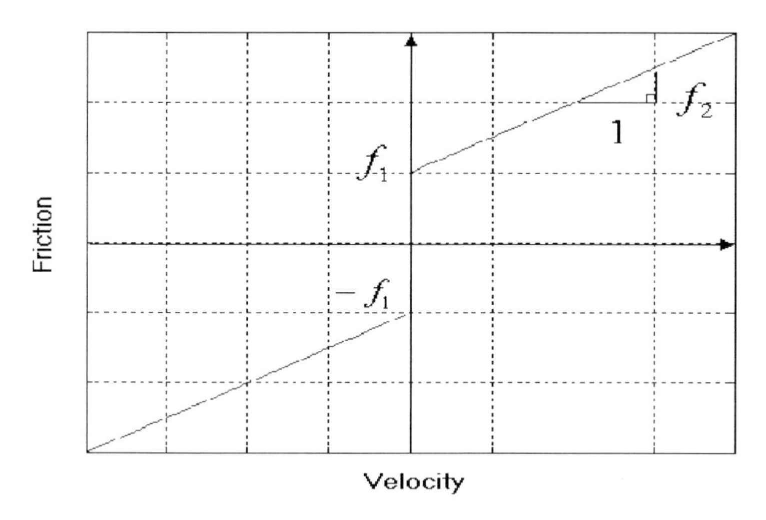

Fig. 3.5. F-$\dot{x}$ characteristics

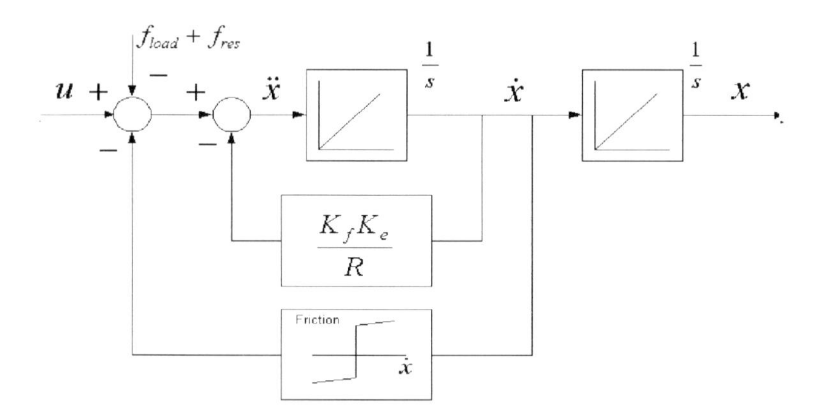

Fig. 3.6. Model of the servo-mechanical system under the influence of friction

The DF of the equivalent relay (N_{ER}) is simply the sum of the individual DFs due to the feedback relays (N_{FR1}), (N_{FR2}) and the inherent system relay (N_{SR}), *i.e.*,

$$N_{ER} = N_{FR1} + N_{FR2} + N_{SR}.$$

According to Gelb and Vander Velde (1968),

$$N_{FR1}(a) = \frac{4h_1}{\pi a},$$
$$N_{FR2}(a) = -j\frac{4h_2}{\pi a},$$

$$N_{SR}(a) = j\left(\frac{4f_1}{\pi a} + wf_2\right),$$
$$N_{ER}(a) = \frac{4h_1}{\pi a} + j\left(\frac{4(f_1 - h_2)}{\pi a} + wf_2\right).$$

For DF analysis, it is more convenient to work with the transfer function of the linear system. The transfer function of the linear system from u to x is assumed as

$$G_p(s) = \frac{K_p}{s(T_p s + 1)}, \tag{3.11}$$

where $K_p = 1/a$ and $T_p = b/a$. Under the relay feedback, the amplitude and oscillating frequency of the limit cycle is thus given approximately by the solution to

$$G_p(j\omega) = -\frac{1}{N_{ER}(a)}, \tag{3.12}$$

i.e., the intersection of the $G_p(j\omega)$ and the negative inverse DF of the equivalent relay.

The complex equation at Equation (3.12) will generate the following two real equations:

$$|G_p(j\omega)| = \left|\frac{1}{N_{ER}(a)}\right|,$$
$$argG_p(j\omega) + arg(N_{ER}(a)) = -\pi.$$

Clearly, two unknown parameters can be obtained from the solution of these equations.

The negative inverse DF of the equivalent relay is approximately a ray to the origin in the third quadrant of the complex plane, if $h_2 > f_1$ as shown in Figure 3.9. The angle at which this ray intersects the real axis depends on the relative relay amplitude of h_1 and h_2. In this way, a sustained limit cycle can be induced from servo-mechanical systems, similar to the more conventional single relay set-up for industrial processes.

Note that the choice of $h_1 = 0$ and $h_2 > f_1$ will lead to a double integrator phenomenon, where no sustained oscillation can be obtained from relay feedback.

By varying h_1 and/or h_2, two relay experiments can be conducted, thus deriving equations from which the unknowns T_p, f_1 and f_2 can be computed, assuming the gain K_p is known or estimated from other tests. It is straightforward to show that the parameters can be directly computed from the following equations:

$$T_p = \frac{4h_{1,1}K_p}{\pi a_1 \omega_1^2},$$
$$f_1 = \frac{w_2 a_2 h_{2,1} - w_1 a_1 h_{2,2}}{w_2 a_2 - w_1 a_1},$$

$$f_2 = -\frac{4}{\pi a_2 w_2}\left(\frac{h_{1,2}}{T_p\omega_2} + f_1 - h_{2,2}\right). \tag{3.13}$$

where ω_1, ω_2 are the sustained oscillating frequencies of the limit cycle oscillations from the relay experiments, a_1 and a_2 are the associated amplitudes of the limit cycles, $h_{1,1}$ and $h_{2,1}$ are the amplitudes used in the first experiment for the relays FR1 and FR2 respectively, and $h_{1,2}$ and $h_{2,2}$ are the corresponding relay amplitudes used in the second experiment.

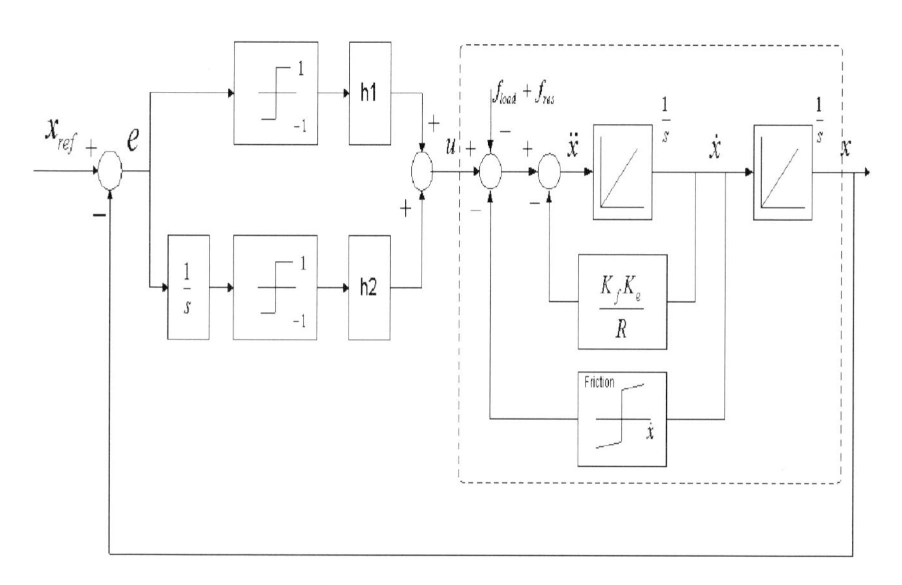

Fig. 3.7. Dual relay set-up

3.2.2 Simulation

To illustrate the accuracy of the estimates of f_1 and f_2 from the relay method, a simulation example is provided.

Consider the process:

$$G_p(s) = \frac{10}{s(0.2685s+1)}, \tag{3.14}$$

with $f_1 = 0.5, f_2 = 0.01$. In the first experiment, the relay parameters are chosen as $h_1 = 2$ and $h_2 = 1.5$. T_p is correctly identified as $T_p = 0.265$.

In the second experiment, the parameters are chosen as $h_1 = 1$ and $h_2 = 0.7$. f_1 and f_2 are correctly identified as $f_1 = 0.5104$ and $f_2 = 0.0065$. The limit cycle oscillations corresponding to the two experiments are shown in Figure 3.10 and 3.11.

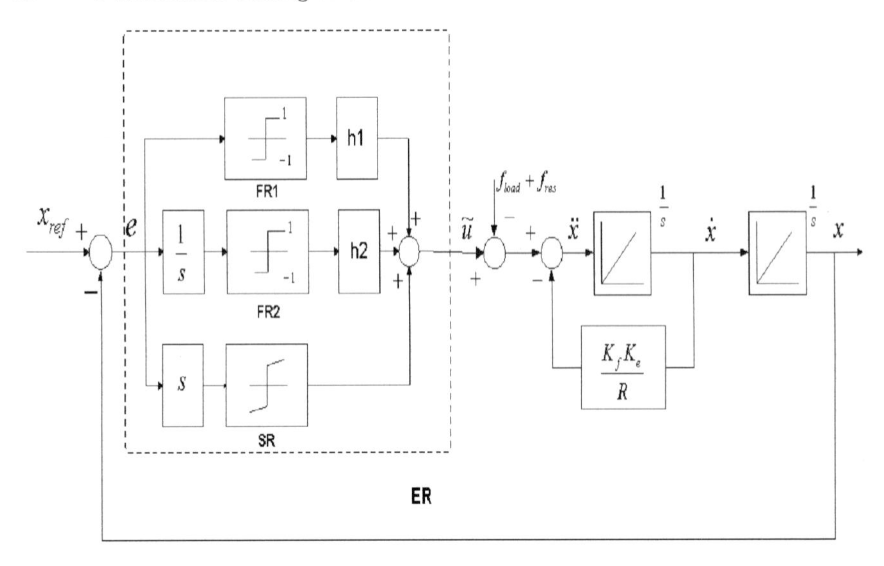

Fig. 3.8. Equivalent system

3.2.3 Initialisation of Adaptive Control

It should be noted that, while the parameter estimation is self-adapting in an adaptive controller (such as the scheme presented in Chapter 2), a good set of initial values provided by the relay experiments is important to ensure good initial transient behaviour and efficient convergence of the parameter estimates. The following simulation example will illustrate this point clearly.

The exact parameters used in the simulation are $a = -10.5$, $b = 0.1429$, $f_1 = 10$ and $f_2 = 10$. The adaptive controller is desired to track a pre-specified trajectory. Figure 3.12 shows the adaptive control performance with zero initial values, *i.e.*, $a = b = f_1 = f_2 = 0$. The convergence rate is slow and the tracking error is large. Figure 3.13 shows the performance when initial values of $a = -5$, $b = 0.05$, $f_1 = 6.9979$ and $f_2 = 6.9979$ are used. The tracking error is reduced and convergence rate is faster. Figure 3.14 shows the performance when good initial values are used with $a = -10$, $b = 0.1$, $f_1 = 9.7971$ and $f_2 = 9.7971$. Both the tracking error and convergence rate exhibit improved characteristics compared to the preceeding two cases.

3.3 Optimal Features Extraction from Relay Oscillations

In many relay feedback applications, it is required to measure the amplitude, frequency and also phase shift quantities from sampled noisy, but periodic, oscillations. Under the influence of measurement noise, it may be difficult to extract these parameters accurately. These parameters are used in the design of the controller, directly or indirectly. Thus, a reliable and accurate

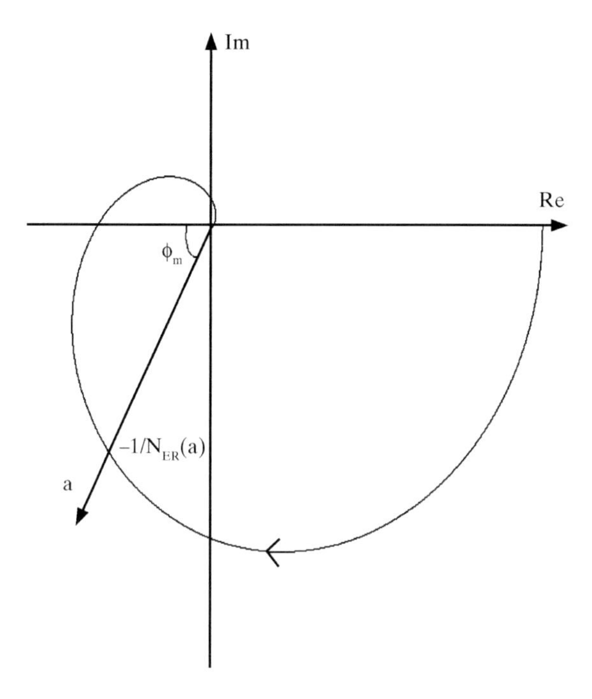

Fig. 3.9. Negative inverse describing function of the modified relay

identification of the key parameters associated with relay oscillations under the influence of noise is important. A non-linear least-squares (LS) method can be applied in a two-stage identification experiment.

Denote by $\{\bar{x}(t) \mid t = t_0, t_0 + T_s, \cdots, t_0 + (N_p - 1)T_s\}$ a data series of a sampled noisy sinusoidal signal where N_p is the total number of point, T_s the sampling period and t_0 is the initial time. The true signal is

$$x(t) = A\sin(\omega t + \theta). \tag{3.15}$$

The optimisation problem is to locate a parameter set which will minimise a performance index such as $J(A, \omega, \theta)$ given by

$$J(A, \omega, \theta) = \sum_{j=0}^{N_p-1} [\bar{x}(t_0 + jT_s) - x(t_0 + jT_s)]^2. \tag{3.16}$$

This is clearly a non-linear least-squares problem. As shown in what follows, the problem can be simplified to a two-stage linear LS identification problem.

Stage 1: Fixed ω

When ω is fixed, Equation (3.16) can be converted to a linear LS problem. Defining

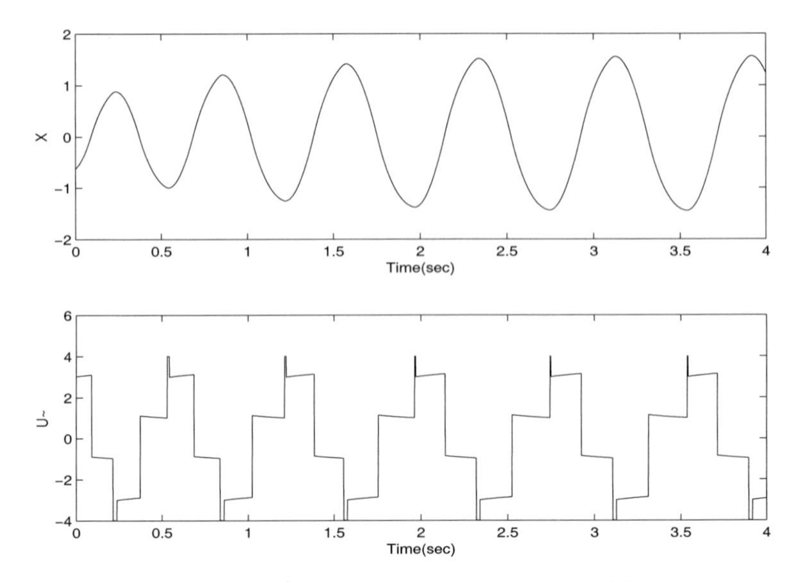

Fig. 3.10. Input/output signals with $h_1 = 2$ and $h_2 = 1.5$

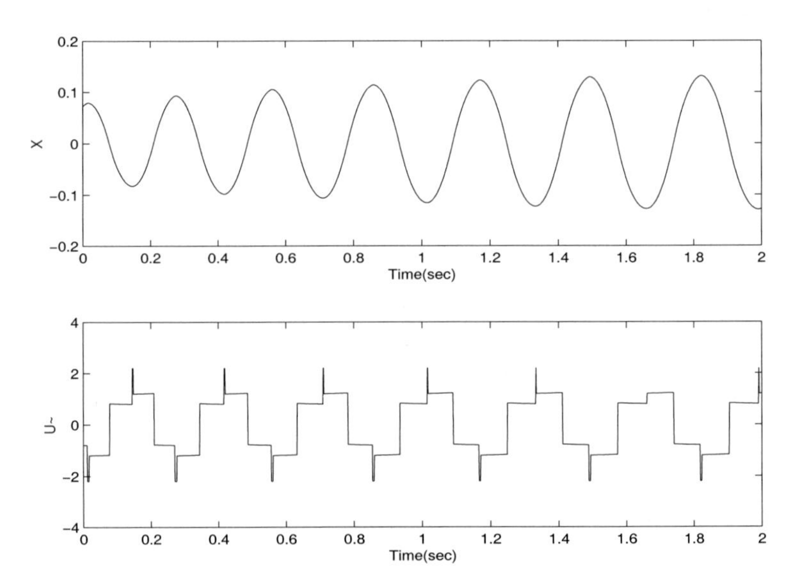

Fig. 3.11. Input/output signals with $h_1 = 1$ and $h_2 = 0.7$

$$\alpha_1 = A\sin(\theta), \quad ,\alpha_2 = A\cos(\theta), \tag{3.17}$$

for a given ω, the optimisation problem is to locate A and θ so that $J_\omega(A, \theta)$ is minimised, where

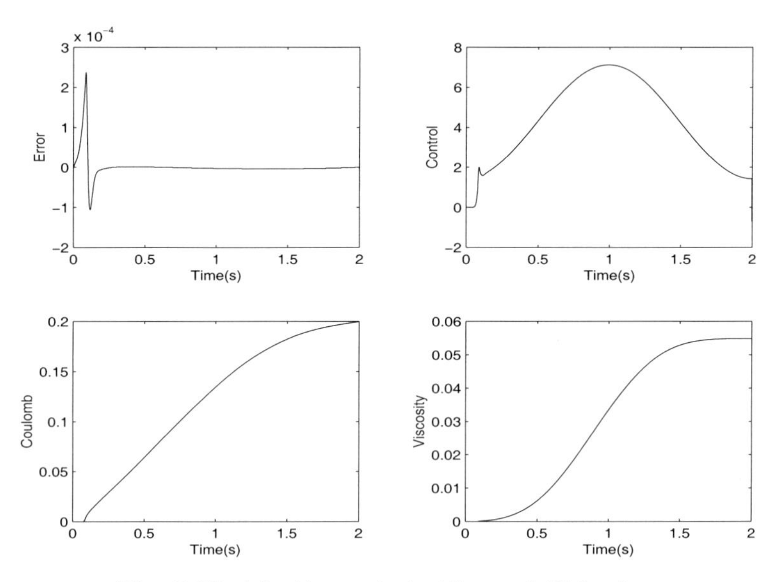

Fig. 3.12. Adaptive control with zero initial values

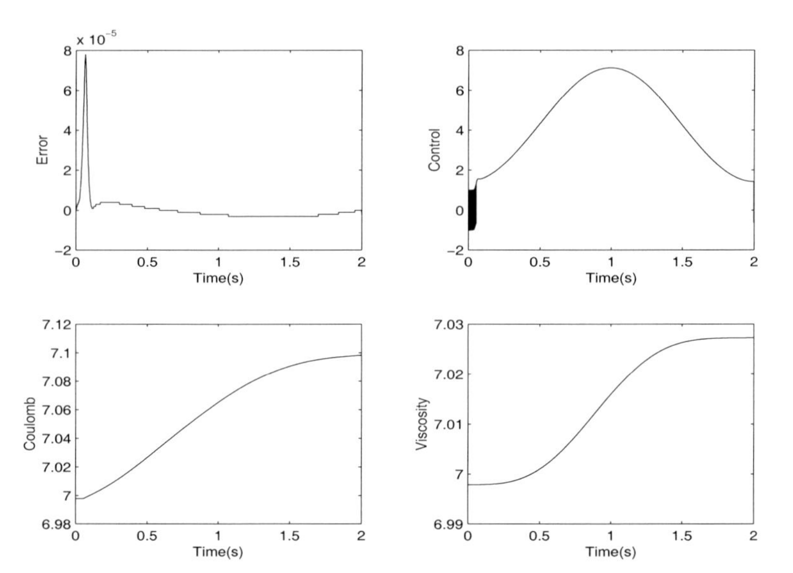

Fig. 3.13. Adaptive control with initial values: $a = -5$, $b = 0.05$, $f_1 = 6.9979$ and $f_2 = 6.9979$

$$J_\omega(A, \theta) = \sum_{j=0}^{N_p-1} [\bar{x}(t_0 + jT_s) - \alpha_1 \sin(\omega(t_0 + jT_s)) - \alpha_2 \cos(\omega(t_0 + jT_s))]^2. \tag{3.18}$$

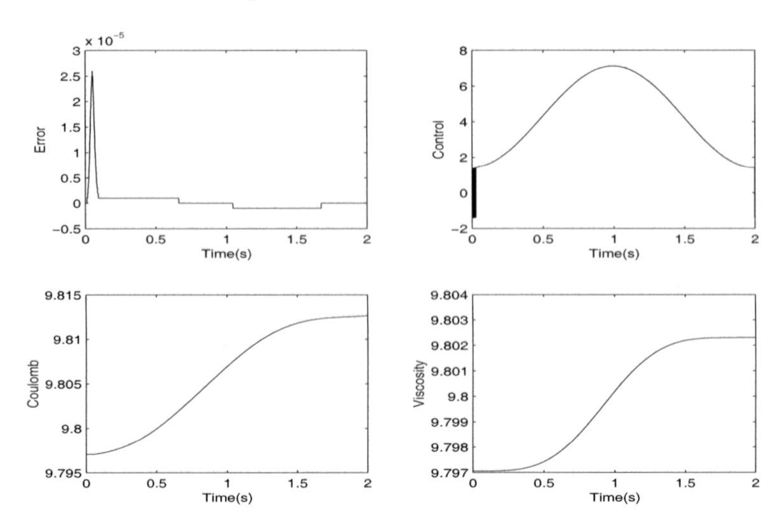

Fig. 3.14. Adaptive control with initial values: $a = -10$, $b = 0.1$, $f_1 = 9.7971$ and $f_2 = 9.7971$

This is clearly a linear LS problem which can be directly solved.

Stage 2: Varying ω

The parameter optimisation process can be repeated for a range of frequency ω in the neighbourhood of the estimated value. It can be defined that

$$\min_{A,\omega,\theta} J(A,\omega,\theta) = \min_{\omega}\{\min_{A,\theta} J_\omega(A,\theta)\}. \tag{3.19}$$

In this way, the complete optimal parameter set (A, ω, θ) can be obtained.

Figure 3.15 shows the extraction of a sinusoidal profile from the noisy oscillation signal of a relay feedback experiment.

3.4 Experiments

In this section, experimental results are provided to illustrate the effectiveness of the relay method. The experimental set-up is similar to that presented in Section 2.2.9.

Two relay experiments are conducted according to the procedures described in Section 3.2. T_p is identified as $T_p = 0.073$. The friction parameters are identified as $f_1 = 0.238$ and $f_2 = 0.001$. The limit cycle oscillations arising from the two experiments are shown in Figure 3.16 and Figure 3.17.

With the model parameters, a PID feedback controller and a feedforward friction compensator can be properly initialised. The overall control system is shown in Figure 3.18. Since the mechanical structure and other components

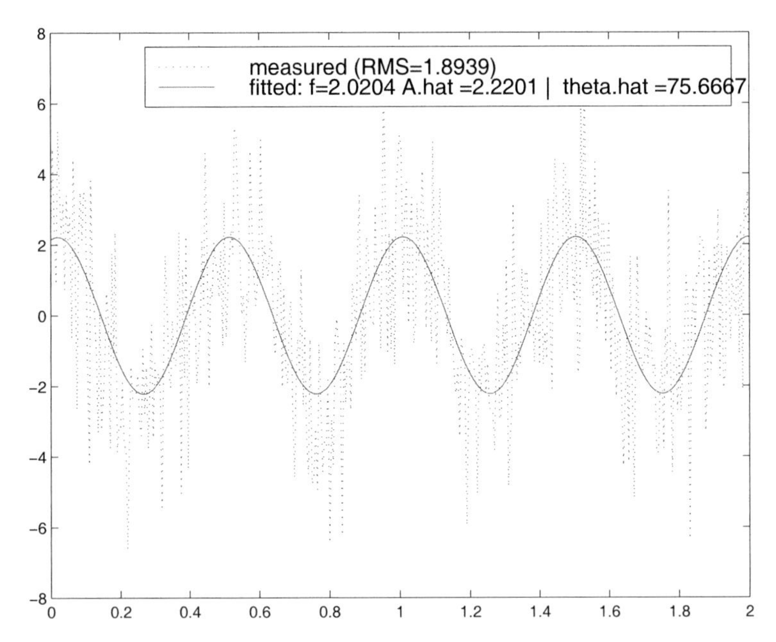

Fig. 3.15. Feature extraction from a noisy sinusoidal signal (*solid*—extracted sinusoid, *dotted*—actual sinusoid)

in the system have inherent and unmodelled high-frequency dynamics which should not be excited, small adaptation gains are used.

Figure 3.19 and Figure 3.20 show the tracking performance to a reference sinusoidal profile with and without the feedforward friction compensator. Clearly, with the friction compensator, the root-mean-square (RMS) value of the tracking error can be drastically reduced from 11.2 μm to around 1.01 μm.

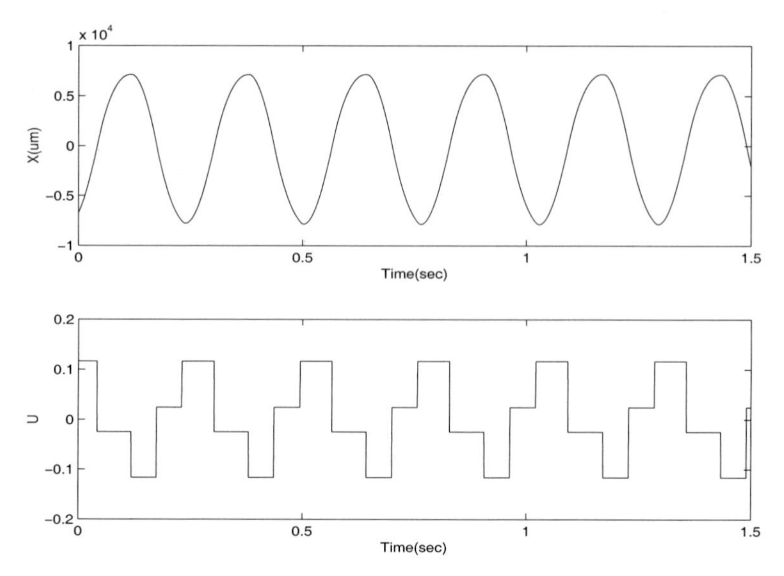

Fig. 3.16. Input-output signals under the first relay experiment

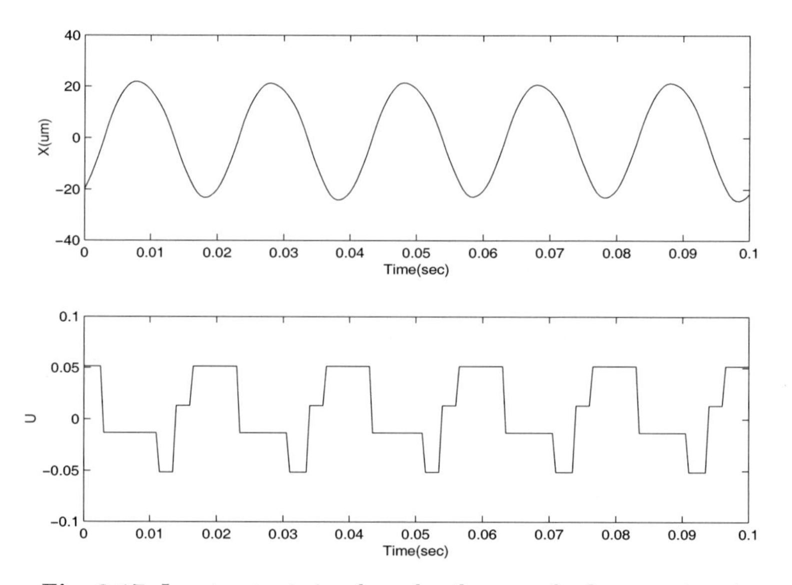

Fig. 3.17. Input-output signals under the second relay experiment

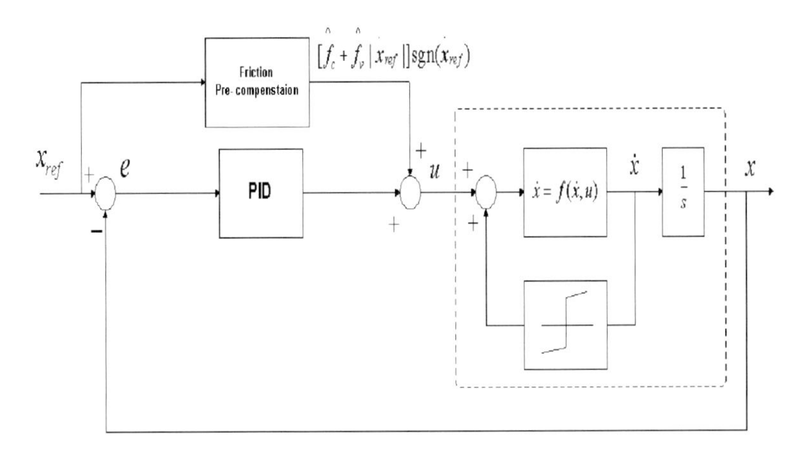

Fig. 3.18. PID with friction pre-compensator

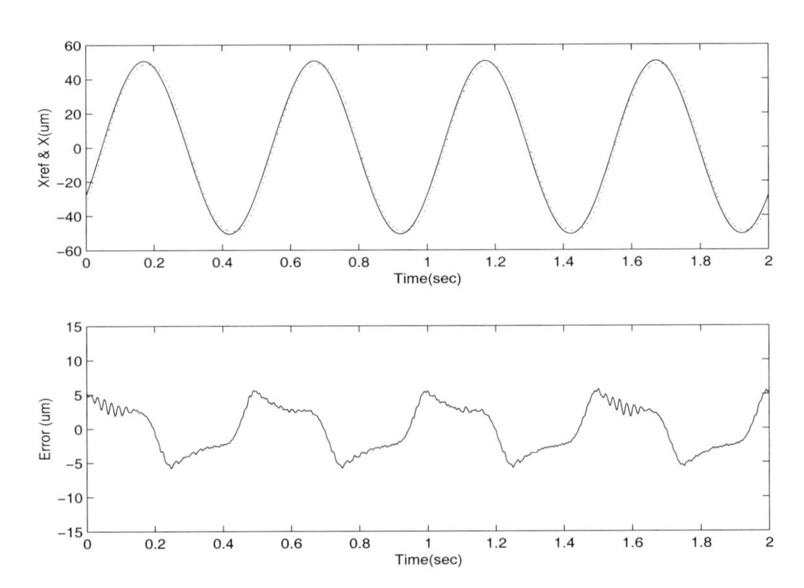

Fig. 3.19. Tracking performance without friction compensation

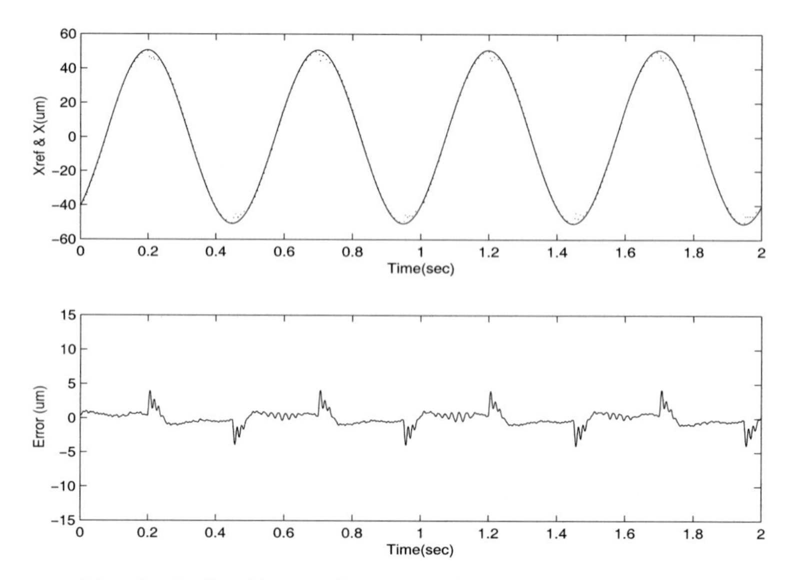

Fig. 3.20. Tracking performance with friction compensation

4

Co-ordinated Motion Control of Gantry Systems

Among the various configurations of long travel and high precision Cartesian robotic systems, one of the most popular is the *H*-type which is more commonly known as the moving gantry system. In this configuration, two motors which are mounted on two parallel slides move a gantry simultaneously in tandem. An example of this stage is shown in Figure 4.1. This gantry system consists of four sub-assemblies, *viz.*, the X and Y-axis sub-assemblies, the planar platform, and two orthogonal guide bars. When positioning precision is of the primary concern, direct drive linear motors are usually used and fitted with aerostatic bearings for optimum performance. Another setup of H-type gantry stage is shown in Figure 4.2. It consists of two X-axis servo motors: SEMs MT22G2-10 and a Y-axis servo motor: Yaskawa's SGML-01AF12. This gantry configuration has been in use for large overhead travelling cranes in ports, rolling mills and flying shear.

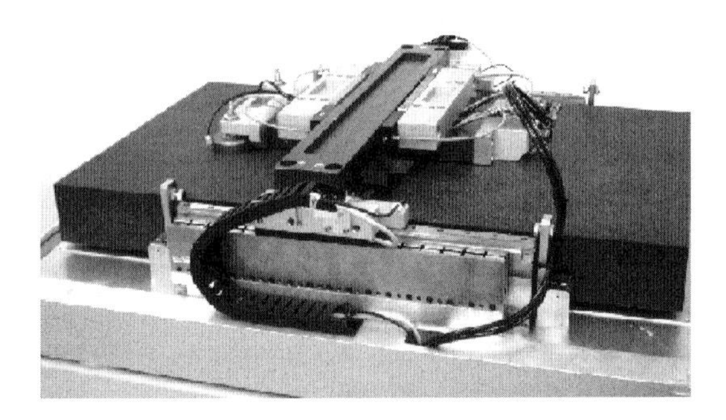

Fig. 4.1. Example of a precision gantry stage

The moving gantry stage is usually designed to provide a high-speed, high-accuracy X, Y and Z motion to facilitate automated processes in flat panel displays, printed circuit board manufacturing, precision metrology, and circuit assembly where high part placement accuracy for overhead access is necessary. The stage is equipped with a high power density due to the dual drives, and it can yield high speed motion with no significant lateral offset when the two drives are appropriately co-ordinated and synchronised in motion. In certain applications, such as in wafer steppers, the dual drives can also be used to produce a small "theta" rotary motion, without any additional rotary actuators. The application domain of the moving gantry stage is rapidly expanding along with the increasingly stringent requirements arising from developments in precision engineering and nanotechnology. To date, it is difficult to find one precision machine manufacturer who does not provide for a moving gantry stage in one form or another.

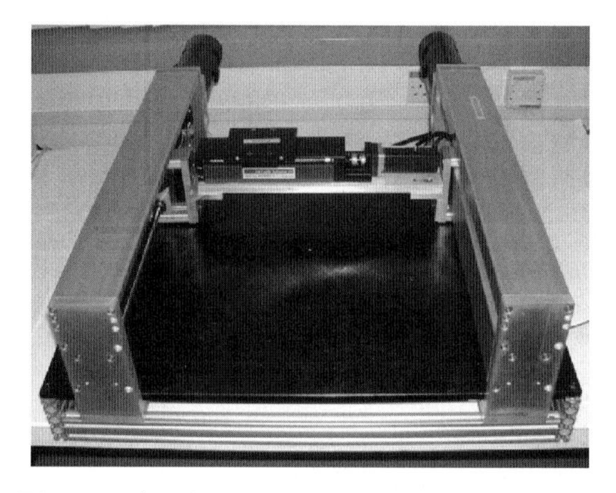

Fig. 4.2. Another structurally-similar gantry stage

The main challenges to address in order to harness the full potential of this configuration of Cartesian stages are mainly in the control system. In addition to precision motion control of the individual motors, efficient synchronisation among them is crucially important to minimise the positional offsets which may arise due to different drive and motor characteristics, non-uniform load distribution of the gantry and attached end-effectors, and possibly time-varying thermo-mechanical properties. In particular, disturbances in the form of dynamical load changes which can be fairly asymmetrical in nature, have to be adequately addressed. This chapter is devoted to address the precision motion control aspects of a moving gantry robotic system, since these are fast becoming important enabling technologies to facilitate the fulfillment of high accuracy processes.

In this chapter, current control approaches in commercial motion control platforms are first surveyed and presented. The performance of these schemes and their limitations will be investigated and compared through simulation and experiments. The first scheme assigns one motor as the master and the other as the slave. The command signal is only transmitted to the master, and the slave simply tracks the motion of the master. The second scheme addresses explicitly the possibly different dynamics of the two motors. An independent control loop under a common command signal is set up for each motor, in a configuration similar to supervisory control. However, similar to the master/slave scheme, there is still no feedback of inter-axis motion offset for a truly co-ordinated control. The third scheme designs a disturbance observer-augmented composite controller to overcome the deficiencies of the aforementioned ones. The main function of the disturbance observer acts as a soft sensor to derive the equivalent disturbance signals based only on prevailing input and output signals, and available system models. The disturbances can then be efficiently and automatically corrected for using the observed disturbance signals.

A recent scheme employing an adaptive control algorithm based on a dynamic Lagrangian model of the stage is given in the chapter to overcome the deficiencies of the aforementioned ones. The methodology employs an adaptive control algorithm based on a dynamic Lagrangian model of the stage. The model is detailed enough to address the main concerns and yet generic enough to cover various types of H gantry stage. Furthermore, only two basic parameters (the length and width of the stage) regarding the stage need to be measured. Both simulation and experimental results are duly provided in this chapter to illustrate the relative performance of the control schemes and their competitive advantages.

4.1 Co-ordinated Control Schemes

Three advanced control schemes suitable for the control of moving gantry stages will be duly described in the following sub-sections. The first two schemes are commonly available in existing industrial motion controllers.

4.1.1 Classical Master/Slave Approach

In the so-called "*classical master/slave*" approach, one motor is chosen as the master motor of the pair. This master motor directly executes the desired trajectories. The encoder of the master motor is also used as the master encoder for the parallel slave motor. Thus, in essence, the slave motor simply follow the motion of the master motor. Figure 4.3 depicts a block diagram of the master/slave control scheme.

This method is relatively very simple and it works just as if a single motor were driving the motion. It is also an effective method to use during jogging and homing moves – the master motor is commanded to jog or home, and the slave motor simply follows along. However, the performance achievable in terms of trajectory tracking can be rather limited, due to the fact that the actual trajectory of the master motor acts as the commanded trajectory of the

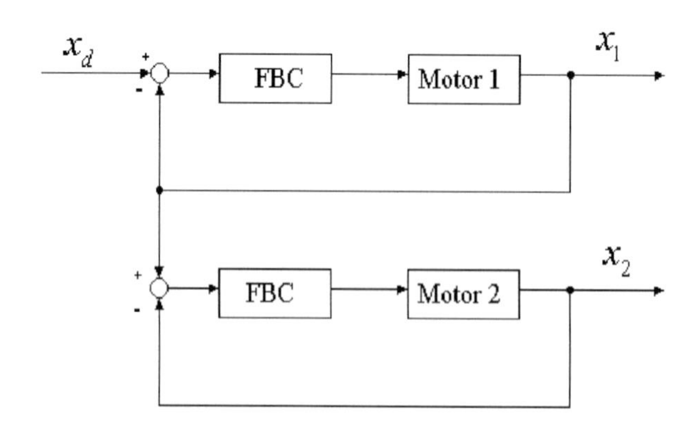

Fig. 4.3. Block diagram of the master/slave control scheme

slave motor. Clearly, the actual trajectory of the master motor can never be as smooth and accurate as the commanded trajectory and if this now becomes the commanded trajectory of the slave motor, the actual trajectory of the slave motor will deviate even more from the desired one. This is particularly true for gantry systems, where both motors are expected to have similar resonant frequencies. Oscillations in the master motor will inevitably be fed into the slave motor and consequently be amplified significantly. In addition, when the slave motor encounters a disturbance, the master will not be able to know and appropriately address it.

4.1.2 Set-point Co-ordinated Control

In the "*set-point co-ordinated control*" method, the two motors are assigned to the same axis in the same co-ordinate system. A supervisory motion program drives the axis through the two motors which share an identical commanded trajectory pre-planned for this axis. Each servo loop of the two motors then has the responsibility of keeping the actual trajectory as closely as possible to the commanded trajectory, since each of them has its own individual servo loop. Presuming all motors have tight servo loops, this method provides a tight and smooth link between the motors for the gantry. Figure 4.4 provides a block diagram of the set-point co-ordinated control scheme.

Since the deficiencies of one motor in following the commanded trajectory do not directly affect the other motor, this scheme is in general superior to using a master/slave technique. The main problem with this form of control is due to differences in the dynamics of the two motors or in their load, in which case there may be a rather considerable difference in the positions of the motors.

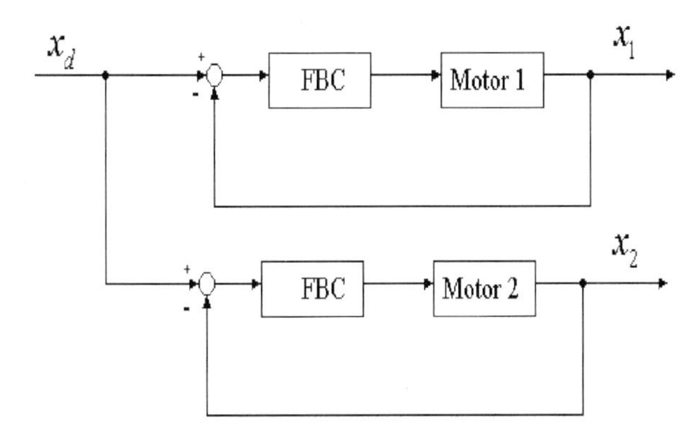

Fig. 4.4. Block diagram of the set-point co-ordinated control scheme

4.1.3 Fully Co-ordinated Control

The overall control structure is shown in Figure 4.5.

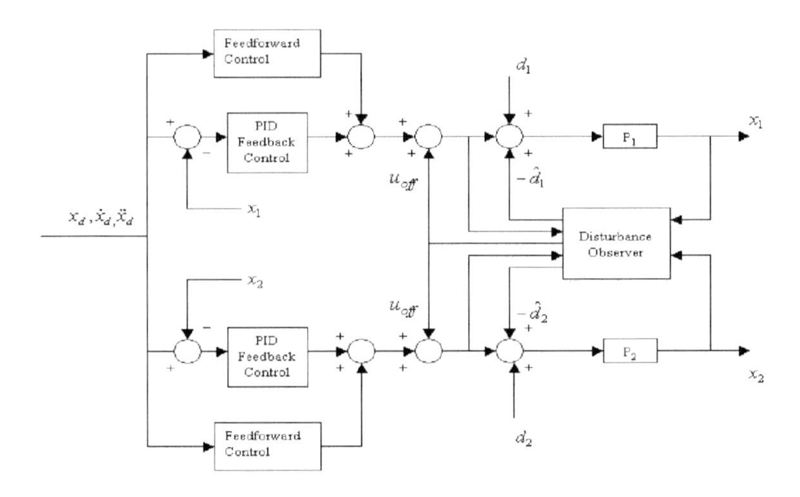

Fig. 4.5. Overall structure of the control system

The designs of the feedforward and feedback control components are similar to those described in Section 2.2.5. As the achievable performance of a precision gantry system is unavoidably and very signficantly limited by the amount of disturbances present, and the uniformity of their distribution among the motors, a disturbance observer is augmented to the composite controller.

Figure 4.6 shows the block diagram of the "Disturbance Observer" part of the control system which uses an estimate of the actual disturbance, deduced from a disturbance observer, to compensate for the disturbances. x_i, u_i, d_i and $\hat{d}_i$ denote the position signal, control signal, actual and observed disturbance signal associated with the i-axis. u_{off} represents a constituent control signal to correct for the inter-axis offset. This can be the output of a P or PI controller. Where the measurement noise is tolerable, an additional derivative action, possibly in conjunction with a filter, may also be installed to provide a derivative offset correction. The disturbance observer estimates the disturbance based on the output x_i and the control signal u_i. P_i denotes the actual system. $P_{n,i}$ denotes the nominal system. The design of the disturbance is similar to that described in Section 2.6. This design possesses several important and useful features. First, it incorporates a feedforward component to facilitate a high speed response. The feedforward component addresses model-based characteristics relating to the servo motors, including specific friction, ripples and possibly other torque-impeding characteristics. Second, an optimal PID feedback controller is designed based on the LQR (Linear Quadratic Regulator) approach. This PID feedback control is intended to provide optimal command response and stability properties. Where additional state variable measurements are available, full-state feedback controllers may be designed along similar design rules to achieve additional enhancement. Third, a disturbance observer is augmented to the composite control structure to provide a fast response to load disturbances and other exogeneous signals acting asymmetrically on the two motors. This feature is especially useful since load disturbances are major factors affecting the control performance, especially when the motors jointly carry a dynamical and asymmetrical load such as an additional servo system running across the gantry.

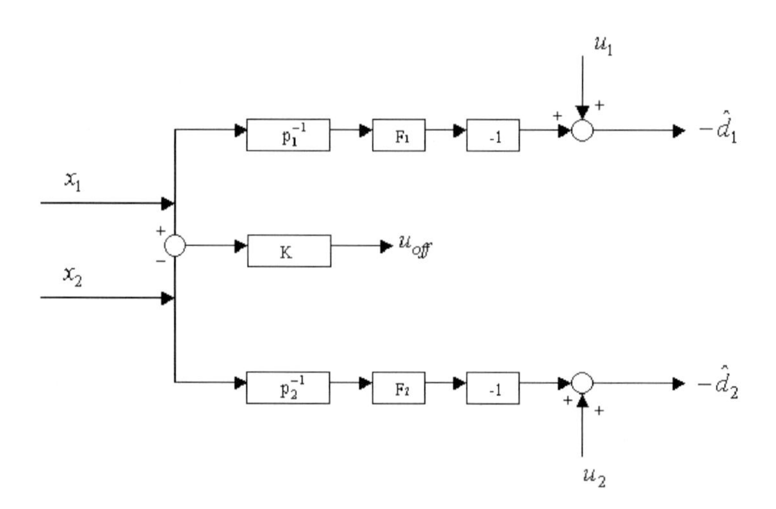

Fig. 4.6. Block diagram of the disturbance observer

4.2 Simulation Study

A simulation study using all three control schemes presented is conducted on a moving gantry stage based on permanent magnet linear motors (PMLM). In the simulation, the inevitably different dynamical properties of the two motors are reflected by using different model parameters. In addition, dynamical load changes are simulated according to Figure 4.7.

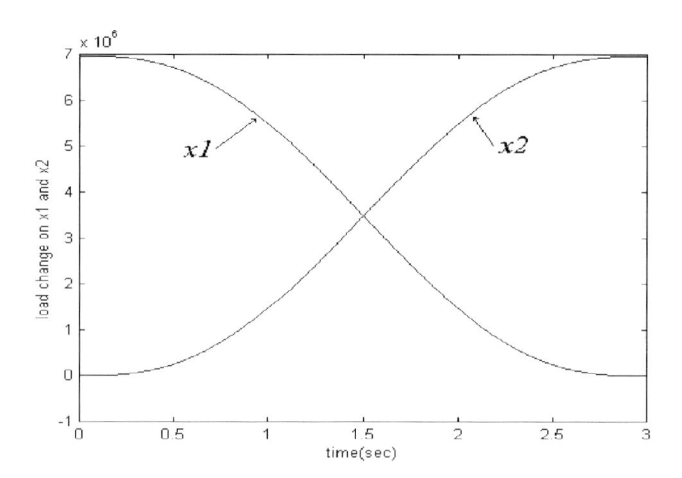

Fig. 4.7. Dynamic load changes

4.2.1 Control Task

The desired trajectory of the gantry for the simulation study is chosen to be a fifth-order polynomial as shown in Figure 4.8. The control performance will be evaluated with respect to the tracking errors of the individual servo loops and the inter-loop motion offset.

4.2.2 Results

Case-1. Master/Slave Control

Two linear motors, x1 and x2, are used as the master and slave motors respectively. The actual position of x1 is used as the motion command signal for the servo loop of x2. Both servo loops use PID controllers which are fine-tuned to achieve optimum performance. The tracking errors for the two loops are shown in Figure 4.9. A maximum tracking error of about 7 μm is registered. The positional offset between the two axes throughout the trajectory is around 7 μm, reflecting the poor inter-axis coordination in this control scheme.

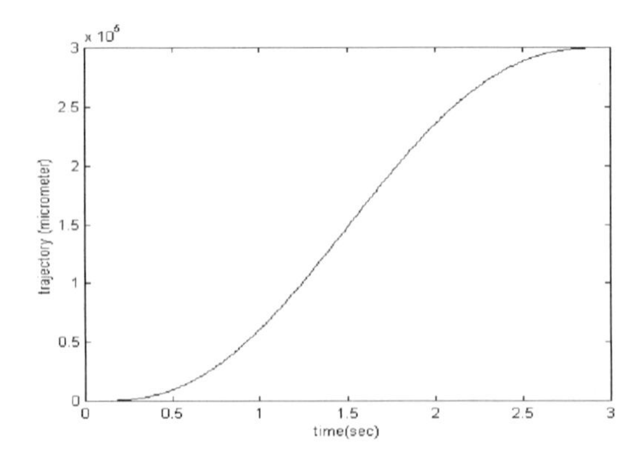

Fig. 4.8. The desired trajectory

Case-2. Set-point Co-ordinated Control
Under the set-point co-ordinated control scheme, each of the two axes has its own feedback control loop, but there is no feedback of the inter-axis positional offset to either controller. The same PID controller in Case-1 is used in this case. The results are shown in Figure 4.10. A maximum tracking error of 6.2 μm is registered for the individual control loop, with a position inter-axis offset of 3.2 μm.

Case-3. Fully Co-ordinated Control
In this case, apart from the individual control loop, the inter-axis positional offset is utilised to construct an additional control input for co-ordination purposes. The results are shown in Figure 4.11. A maximum tracking error of 3.2 μm is registered for the individual control loop, with a position inter-axis offset of less than 2.7 μm. The average offset is lower than the case of set-point co-ordinated control.

From the simulation study, the fully co-ordinated control clearly exhibits a better performance with respect to both tracking errors for each axis and the inter-axis positional offset.

4.3 Experiments

Real-time experiments are carried out on two configurations of X-Y table. Both set-ups use a 2.5 μm resolution digital encoders installed on the x and y motors. The specifications for the motors are listed in Table 4.3. One table is configured to a moving gantry type, with two motors driving the load along the x direction. Figure 4.12 shows a photograph of the table.

The other table is a more conventional one with one motor each along the x and y direction, the photograph of which is as shown in Figure 4.13.

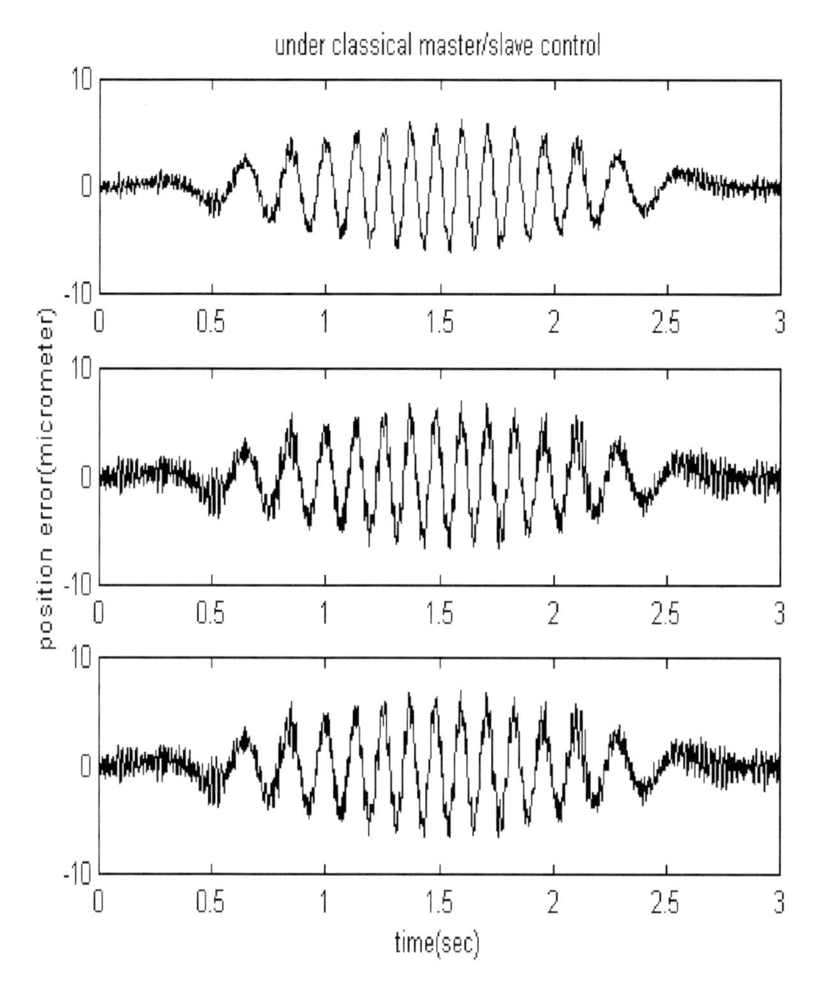

Fig. 4.9. Tracking error for x1-axis under classical master/slave control(*top*); Tracking error for x2-axis under classical master/slave control(*middle*); Positional inter-axis offset under classical master/slave control(*bottom*)

For this second configuration, the control task in the experiment is to execute an XY diagonal as straightly and precisely as possible. Such requirements on precise diagonal motion are essential for the calibration of machine geometrical properties. Clearly, in this application, a tight co-ordination between the X and Y motors is as important as the requirement for the moving gantry stage.

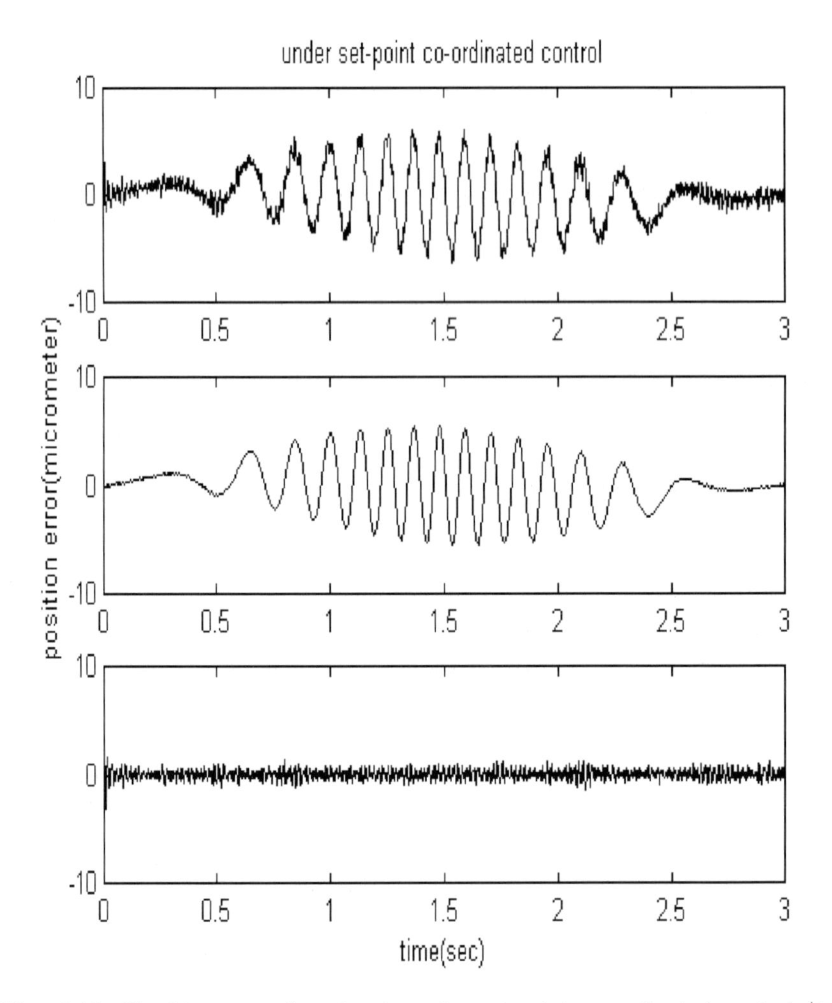

Fig. 4.10. Tracking error for x1-axis under set-point co-ordinated control (*top*); tracking error for x2-axis under set-point co-ordinated control (*middle*); positional inter-axis offset under set-point co-ordinated control (*bottom*)

4.3.1 XY Table–Configuration I

Case-1. Master/Slave Control

Experimental results are shown in Figure 4.14. A maximum tracking error of 15 μm is registered for the individual control loop with a position inter-axis offset of as much as 16 μm. In high precision applications, this offset may pose problems both in the mechanical alignment as well as in the achievements of the desired motion precision.

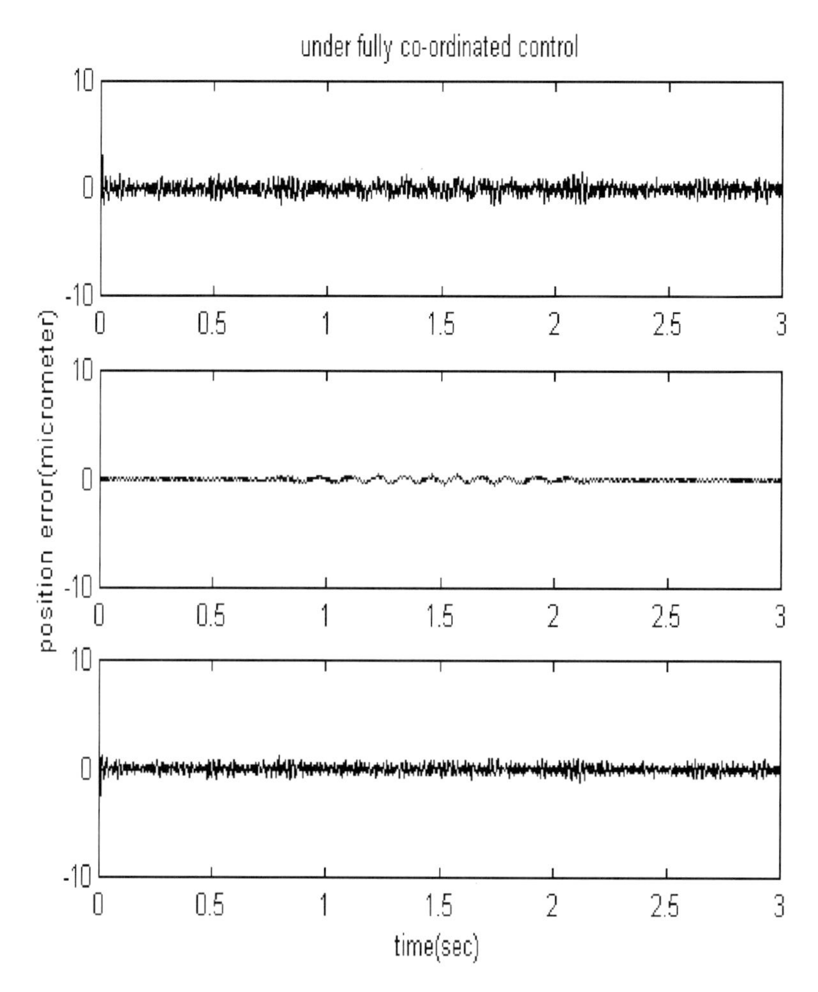

Fig. 4.11. Tracking error for x1-axis under fully co-ordinated control(*top*); Tracking error for x2-axis under fully co-ordinated control(*middle*); Positional inter-axis offset under fully co-ordinated control(*bottom*)

Case-2. Set-point Co-ordinated Control
Experimental results are shown in Figure 4.15. A maximum tracking error of 13 μm is registered for the individual control loop. The inter-axis position offset between the two axes x1 and x2 reachs 8 μm, which is better than that under the master/slave control.

Case-3. Fully Co-ordinated Control
Experimental results are shown in Figure 4.16. A maximum tracking error of 12 μm is registered for the individual control loop. The maximum position

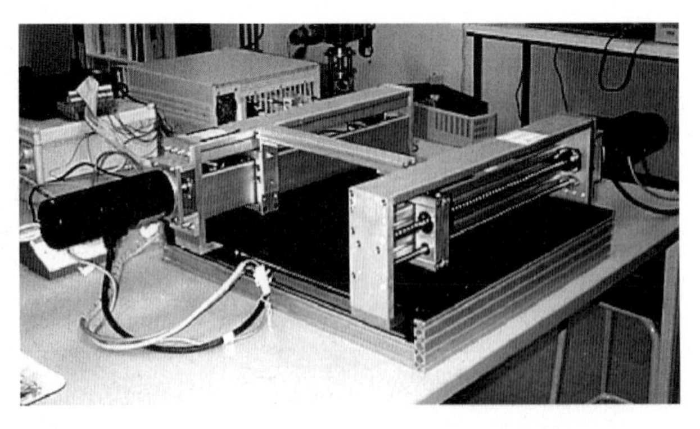

Fig. 4.12. XY Table-Configuration I

Fig. 4.13. XY Table-Configuration II

Table 4.1. Parameters of MT22G2 DC servomotor

Contents	Units	MT22G2-10
Torque constant(K_f)	N/Amp	0.10
Armature resistance less brushes (R)	ohms	0.63
Voltage constant EMF(K_e)	volt/rad/sec	0.10
Armature inductance(L)	mh	2.1
Max. velocity	RPM	5000

inter-axis offset between the two axes x1 and x2 is only 5 μm, which is the best results among the three control schemes.

4.3.2 XY Table-Configuration II

Case-1. Master/Slave Control
Experimental results are shown in Figure 4.17. A maximum tracking error of 12 μm is registered for the individual control loop.

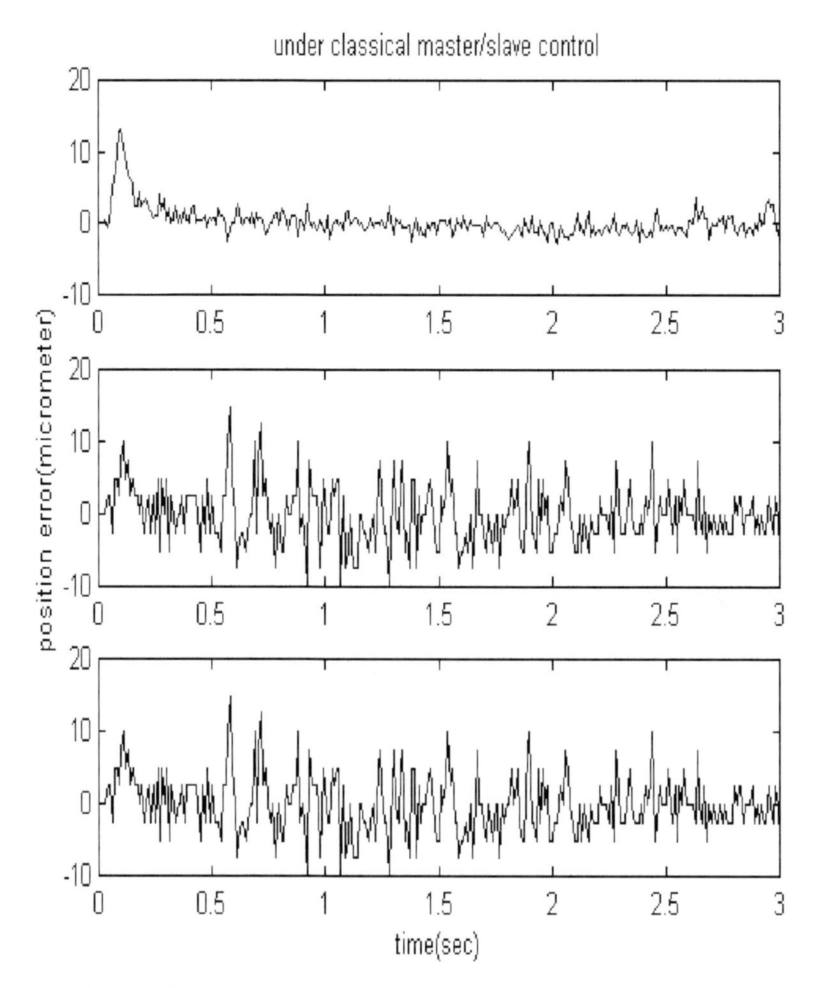

Fig. 4.14. Tracking error for the x1-axis under classical Master/Slave control (*top*); tracking error for the x2-axis under classical Master/Slave control (*middle*); positional inter-axis offset under classical Master/Slave control (*bottom*)

Case-2. Set-point Co-ordinated Control
Experimental results are shown in Figure 4.18. A maximum tracking error of 8 μm is registered for the individual control loop.

Case-3. Fully Co-ordinated Control
Experimental results are shown in Figure 4.19. A maximum tracking error of 6 μm is registered for the individual control loop. Figure 4.20 shows a comparison of the inter-axis offset arising from the use of each of the three

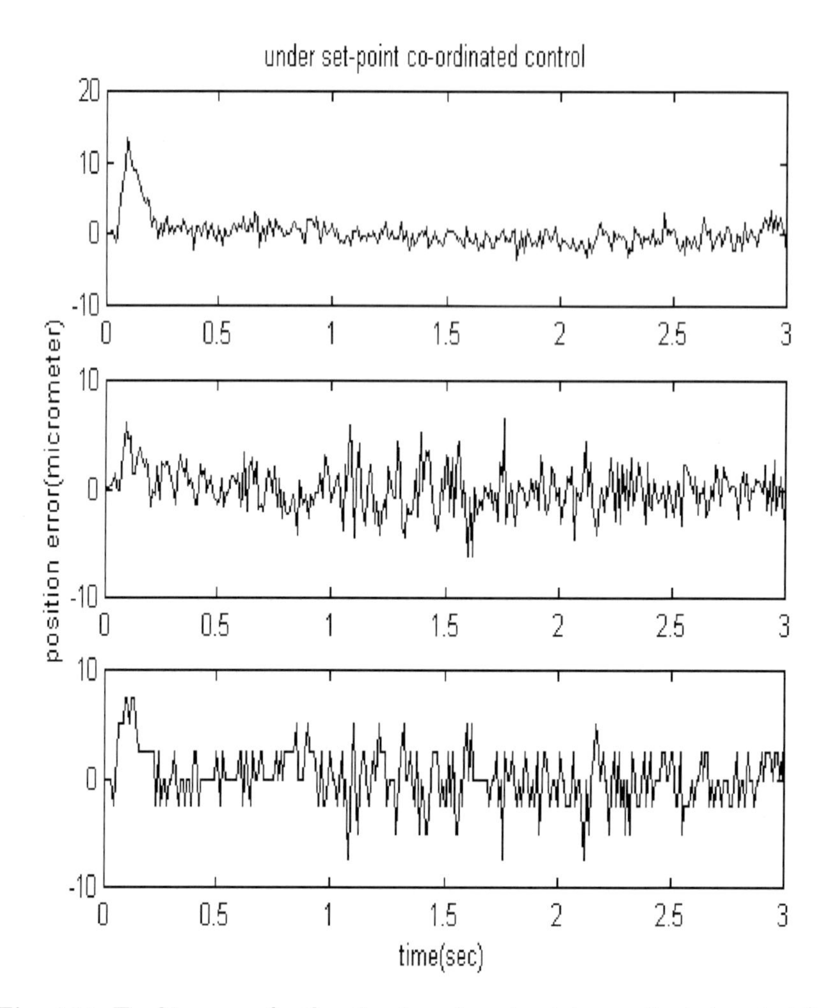

Fig. 4.15. Tracking error for the x1-axis under set-point co-ordinated control (*top*); tracking error for the x2-axis under set-point co-ordinated control (*middle*); positional inter-axis offset under set-point co-ordinated control (*bottom*)

control schemes. Clearly, a significant improvement in both tracking errors of individual loops and the inter-axis positional offset is achieved in the fully co-ordinated control.

4.4 Adaptive Co-ordinated Control Scheme

In the preceding sections, control methods presented are essentially non-parametric schemes, or partially parametric ones based on linear dominant

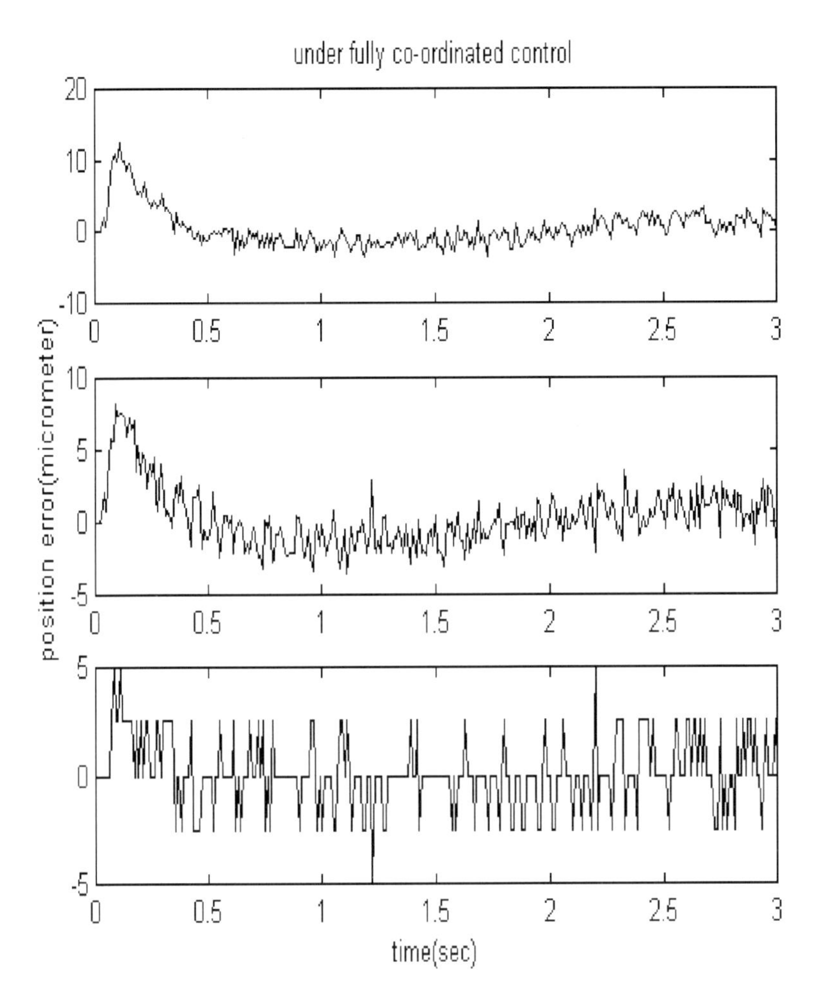

Fig. 4.16. Tracking error for the x1-axis under fully co-ordinated control (*top*); tracking error for the x2-axis under fully co-ordinated control (*middle*); positional inter-axis offset under fully co-ordinated control (*bottom*)

linear models without explicitly modeling the cross-axis effects. In more demanding applications, coupling and disturbances along the X and Y direction and load change which can be fairly asymmetrical in nature, may have to be adequately addressed. In this section, an adaptive control scheme is designed based on a physical model, which is able to adaptively estimate the model parameters without much *a priori* information assumed.

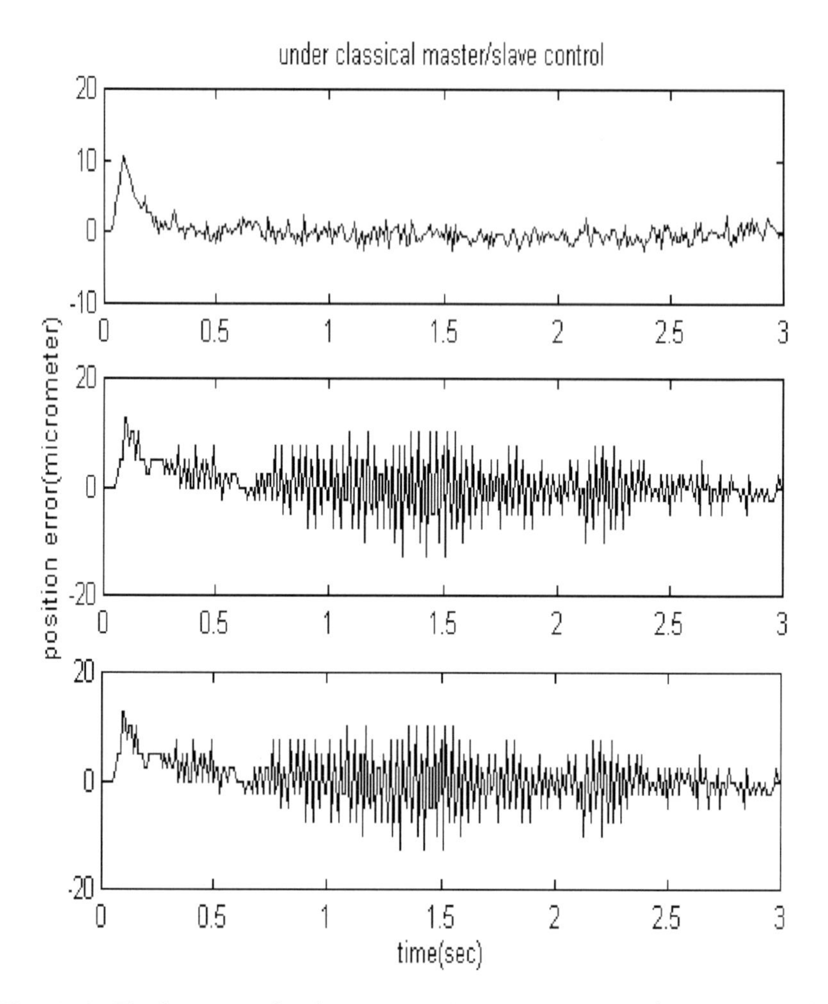

Fig. 4.17. Tracking error for the x-axis under classical Master/Slave control (*top*); tracking error for the y-axis under classical Master/Slave control (*middle*); positional inter-axis offset under classical Master/Slave control (*bottom*)

4.4.1 Dynamic Modelling of Gantry Stage

Although there are various configurations of H-type gantry stages, many of them are intrinsically similar. A typical gantry stage may be considered as a three-degree-of-freedom servo-mechanism, which can be adequately described by the schematics in Figure 4.21. Two servomotors carry a gantry on which a slider holding the load (*e.g.*, the tool) is mounted. One motor yields a linear displacement x_1 (measured from origin O), while the other yields a linear displacement x_2. Ideally $x_1 = x_2$, but they may differ in practice owing to

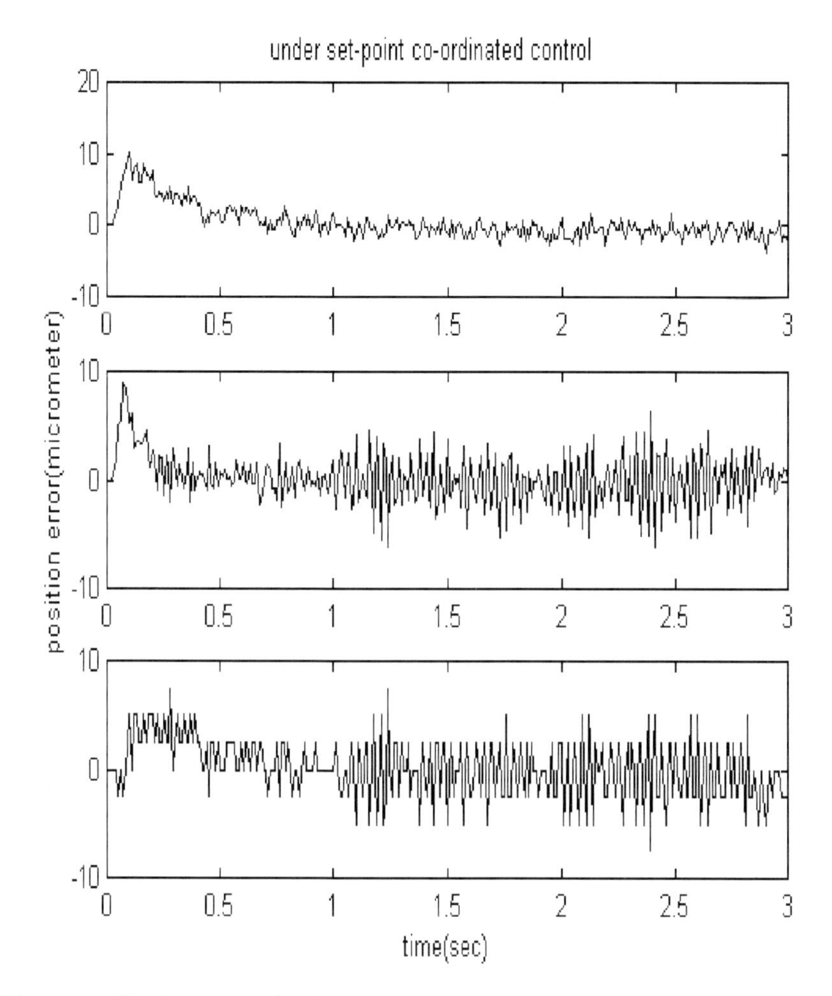

Fig. 4.18. Tracking error for the x1-axis under set-point co-ordinated control (*top*); tracking error for the y-axis under set-point co-ordinated control (*middle*); positional inter-axis offset under set-point co-ordinated control (*bottom*)

different dynamics exhibited by each of the motors, and also the dynamic loading present due to the translation of the slider along the gantry. The central point C of the gantry is thus constrained to move along the dashed line with two degrees of freedom. The displacement of this central point C from the origin O is denoted by x. The gantry may also rotate about an axis perpendicular to the plane of Figure 4.21 due to the deviation between x_1 and x_2, and this rotational angle is denoted by θ. The slider motion relative to the gantry is represented by y. It is also assumed that the gantry is symmetric and the distance from C to the slider mass center S is denoted by $d = w+v$.

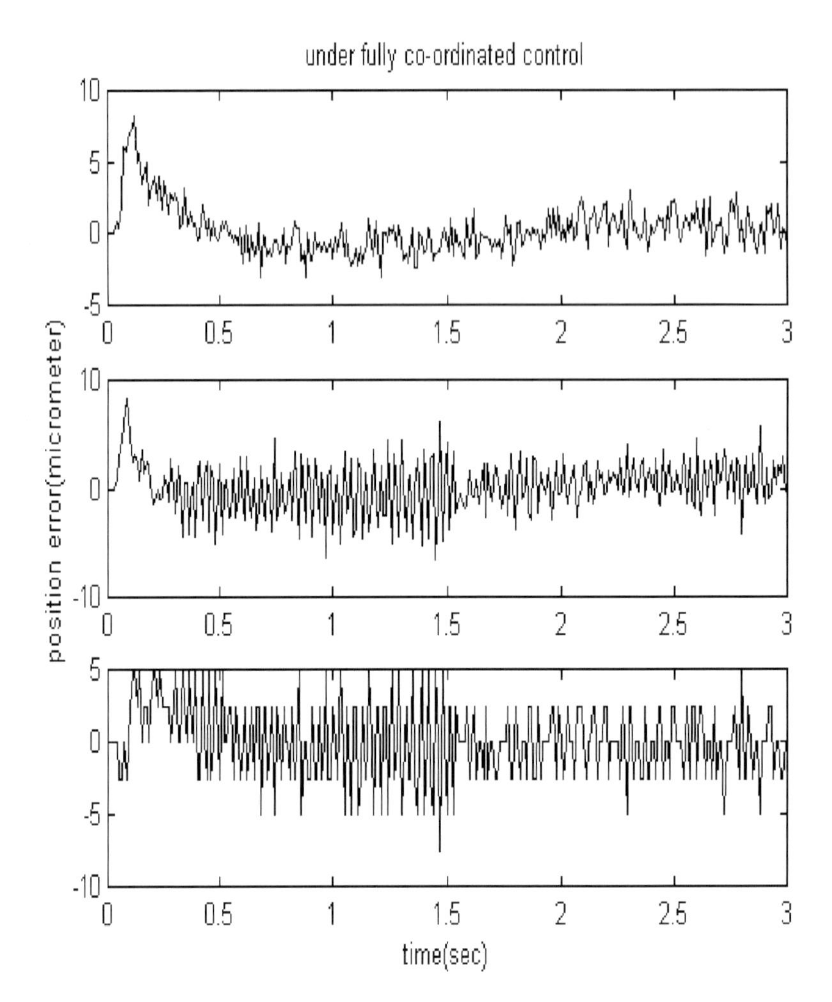

Fig. 4.19. Tracking error for the x-axis under fully co-ordinated control(*top*); Tracking error for the y-axis under fully co-ordinated control(*middle*); Positional inter-axis offset under fully co-ordinated control(*bottom*)

With this formulation of the gantry stage, it is imminent to proceed with the dynamic modeling of the gantry stage.

Let m_1, m_2 denote the mass of the gantry and slider respectively, l denotes the length of the gantry arm, I_1, I_2 denote the moment of inertia of the gantry arm and slider respectively, (we assume that $I_1 = m_1(l/2)^2, I_2 = m_2(\frac{l}{2} + y)^2$) and $X = [x\ \theta\ y]^T$, where $x = x_1 + \frac{x_2 - x_1}{2}$ (refer to Figure 4.21).

The positions of m_i, $i = 1, 2$ are given by

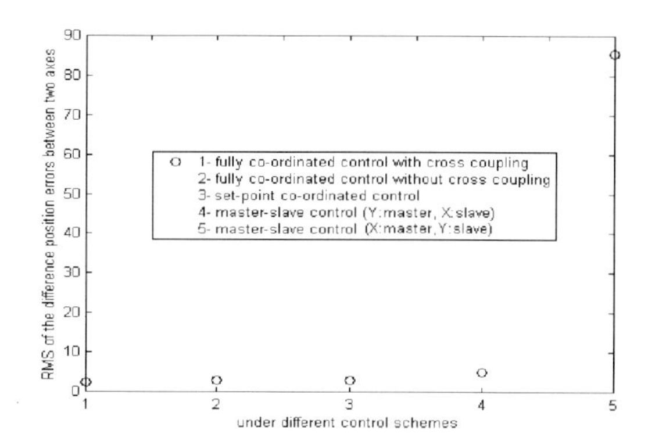

Fig. 4.20. RMS comparison of positional inter-axis offset under different control schemes

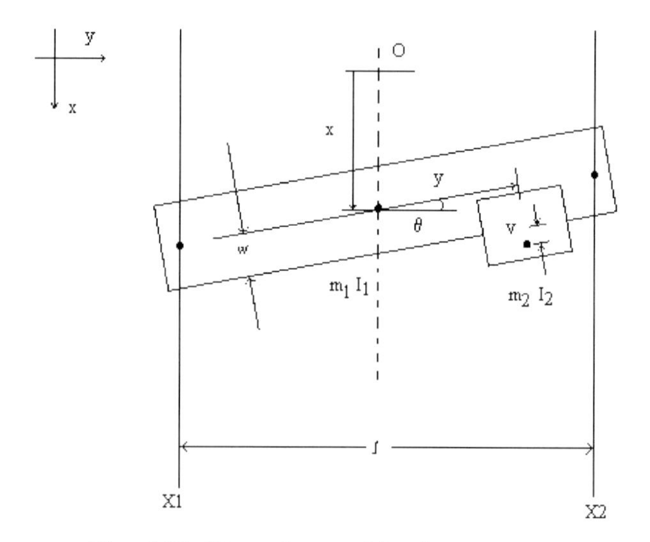

Fig. 4.21. Three-degree-of-freedom structure

$$x_{m1} = x, \tag{4.1}$$

$$y_{m1} = 0, \tag{4.2}$$

$$x_{m2} = x + dcos\theta - ysin\theta, \tag{4.3}$$

$$y_{m2} = ycos\theta + dsin\theta, \tag{4.4}$$

which lead to the corresponding velocities as

$$v_{m1} = \begin{bmatrix} \dot{x} \\ 0 \end{bmatrix}, \tag{4.5}$$

$$v_{m2} = \begin{bmatrix} \dot{x} - d\dot{\theta}sin\theta - \dot{y}sin\theta - y\dot{\theta}cos\theta \\ \dot{y}cos\theta - y\dot{\theta}sin\theta + d\dot{\theta}cos\theta \end{bmatrix}. \quad (4.6)$$

Thus, the total kinetic energy may be computed as

$$\begin{aligned} K &= \frac{1}{2}m_1 v_{m1}^T v_{m1} + \frac{1}{2}m_2 v_{m2}^T v_{m2} + \frac{1}{2}(I_1 + I_2)\dot{\theta}^2 \\ &= \frac{1}{2}(m_1 + m_2)\dot{x}^2 + \frac{1}{2}(I_1 + I_2 + m_2 y^2 + m_2 d^2)\dot{\theta}^2 \\ &+ \frac{1}{2}m_2\dot{y}^2 - \dot{x}\dot{\theta}m_2[dsin\theta + ycos\theta] - \dot{x}\dot{y}m_2 sin\theta \\ &+ \dot{\theta}\dot{y}m_2 d, \end{aligned} \quad (4.7)$$

which can be further written as

$$K = \frac{1}{2}\dot{X}^T D\dot{X}, \quad (4.8)$$

where D is the inertia matrix given by

$$D = \begin{bmatrix} m_1 + m_2 & -m_2 dsin\theta - m_2 ycos\theta & -m_2 sin\theta \\ -m_2 dsin\theta - m_2 ycos\theta & I_1 + I_2 + m_2 y^2 + m_2 d^2 & m_2 d \\ -m_2 sin\theta & m_2 d & m_2 \end{bmatrix}. \quad (4.9)$$

Next, the elements of the Coriolis and centrifugal matrix C can be derived from

$$C_{ij} = \sum_{k=1}^{3}(c_{ijk}\dot{q_k}), \quad (4.10)$$

where $\dot{q_1}$, $\dot{q_2}$ and $\dot{q_3}$ represents the derivative of x, θ and y respectively, and c_{ijk}, the Christoffel symbols, are computed as

$$c_{ijk} = \frac{1}{2}[\frac{\partial d_{ij}(q)}{\partial q_k} + \frac{\partial d_{ik}(q)}{\partial q_j} + \frac{\partial d_{jk}(q)}{\partial q_i}], \quad (4.11)$$

where d_{ij} represents the element in the ith row and jth column of the inertia matrix D. Substituting the assumed inertia equation I_1 and I_2 into Equation (4.9) and computing Equation (4.11), matrix C can be expressed as

$$C = m_2 \begin{bmatrix} 0 & C_{01} & -\dot{\theta}cos\theta \\ C_{01} & C_{02} & (\frac{l}{2} + 2y)\dot{\theta} - \dot{x}cos\theta \\ -\dot{\theta}cos\theta & (\frac{l}{2} + 2y)\dot{\theta} - \dot{x}cos\theta & 0 \end{bmatrix}, \quad (4.12)$$

where

$$\begin{aligned} C_{01} &= y\dot{\theta}sin\theta - d\dot{\theta}cos\theta - \dot{y}cos\theta, \\ C_{02} &= (ysin\theta - dcos\theta)\dot{x} - (\frac{l}{2} + 2y)\dot{y}. \end{aligned}$$

Finally, the dynamic model is expressed as

$$D\ddot{X} + C\dot{X} + BF = BU, \tag{4.13}$$

where

$$B = \begin{bmatrix} 1 & 1 & 0 \\ lcos\theta & -lcos\theta & 0 \\ 0 & 0 & 1 \end{bmatrix}, \tag{4.14}$$

$$F = [F_{x1}, F_{x2}, F_y]^T, \tag{4.15}$$

$$U = [u_{x1}, u_{x2}, u_y]^T. \tag{4.16}$$

F_{x1}, F_{x2}, F_y are the frictional forces, and u_{x1}, u_{x2}, u_y are the generated mechanical forces along x_1, x_2 and y respectively. The frictional forces, F, are assumed to be adequately described by the Tustin model,

$$F_z = d_z\dot{z} + f_z sgn(\dot{z}), \tag{4.17}$$

where z represents x_1, x_2 or y.

4.4.2 Model-based Adaptive Control Design

For the actual real system, it is a challenging and difficult task to obtain the exact values of the parameters of the model m_1, m_2, d_i and f_i $(i = x_1, x_2, y)$ accurately. To this end, an adaptive controller which is designed based on the dynamic model, is constructed.

Define the filtered error $s = \Lambda e + \dot{e}$ where $e = X_d - X$, and X_d, representing the desired trajectories, is twice differentiable; Λ is a user-defined parameter. Thus, Equation (4.13) can be expressed as

$$D\dot{s} = D(\Lambda\dot{e} + \ddot{X}_d) + C\dot{X} + BF - BU. \tag{4.18}$$

The parameters D, C, and F may be further expressed as follows:

$$D = m_1 D_0 + m_2 D_1, \tag{4.19}$$

$$C = m_2 C_0, \tag{4.20}$$

$$F = \sum_{i=1}^{3}(d_i F_{0i} + f_i F_{1i}), \tag{4.21}$$

where the various coefficients (D_0, D_1 *etc.*) are expressed in Equations (4.22)–(4.30):

$$D_0 = \begin{bmatrix} 1 & 0 & 0 \\ 0 & (l/2)^2 & 0 \\ 0 & 0 & 0 \end{bmatrix}, \tag{4.22}$$

$$D_1 = \begin{bmatrix} 1 & -dsin\theta - ycos\theta & -sin\theta \\ -dsin\theta - ycos\theta & (\frac{l}{2}+y)^2 + y^2 + d^2 & d \\ -sin\theta & d & 1 \end{bmatrix}, \tag{4.23}$$

$$C_0 = \begin{bmatrix} 0 & C_{01} & -\dot{\theta}cos\theta \\ C_{01} & C_{02} & (\frac{l}{2}+2y)\dot{\theta} - \dot{x}cos\theta \\ -\dot{\theta}cos\theta & (\frac{l}{2}+2y)\dot{\theta} - \dot{x}cos\theta & 0 \end{bmatrix}, \tag{4.24}$$

$$F_{01} = [\dot{x}_1, 0, 0]^T, \tag{4.25}$$

$$F_{02} = [0, \dot{x}_2, 0]^T, \tag{4.26}$$

$$F_{03} = [0, 0, \dot{y}]^T, \tag{4.27}$$

$$F_{11} = [sgn(\dot{x}_1), 0, 0]^T, \tag{4.28}$$

$$F_{12} = [0, sgn(\dot{x}_2), 0]^T, \tag{4.29}$$

$$F_{13} = [0, 0, sgn(\dot{y})]^T. \tag{4.30}$$

Let

$$V = \frac{1}{2}\dot{D} = m_2 V_0. \tag{4.31}$$

Thus, V_0 may be expressed as

$$V_0 = \frac{1}{2}\begin{bmatrix} 0 & V_{01} & -\dot{\theta}cos\theta \\ V_{01} & 2(\frac{l}{2}+y)\dot{y} + 2y\dot{y} & 0 \\ -\dot{\theta}cos\theta & 0 & 0 \end{bmatrix}, \tag{4.32}$$

with

$$V_{01} = d\dot{\theta}cos\theta - \dot{y}cos\theta + y\dot{\theta}sin\theta.$$

Now the filtered error Equation (4.18) can be re-written as

$$D\dot{s} = -Vs + m_1 D_0(\Lambda\dot{e} + \ddot{X}_d) + m_2[V_0 s + D_1(\Lambda\dot{e} + \ddot{X}_d) + C_0\dot{X}] + \sum_{i=1}^{3}(d_i BF_{0i} + f_i BF_{1i}) - Bu. \tag{4.33}$$

The proposed adaptive controller is given by

$$U = B^{-1}Ks + \hat{m}_1 B^{-1} D_0(\Lambda\dot{e} + \ddot{X}_d) + \hat{m}_2 B^{-1}[V_0 s + D_1(\Lambda\dot{e} + \ddot{X}_d) + C_0\dot{X}] + \sum_{i=1}^{3}(\hat{d}_i F_{0i} + \hat{f}_i F_{1i}), \tag{4.34}$$

along with the following adaptation rules:

$$\dot{\hat{m}}_1 = \gamma_1 s^T D_0(\Lambda\dot{e} + \ddot{X}_d), \tag{4.35}$$

$$\dot{\hat{m}}_2 = \gamma_2 s^T [V_0 s + D_1(\Lambda\dot{e} + \ddot{X}_d) + C_0\dot{X}], \tag{4.36}$$

$$\dot{\hat{d}}_i = \gamma_{3i} s^T BF_{0i}, \tag{4.37}$$

$$\dot{\hat{f}}_i = \gamma_{4i} s^T BF_{1i}, \tag{4.38}$$

where $K > 0$ is positive definite, and $\hat{m}_1, \hat{m}_2, \hat{d}_i, \hat{f}_i$ are estimates of m_1, m_2, d_i, f_i, respectively.

4.4.3 Stability Analysis

Define the following Lyapunov function:

$$v = s^T Ds + \frac{1}{\gamma_1}\tilde{m}_1^2 + \frac{1}{\gamma_2}\tilde{m}_2^2 + \sum_{i=1}^{3}(\frac{1}{\gamma_{3i}}\tilde{d}_i^2 + \frac{1}{\gamma_{4i}}\tilde{f}_i^2), \tag{4.39}$$

where $\tilde{m}_1, \tilde{m}_2, \tilde{d}_i, \tilde{f}_i$ are the estimation error of m_1, m_2, d_i, f_i respectively. Differentiating v and substitute in Equation (4.33) and the control law at Equation (4.34)

$$\begin{aligned}\dot{v} = &-2s^T Ks + 2\tilde{m}_1 s^T D_0(\Lambda\dot{e} + \ddot{X}_d) + 2\tilde{m}_2 s^T[V_0 s + D_1(\Lambda\dot{e} + \ddot{X}_d) + C_0\dot{X}] \\ &+2\sum_{i=1}^{3} s^T(\tilde{d}_i BF_{0i} + \tilde{f}_i BF_{1i}) - 2\frac{1}{\gamma_1}\tilde{m}_1\dot{\hat{m}}_1 - 2\frac{1}{\gamma_2}\tilde{m}_2\dot{\hat{m}}_2 \\ &-2\sum_{i=1}^{3}(\frac{1}{\gamma_{3i}}\tilde{d}_i\dot{\hat{d}}_i + \frac{1}{\gamma_{4i}}\tilde{f}_i\dot{\hat{f}}_i).\end{aligned} \tag{4.40}$$

Incorporating the adaptive laws at Equations (4.35)–(4.38), $\dot{v}$ becomes

$$\dot{v} = -2s^T Ks. \tag{4.41}$$

This implies that $s, \hat{m}_1, \hat{m}_2, \hat{d}_i, \hat{f}_i$ are bounded. Based on the defined filtered error equation, since Λ is positive definite and s is bounded, it follows that e is bounded. This also implies that $\dot{e}$ is bounded, and in turn, that $X, \dot{X}$ are bounded. Furthermore, from Equation (4.33), it can be concluded that $\dot{s}$ is bounded, and Equation (4.41), together with the definition of v, jointly imply that

$$\lim_{t\to\infty} Ks^2 = V(0) - \lim_{t\to\infty} V(\infty). \tag{4.42}$$

Finally, applying Barbalat's lemma, $\lim_{t\to\infty} s(t) = 0$.

4.4.4 Software Simulation

To verify the effectiveness of the present approach, the results of using three decoupled PID controllers on each individual axis is compared with the developed adaptive controller applied to a software version of the dynamic gantry model. A MATLAB® simulation study is set up in each case. The gantry's parameters are selected as follows: masses: m_1=1 kg, m_2=1 kg, length l=0.415 m, distance d=0.015 m and the friction parameters are $d_1 = d_2 = d_3 = 1$, and

$f_1 = f_2 = f_3 = 1$. The desired trajectories (position, velocity and acceleration) are as depicted in Figure 4.22. The trajectory would span a distance of 0.01 m, periodically in 4 s. The maximum velocity and acceleration attained are 0.094 m/s and 0.145 m/s^2 respectively.

For the PID controllers, the following PID control law is used

$$u = K_p e + K_i \int e + K_d \frac{\partial e}{\partial t}. \tag{4.43}$$

Using independent axis control, and assuming identical dynamics for each axis; in this simulation, all three PID controller are tuned as K_p=400, K_i=50 and K_d=30. The adaptive controllers' parameters are configured as: $\gamma_1 = 45,000$, $\gamma_2 = 2,800$, $\gamma_{31} = 4000$, $\gamma_{32} = 4000$, $\gamma_{33} = 28,000$, $\gamma_{41} = \gamma_{42} = \gamma_{43} = 100$, K=diag(100 10 10), and Λ equates the identity matrix, *i.e.* Λ=diag(1 1 1).

The simulation results showing the error responses for individual axes are depicted in Figure 4.23, whilst the inter-axis offset error is shown in Figure 4.24. The control signals coming from the controller are recorded in Figure 4.25. The data collated from the PID-based simulations are represented by dotted lines whilst the adaptive-based simulations are represented by solid lines.

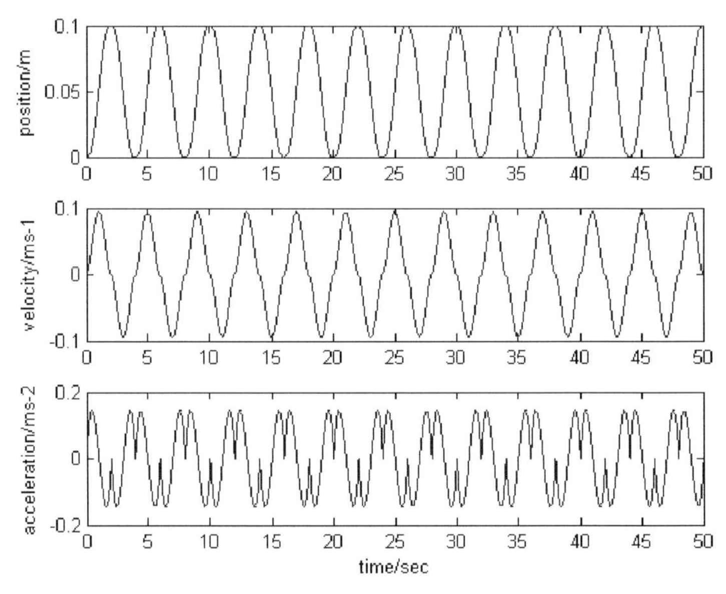

Fig. 4.22. Desired position, velocity and acceleration trajectories for x_1, x_2 and y

For a short time duration from $t = 0$ to $t = 3$, the PID control outperforms the adaptive controller. This is expected as the learning parameters have been initialized to zero with no *a priori* knowledge assumed. Subsequently, after

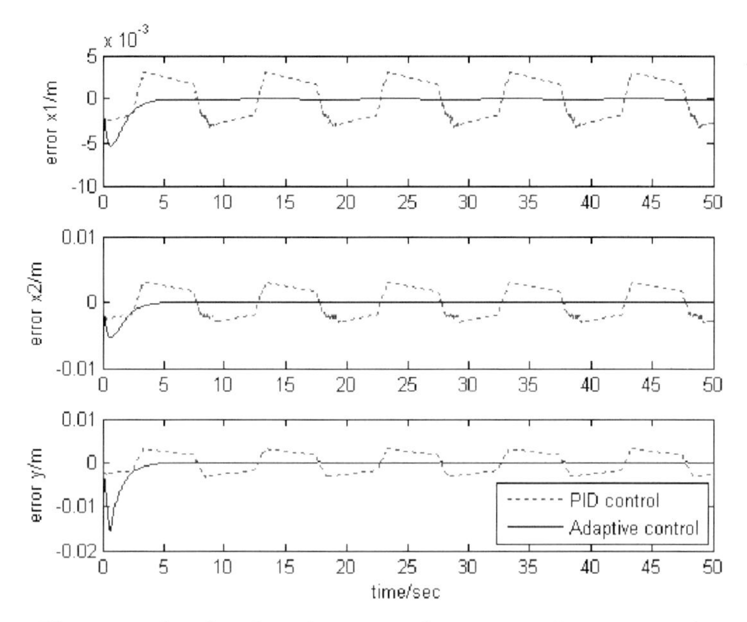

Fig. 4.23. Simulated tracking error for x_1, x_2 and y respectively

some parameter adaptation, the proposed approach quickly yielded significantly improved performance over PID control. On a further note, it can be seen that prior to attaining steady state, the tracking errors (for x_1, x_2 and y) has achieved reasonable performances.

4.4.5 Implementation Results

The stage used for the experimental setup is the gantry stage as mentioned earlier in Figure 4.2. The motor specifications are listed in Table 4.2.

Table 4.2. Specifications of gantry motors

Content	X-Axis servo motor	Y-Axis servo motor
	SEM MT22G2-10	Yaskawa SGML-01AF12
Power	350W	100W
Torque	0.70Nm	0.318Nm
Velocity	5000RPM	3000RPM
Resolution	10μm	10μm

For the PID-controlled implementation, PID controllers are tuned as K_p=90, K_i=5 and K_d=1, for the two X-axes (X_1 and X_2) whilst the Y-axis is tuned as K_p=30, K_i=1 and K_d=0. As noted in Table 4.2, the X-axes motors

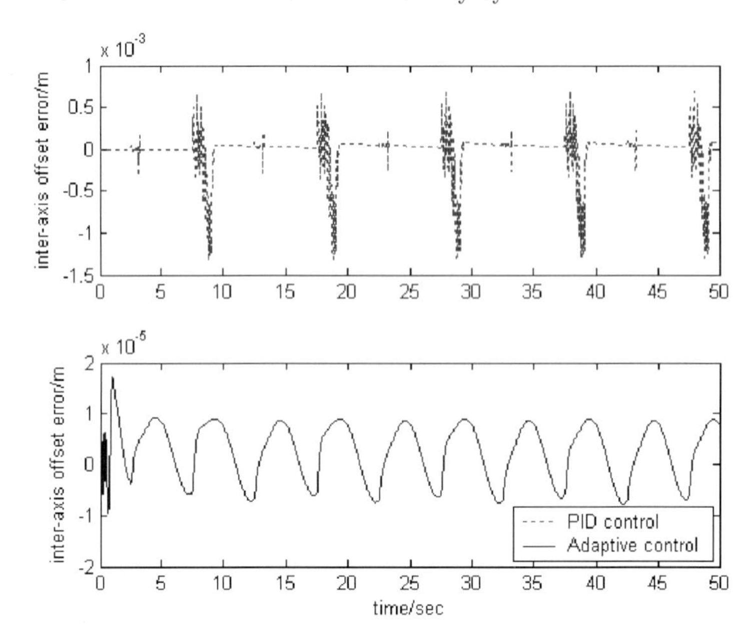

Fig. 4.24. Simulated inter-axis offset error between x_1 and x_2 using (a) PID Control and (b) Adaptive Control

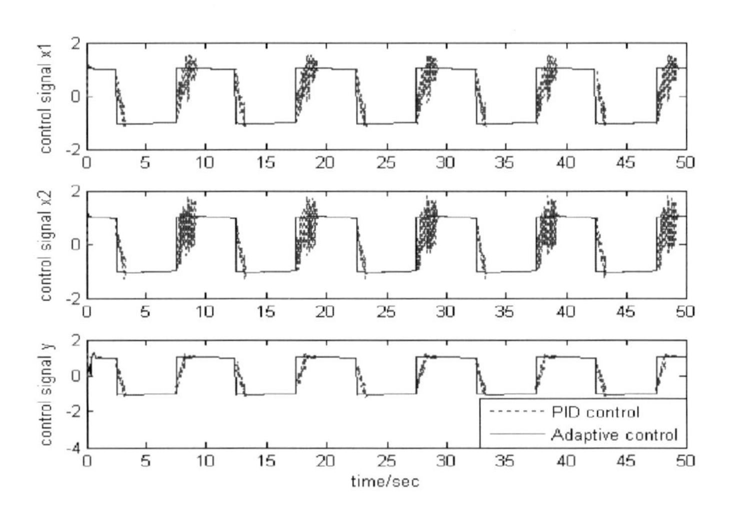

Fig. 4.25. Simulated control Signal for x_1, x_2 and y respectively

are in the same class and different from the Y-axis motor, hence the X-axes and Y-axis need to be tuned differently. The adaptive controllers' parameters are configured as $\gamma_1 = \gamma_2 = \gamma_{31} = \gamma_{32} = \gamma_{33} = \gamma_{41} = \gamma_{42} = \gamma_{43} = 1.8$, K=diag(40 3 5), and Λ=diag(1 1 1).

Trajectories similar to the software simulations are used and the results are shown in Figures 4.26, 4.27, and 4.28. These figures present similar characteristics to those obtained from software simulations. The adaptive controller is able to yield individual axis error of under 0.38 mm at steady state as compared to the PID performance of 0.96 mm for both x_1 and x_2 axes, whilst the y-axis error is kept under 2 mm for both controllers (refer to Figure 4.26). In addition, the adaptive controller is able to minimize the inter-axis offset error (by manipulation of the parameter K), whilst the decoupled PID controllers were only able to track individual trajectories independently. This performance is reflected by the resultant inter-axis offset error of 0.32mm using the adaptive controller vs 0.81 mm for the decoupled PID controller (refer to Figure 4.27).

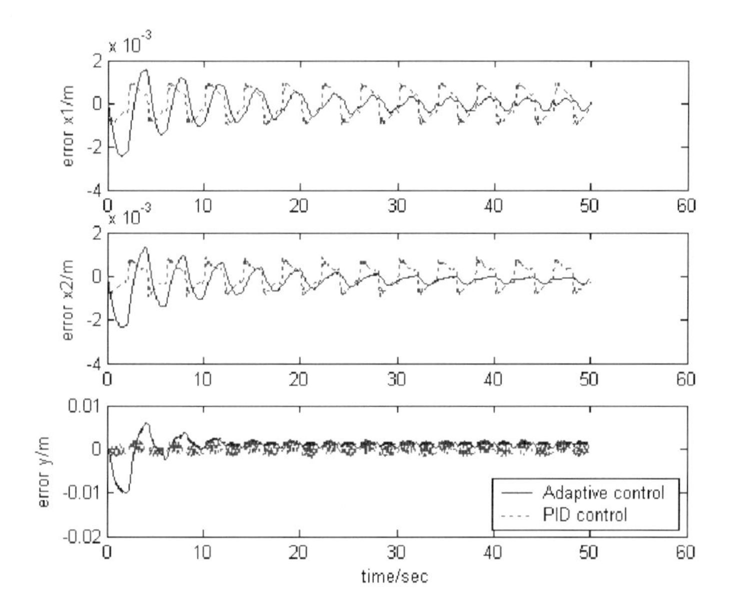

Fig. 4.26. Tracking error for x_1, x_2 and y respectively

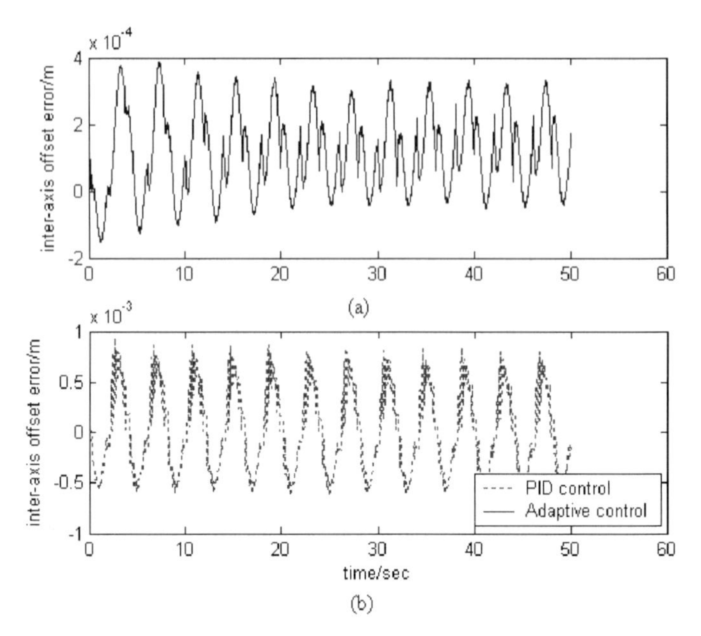

Fig. 4.27. Inter-axis offset error between x_1 and x_2 using (**a**)adaptive control and (**b**) PID control

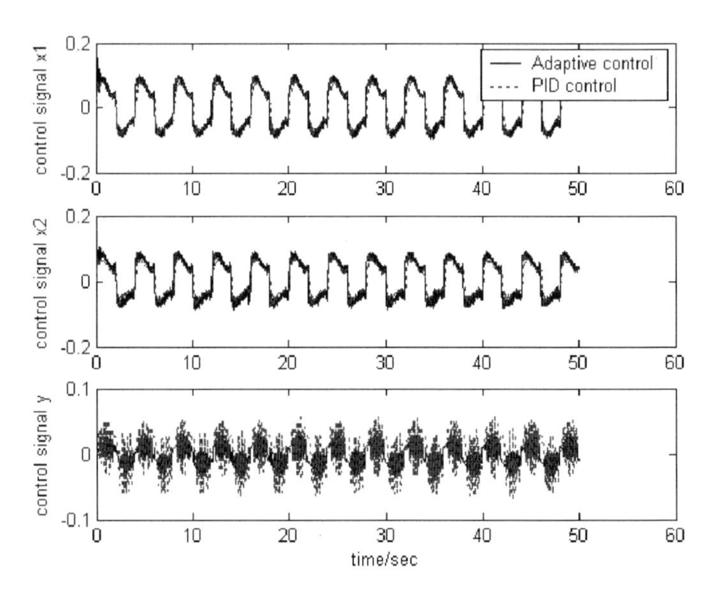

Fig. 4.28. Control signal for x_1, x_2 and y respectively

5

Geometrical Error Compensation

In automated positioning machines such as Co-ordinate Measuring Machines (CMM) and machine tools, the relative position errors between the end-effector of the machine and the workpiece directly affect the quality of the final product or the process of concern. These positioning inaccuracies arise from various sources, including static/quasi-static sources such as geometrical errors from the structural elements, tooling and fixturing errors, thermally induced and load induced errors, and also dynamic ones due to the kinematics of the machine (Hayati 1983; Zhang *et al.* 1985; Duffie and Maimberg 1987; Weekers and Schellekens 1995). These errors may be generally classified under two main categories:

- Systematic errors which are completely repeatible and reproducible, and
- Apparently random errors which vary under apparently similar operating conditions.

Although a complete elimination of machine errors is physically unachievable, these errors may be reduced to a level which is adequate for the particular application of the machine with a sufficiently high investment in the machine design and construction. It is widely reckoned that for an increase in the precision requirements, the corresponding increase in cost will be far steeper. Thus, rather than relying solely on the precision design and construction of the machine, which is expensive, this performance-cost dilemma sets the motivation for an alternative corrective approach in the form of an appropriate error compensation in the machine control to achieve comparable machine precision at a much reduced cost. Several CMM manufacturers reported that geometric error compensation by software techniques has produced a reduction in production costs, which they estimated to be between 5% and 50% (Satori 1995). While widespread incorporation of error compensation in machine tools remains to be seen, the application in CMM is tremendous and today it is difficult to find a CMM manufacturer who does not use error compensation in one form or another (Hocken 1980).

Compensation for errors in machines is not new. The development in error compensation is well-documented by Evans (1989). Early compensation methods mainly utilised mechanical correctors in the form of leadscrew correctors, cams, reference straightedges *etc.* Maudslay and Donking, for example, used leadscrew correction to compensate for the errors in their scales producing machine. Compensation *via* mechanical correction, however, inevitably increases the complexity of the physical machine. Furthermore, mechanical corrections rapidly cease to be effective due to mechanical wear and tear. The corrective components have to be serviced or replaced on a regular basis, all of which contribute to higher machine downtime and costs.

The evolution of control systems from mechanical and pneumatic-based subsystems to microprocessor-based systems has opened up a wide range of new and exciting possibilities. Many operations which used to be the result of complex linkages of levers, cams and bailing wires can now be carried out equivalently and more efficiently with program codes residing in the memory of an electronic computer. Software-based error compensation schemes thus blossomed in the 1970s. The first implementation was on a Moore N.5 CMM, a pioneering piece of work which earned Hocken the CIRP Taylor Medal in 1977. Since then, there has been an explosion of interests in soft compensation of machine errors with new methods developed and implemented (Love and Scarr 1973; Hocken *et al.* 1977; Bush *et al.* 1984).

This chapter addresses the principles and approaches of geometrical error compensation and related issues.

5.1 Overview of the Laser Measurement System

The laser measurement system provides machine tool manufacturers and users with the basic components necessary to measure machine tool positioning accuracy and to use compensation data to correct the machine positioning errors. The basic system includes a laser source, compensation electronics, optics, cables and accessories. A PC is usually necessary to control the system, and specific optics are required to make specific measurements of the machine geometrical properties. The complete system is able to collect and analyse different measurement data for calibration purposes, including linear, angular, straightness and squareness errors. The resolution of a linear laser measurement can reach as high as 1 nm. This high precision is achievable as the measurement process uses the precise wavelength of a laser as a basis for the computation of a distance measurement, thus achieving higher accuracy compared to other measurement systems. A two-frequency laser technique is also frequently incorporated to eliminate the problems resulting from changes in beam intensity, thus achieving better robustness and reliability. The basic measurement made by most laser measurement systems is a linear measurement of the relative movement between an interferometer and its associated retroreflector, along the path of the laser beam. In most cases, the interferometer is the fixed optic and the retroreflector is the moving one. Other measurements, such as angular, straightness and squareness measurements, are really special applications of the basic linear measurement. In many laser measurement systems, error compensation tables (look-up tables) can be automatically generated after the laser measurements and control inputs are obtained. These can be used in the servo systems for compensation purposes.

In this chapter, the presentation will focus mainly on the HP5529A laser measurement system. The laser head incorporates a helium-neon laser with a beam diameter of 6 mm (0.24 in.) and a vacuum wavelength of 632.491354 nm which is accurate to ± 0.1 ppm (parts per million). The laser head uses a proven long-life laser tube. The HP5529A is configurable with various optics and system electronics, resulting in a laser calibration system which can meet the unique physical layout and measurement requirements of many applications, and which is ideal for machine calibration and compensation purposes.

5.2 Components of the Laser Measurement System

The laser measurement system usually consists of:

- The laser head,
- The interferometer and its associated retroreflector,
- The measurement receiver, and
- The measurement and control electronics.

For certain measurements, it is not possible to align the laser head directly with the interferometer input aperture. Thus, the system will also include various beam-directing optics. In addition, the system may also include environmental sensing and/or wavelength tracking devices for further measurement compensation due to changes in the laser wavelength, consequent of the changes in the environmental parameters. The laser head serves as the light beam and reference frequency source. The optics and measurement receiver uses the laser beam to generate the measurement signal. The reference and measurement signals, along with the environment sensor signals, are used by the measurement electronics to generate the linear displacement information. The system controller can read and display this displacement information. In addition, the measurement electronics outputs a real-time signal representing the difference between the destination and the actual position. This error signal can be used in servo electronics to directly drive the servo motors. A schematic of the whole set-up is shown in Figure 5.1.

5.2.1 Laser Head

The wavelength of the laser emitting from the laser head is used as the length standard for the laser measurement system. The laser head generates a coherent light beam composed of two orthogonally polarised frequency components

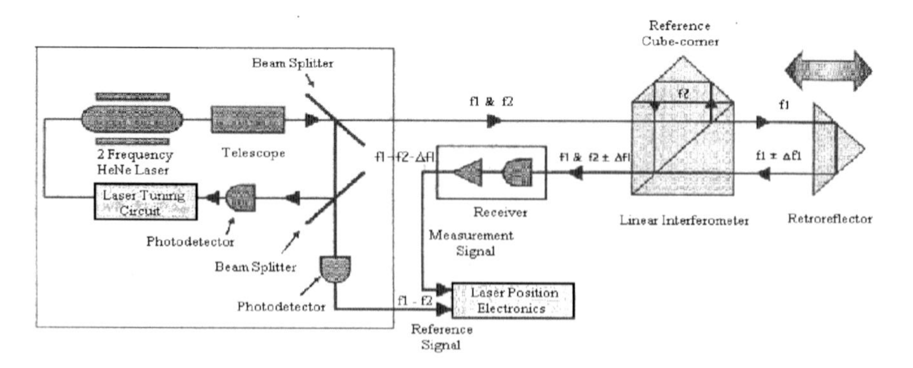

Fig. 5.1. Schematic of the laser measurement system

at slightly different frequencies, f_1 and f_2. Before emerging from the laser head assembly, the beam passes through a beam splitter where a small fraction of the beam is sampled. This portion of the beam is used to generate a reference frequency (by the reference receiver) and to provide an error signal to the laser cavity tuning system. The difference in the amplitudes of f_1 and f_2 is used for tuning purposes while the difference in frequency between f_1 and f_2 is used as the reference frequency signal which is the basis to compute the linear displacement. Figure 5.2 shows a schematic of the laser head.

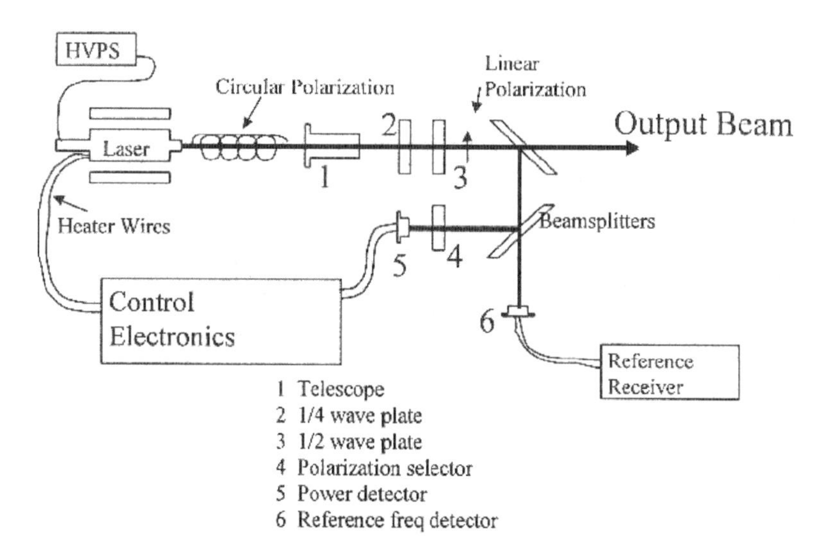

Fig. 5.2. Schematic of the laser head

5.2.2 Interferometer and Reflector

The interferometer and the reflector are the key optical components of a laser measurement system. The major portion of the beam emitting from the laser head is transmitted to an interferometer. The interferometer is a polarising beam splitter that reflects one polarisation and transmits the other. The beam splitter is oriented such that the reflected and transmitted beams are at right angles to each other. The beam (f_2) is reflected off a fixed retroreflector which is usually mounted on the interferometer. The transmitted frequency (f_1) passes through the interferometer and is reflected back by a movable retroreflector. If the distance between the interferometer and the movable retroreflector remains fixed at a certain length corresponding to the zero position, the offset frequency (f_2–f_1) will be identical to the reference signal. Under these conditions, the system will detect no change in the relative position of the interferometer and the movable retroreflector. When the movable retroreflector changes position relative to the fixed interferometer, a Doppler frequency shift occurs. This Doppler-shifted frequency, denoted as $f_1 \pm \Delta f_1$, is dependent on the direction of reflector movement. The two frequency components, $f_1 \pm \Delta f_1$ and f_2, exit the interferometer as a coincident beam. A schematic of the interferometer is shown in Figure 5.3

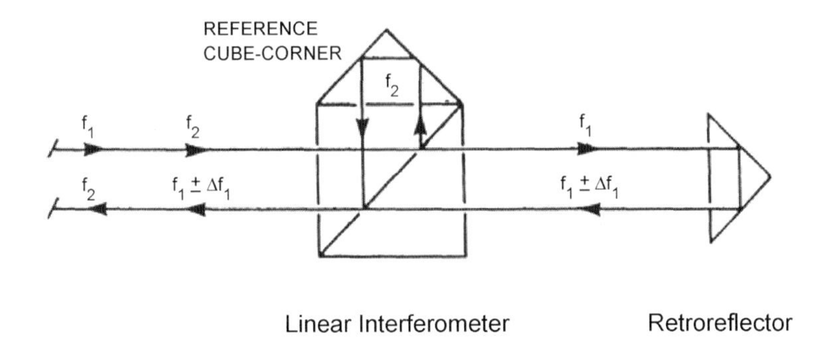

Fig. 5.3. Schematic of the interferometer and reflector

5.2.3 Measurement Receiver

The two orthogonally polarised frequencies of the laser, exiting from the interferometer, enter the measurement receiver. The receiver photodetector circuitry then amplifies and processes the signal to yield the measurement frequency. The displacement information is subsequently derived from the measurement electronics *via* a comparison of the measurement and reference signals. The schematic of the measurement receiver is given in Figure 5.4

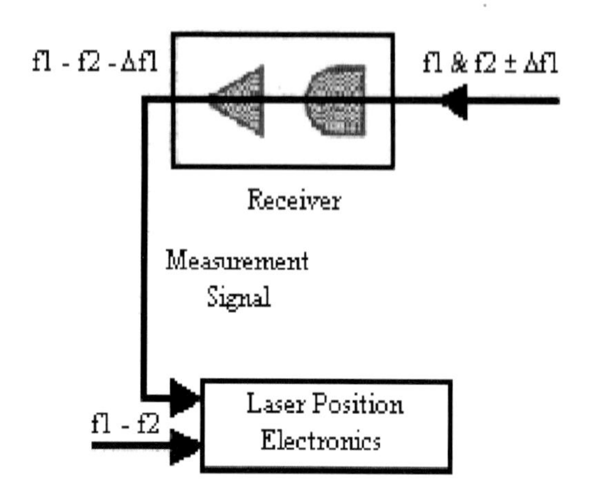

Fig. 5.4. Schematic of the measurement receiver

5.2.4 Measurement and Control Electronics

The balanced signals from the reference and measurement receivers are acquired *via* a measurement and control card. One function of the card is to reduce external noise by reading the signals in common mode and appropriately condition them. Additional signal interpolation may be carried out to yield improved resolution. The card will typically include a wavelength compensation function. The accuracy of the laser measurement system can be affected by changes in the parameters relating to the environment, such as changes in ambient temperature, pressure and relative humidity. This is because the laser wavelength, which is the length standard for measurements, is sensitive to these characteristics. The wavelength deviation can be estimated based on the measurements from the various environmental sensors. It can then be compensated for either manually or automatically.

5.3 Overview of Laser Calibration

The accuracy and precision of a multi-axis machine is determined primarily by the geometrical properties of the machine. Thus, to analyse fully the machine's positioning accuracy, it is necessary to measure the following geometrical characteristics (each of which contributes to positioning accuracy and precision at any point within the workzone of the machine):

- The six degrees of freedom for each measurement axis,
- Squareness between measurement axes,
- Parallelism between measurement axes.

The six degrees of freedom for each motion axis are depicted clearly in Figure 5.5.

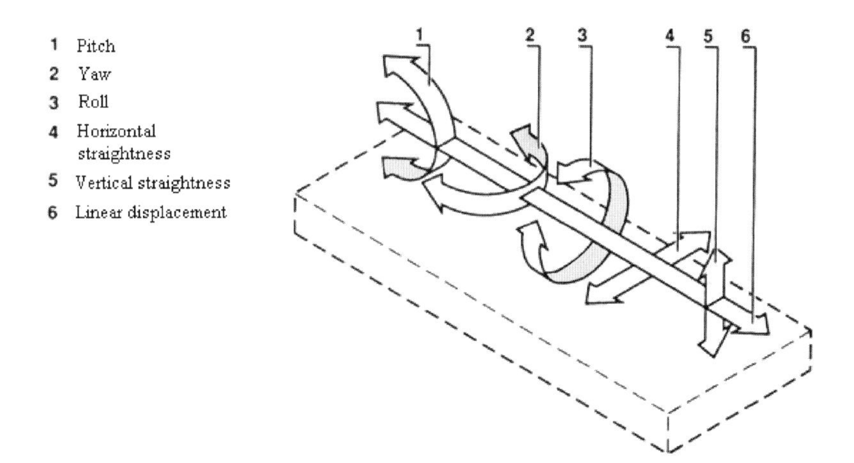

Fig. 5.5. Six degrees of freedom for a machine axis

The squareness and parallelism of travel between two or more axes characterise the relative orientation among the axes. Both measurements can be accomplished by performing two straightness meausrements, with the squareness measurement approach requiring more optics such as the 90^0 reference (the optical square). Most of these geometrical characteristics can be duly obtained using a laser measurement system.

5.3.1 Linear Measurement

Linear measurements refers to the actual distance translated by the moving part when it is controlled to move in a straight line. The retroreflector is mounted on the moving part to allow this measurement. In servo control systems, a position measurement is usually inferred directly from the encoder (or equivalent position measurement device) for the motor. However, due to inherent encoder calibration errors, there will inevitably exist some mismatch between the encoder measurement and the actual position. The laser system would be able to address this situation by giving the end user an assessment of the linear profile of the motor performance. The optics required to obtain the linear measurements are given in Figure A.1 (Appendix A). The set-up for a linear measurement is as shown in Figure 5.6.

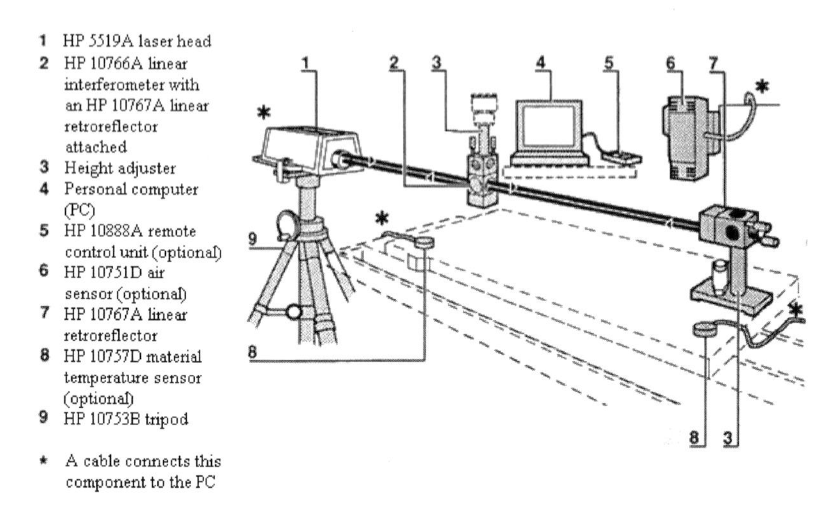

Fig. 5.6. Set-up for a typical linear measurement

5.3.2 Angular Measurement

An angular measurement is concerned with the measurement of the angular displacement (tilt) of the moving part (on which the angular reflector is mounted) from the ideal position. This angular displacement may vary with the linear travel distance of the moving part. The primary causes of an angular deviation include the physical guide imperfections and possibly cogging related effects. The optics and accessories used for the angular measurement are rather similar to those used for linear measurements. A breakdown of these devices and accessories is given in Figure A.2. The set-up for pitch and yaw measurements are given respectively in Figure A.3 and Figure A.4. A closed-up view of the traverse path of the laser beams is given in Figure 5.7 which illustrate that the angular measurement is comprised of two linear measurements at a precisely known separation. Roll measurement is addressed separately in the next section as this measurement will typically require a level-sensitive device to be used.

5.3.3 Straightness Measurement

The objective of a straightness measurement is to determine whether the moving part is moving along a straight path. The main source for a straightness

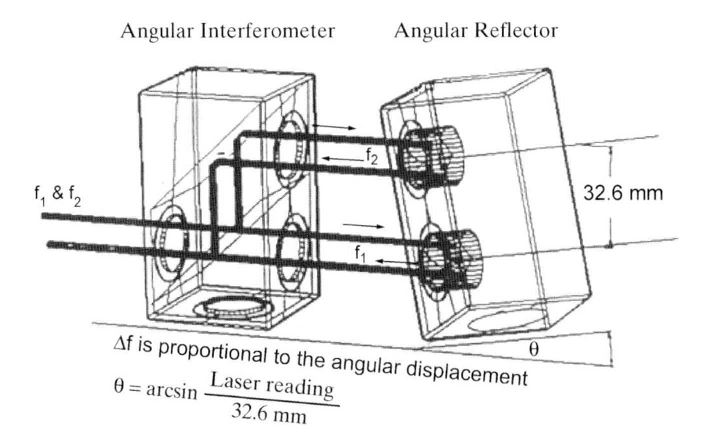

Fig. 5.7. Angular measurement

error is the straightness profile of the guiding mechanisms which guide the motion of the moving part. The optics required for straightness measurement is given in Figure A.5. The straightness profile can be divided into two components: namely the horizontal and vertical straightness. The schematic of the set-up to carry out these measurements is given in Figure A.6. Figure 5.8 illustrates the two light paths of travel within the interferometer. The mirror axis serves as an optical straight edge to provide a reference for the straightness meaurements.

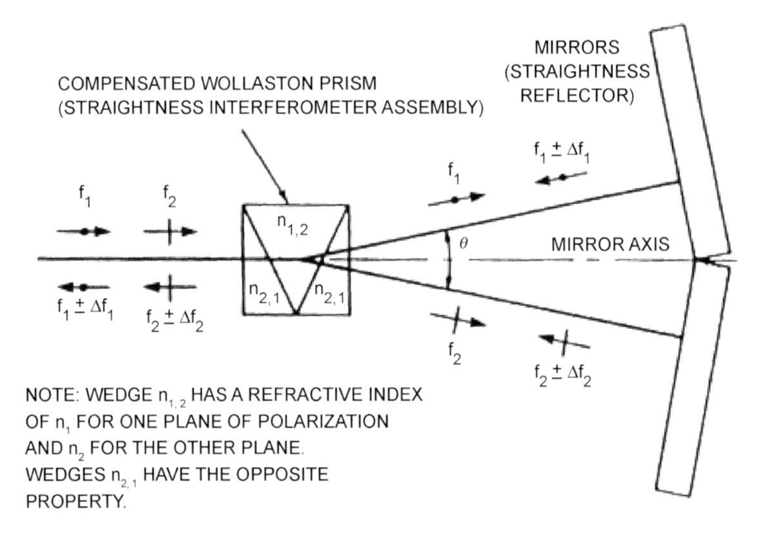

Fig. 5.8. Straightness measurement

5.3.4 Squareness Measurement

Straightness and squareness measurements are usually done concurrently, since a squareness measurement consists of two straightness measurements carried out perpendicularly to each other. These measurements allow the user to determine whether two machine axes are oriented perpendicularly to each other. A milling machine with a horizontal spindle and a bed which moves perpendicularly to the spindle is an example of a machine with two perpendicular axes. A CMM with a probe that moves vertically and mounted on a bridge which moves horizontally is another example. The main cause of a squareness deviation is probably the constraints during the manufacture or assembly of the machine to fix two axes exactly perpendicular to each other. The squareness measurement will be useful to allow the small angular difference to be measured and compensated for. The optics required for squareness measurements are given in Figure A.7. The main procedure for squareness measurement on a horizontal plane is to carry out a measurement along the first axis as shown in Figure A.8 using an optical square, and subsequently to carry out a measurement along the second axis according to the set-up in Figure A.9. The second axis measurement is simply a horizontal straightness measurement along the axis on which the reflector was earlier mounted during the first measurement. Figure 5.9 illustrates the concept of obtaining the squareness error from the two straightness measurements.

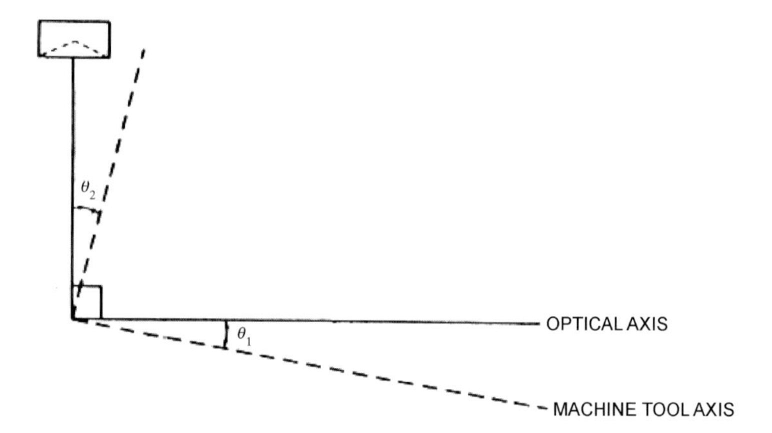

Fig. 5.9. Squareness measurement

The procedure to execute a squareness measurement in the vertical plane is similar to that of the horizontal plane, except for additional requirements in terms of optics. The required set of devices is shown in Figure A.10. The set-up for measurements along the vertical axis (*i.e.*, the z-axis) is given in Figure A.11.

5.4 Roll Measurement Using a Level-sensitive Device

Roll measurement refers to the measurement of rotation about its own axis. The measurement of roll tilt about its own axis is quite tedious even with a full set of laser interferometer equipment. Therefore, an electronic level measurement system is usually applied to facilitate this particular measurement. The principles of operation are straightforward: it make use of a pendulum in conjunction with an electronic detection system to sense precisely the attitude of the pendulum with respect to a reference. The equipment consists of two components, namely the level unit which is to be secured onto the moving part, and the display unit which shows the angular deviation. An accuracy of 0.2 arcsec can readily be achieved by commercial level measurement systems.

For a roll measurement, the level sensor will be fixed onto the moving part. When the axis moves to a designated position, the swing in angle on the pendulum will be reflected *via* the display. There is usually a PC interface provided to acquire the measurements into a PC. Figure 5.10 shows an illustration of the working principles of the level sensor.

5.5 Accuracy Assessment

The main objective behind the calibration of a machine tool or Co-ordinate Measuring Machine (CMM) is to determine its positioning accuracy, *i.e.*, to improve the positioning accuracy of the tool within the work zone. This calibration cannot be done directly, but it can be achieved by measuring the six degrees of freedom for each of the three axes, and the squareness between X, Y and Z, for a 3D Cartesian workzone. Thus, a total of 21 sources of error needs to be calibrated as shown in Figure 5.11. This can be a time-consuming process. An assessment of the accuracy, before and after compensation, is usually done *via* diagonal measurements. As the tool is traversed along a body diagonal of the work zone, all axes must move in concert in order to position accurately along the diagonal. Diagonal measurements are useful in machine tool acceptance testing or in a periodic maintenance program to assess quickly the condition of a machine. Therefore, linear measurements along the work zone diagonals can provide a quick assessment of the overall positioning accuracy. The HP 10768A diagonal measurement kit is an optical accessory to the HP5529A laser measurement system. A schematic of the accessories is shown

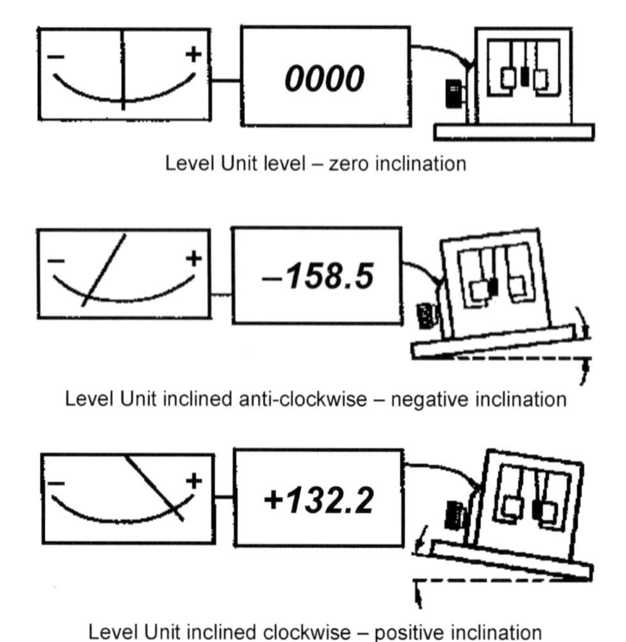

Fig. 5.10. Operation of a level-sensitive device

in Figure A.12. Figure A.13 shows the typical optics used for a diagonal measurement. A typical set-up for a diagonal measurement is shown in Figure A.14.

5.6 Factors Affecting Measurement Accuracy

The accuracy associated with laser measurements are also affected by several factors usually relating to the set-up, optical deformation and also environmental conditions. The main factors will be described.

5.6.1 Linear Measurement Errors

Abbé Errors

The perpendicular distance between the measurement axis of a machine (the scales) and the actual displacement axis is called the Abbé offset. As a result of the Abbè offset which is inevitably existent, an Abbé error occurs when there is an angular displacement of the moving part during its translation. As depicted in Figure 5.12, when the moving part <2> has moved a distance which is measured to be <5>, the corresponding actual distance moved is actually given by <6>.

SOURCES OF ERROR OF A TYPICAL 3 AXIS, FIXED SPINDLE (X AND Y), STACKED STAGE MACHINE DUE TO GEOMETRY AND POSITION FEEDBACK ONLY (SHOWING 21 SOURCES OF ERROR)
Z AXIS
X AXIS
Y AXIS
X AXIS
Y AXIS
Z AXIS
S
P'
R
P
Y'
θYZ
θXZ
θXY
θ
Y

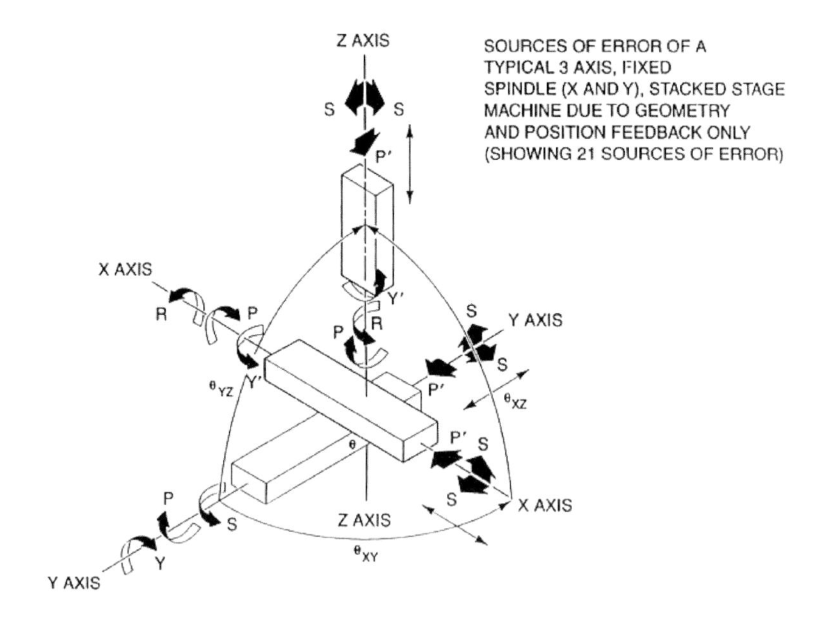

Fig. 5.11. Sources of error for a typical 3D machine

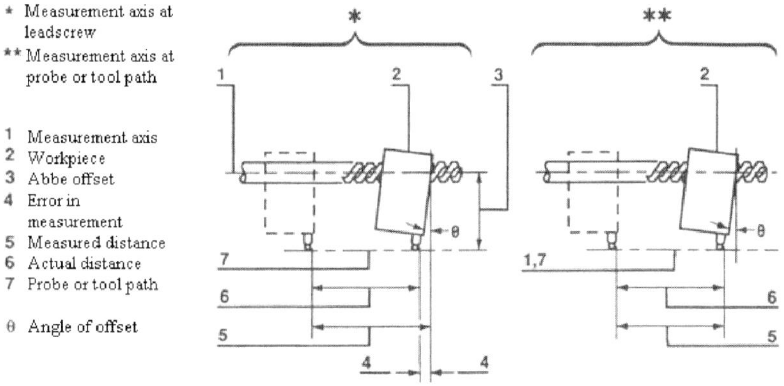

* Measurement axis at leadscrew
** Measurement axis at probe or tool path
1 Measurement axis
2 Workpiece
3 Abbe offset
4 Error in measurement
5 Measured distance
6 Actual distance
7 Probe or tool path
θ Angle of offset
*
**
1,7

Fig. 5.12. Abbé error

The Abbé error increases in proportion to the size of both the angular and Abbé offsets. Thus, to minimise this error, measurements should be taken as close as possible to the moving part, *i.e.*, the Abbé offset should be as small as possible.

Deadpath Errors

Deadpath is the part of the measurement path between the interferometer and the reflector when the reflector is at the zero point. It is ideally zero, so that a Doppler frequency shift is only associated with a translation. Otherwise, the linear measurement may include an additional part which arises due to the deadpath. When there is a variation in the air refractive index, a deadpath error may manifest in an apparent shift of the zero point, resulting in poor machine repeatability. Figure 5.13 shows an example of a deadpath error. To minimise this error, the interferometer optics <2> should be placed as close as possible to the retroreflector <5> without allowing them to touch.

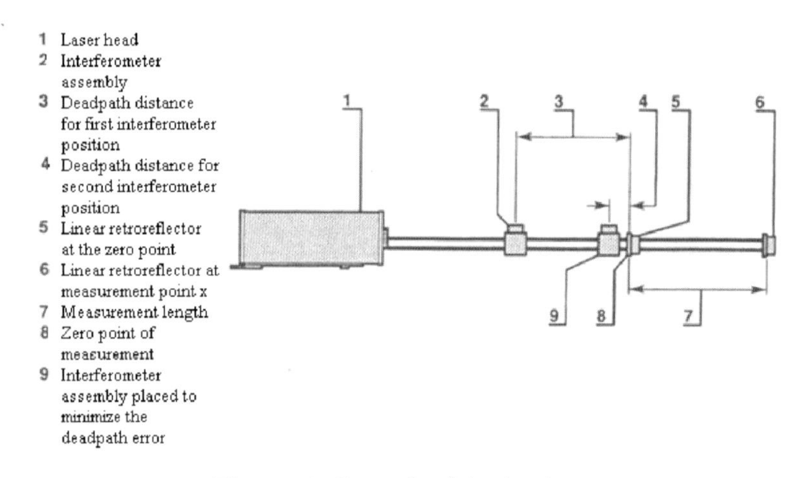

Fig. 5.13. Example of deadpath error

Cosine Errors

Cosine error arises when the laser and the desired measurement axis are not straightly aligned, so that the recorded measurement is shorter than the actual travel of the machine. The error increases with the travel distance and the misalignment. An exaggerated illustration is given in Figure 5.14

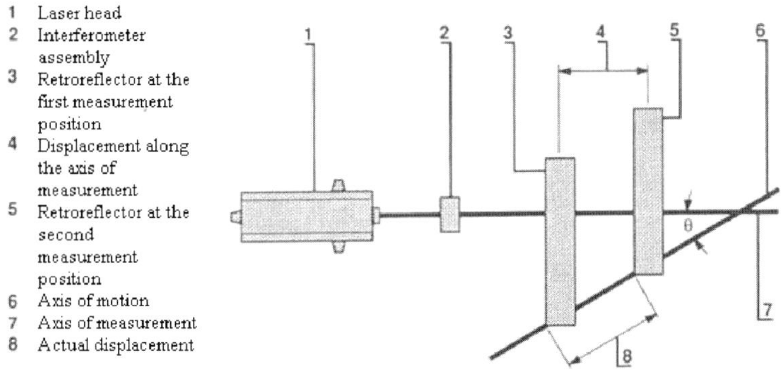

1 Laser head
2 Interferometer assembly
3 Retroreflector at the first measurement position
4 Displacement along the axis of measurement
5 Retroreflector at the second measurement position
6 Axis of motion
7 Axis of measurement
8 Actual displacement

Fig. 5.14. Cosine error

5.6.2 Angular Measurement Errors

The accuracy of an angular measurement can be affected even by a small change in the distance between the retroreflectors. The distance between the retroreflectors needs to be known precisely in order to convert the two linear measurements into an angular one. This change can occur due to variation in the temperature of the angular reflector housing. To minimise this error, excessive handling of the angular reflector, or contact with temperature varying medium should be avoided or minimised.

5.6.3 Straightness Measurement Errors

The accuracy of the straightness measurement depends significantly on the two plane mirrors in the straightness reflector having ideal surface characteristics. If either of the plane mirrors is convex or concave, these characteristics will be misinterpreted as a straightness error, even if the moving part is translating along a perfectly straight line. One way to reduce this error is to rotate the straightness reflector by 180° and obtain a second set of measurements. With at least two sets of results, the average can be taken to calculate the actual deviation. Although this procedure might be time consuming, it is a useful method to compensate for the flatness disparity of the straightness optics.

Straightness related errors can also occur when the reference bisector of the straightness reflector is not aligned with the laser and interferometer, leading to the misalignment (slope) being erroneously interpreted as a straightness error as shown in Figure 5.15.

5.6.4 Environmental Conditions

Environmental factors (especially the temperature factor) can affect the reference laser wavelength, and thus the measurement accuracy. Ambient tem-

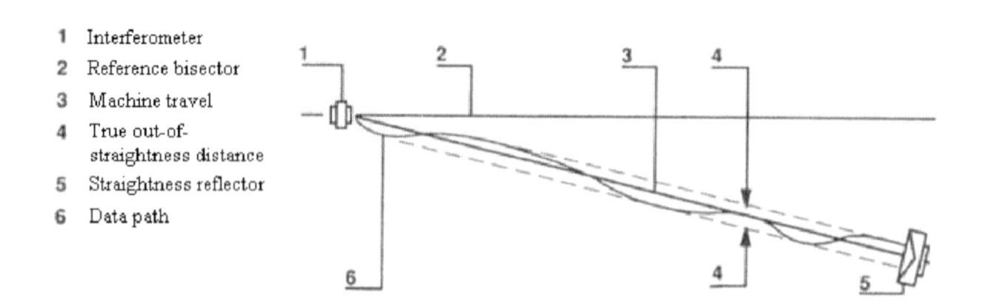

Fig. 5.15. Error in straightness measurement

perature, pressure and relative humidity are among the more common environmental factors which should be addressed where possible. In addition, changes in environmental parameters may also result in an inadequate geometrical error compensation due to different environmental conditions during the calibration and compensation phases, and therefore different geometrical properties of the machine. Although they may appear to be random influences on the machine, they can be compensated to a certain extent.

Thermal effects in CMM and machine tools can produce very complex behaviour in the physical structure of the machine. They arise from a wide variety of sources, including sources not directly related to the operations of the machine such as room temperature, lights, hydraulics, *etc.* A very comprehensive description of these effects was given by Bryan 1990. To minimise the effect of thermal expansion, it is imperative to allow the machine and optics temperatures to stabilise before making measurements. Good environmental temperature regulation may be necessary, depending on the applications and requirements. Full thermal compensation by computer is extremely difficult, although approximations can be made. First-order thermal effects modelling and compensation are more realistic and amenable to practical implementation. At least one sensor *per* axis is required for first-order thermal correction. The thermal effects on geometry are reflected in the use of multiple geometrical error models, each valid for a particular range of operating temperatures.

Excessive vibration and air turbulence can be identified by random drifting of measurement results when the optics are at rest. They can also be identified by the increase of drift or the distance between successive runs as optics distance increases. Preventive methods are to ensure all equipment is rigidly secured and supported, and to use sufficient fans to allow adequate air circulation.

5.7 Overall Error Model

Common to all works on geometric error compensation and more is a model of the machine errors, which is either implicitly or explicitly used in the compensator. The geometrical machine model is designed to compensate for the systematic part of the geometric errors in the machine based on a rigid-body assumption. Consider a 2D meaurement machine as shown in Figure 5.16. Three independent co-ordinate systems, as shown in Figure 5.16, are used in the model with respect to the table (O, X, Y), the bridge (O_1, X_1, Y_1), and the X-carriage (O_2, X_2, Y_2) respectively. It is assumed, as initial conditions, that all three origins coincide and the axes of all three systems are properly aligned.

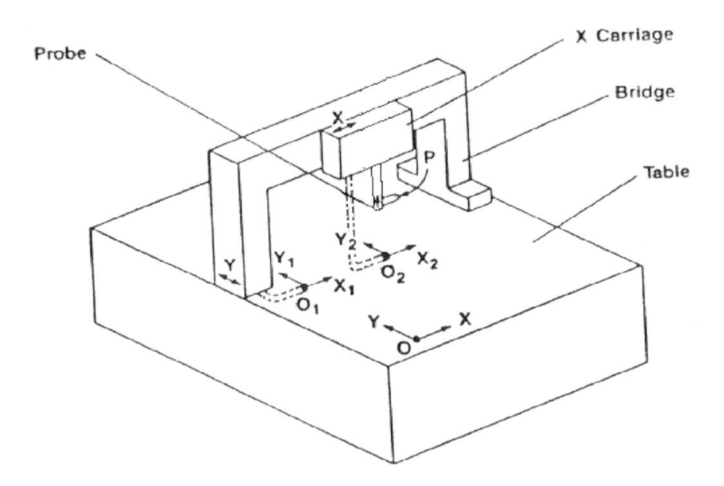

Fig. 5.16. 2D measurement machine

Thus, when the bridge moves a nominal distance Y, the actual position of the bridge origin O_1, with respect to the table system, is given by the vector

$$\overrightarrow{OO_1} = \begin{bmatrix} \delta_x(y) \\ y + \delta_y(y) \end{bmatrix}. \tag{5.1}$$

At the same time, the bridge co-ordinate system rotates with respect to the table system due to the angular error motion. This rotation can be represented by the matrix

$$R_1 = \begin{bmatrix} 1 & \epsilon_y \\ -\epsilon_y & 1 \end{bmatrix}. \tag{5.2}$$

Similarly, when the X carriage moves a nominal distance X, it follows that

$$\overrightarrow{O_1O_2} = \begin{bmatrix} x + \delta_x(x) \\ \delta_y(x) - \alpha x \end{bmatrix}, R_2 = \begin{bmatrix} 1 & \epsilon_x \\ -\epsilon_x & 1 \end{bmatrix}, \tag{5.3}$$

$$\overrightarrow{O_2P} = \begin{bmatrix} x_p \\ y_p \end{bmatrix}, \tag{5.4}$$

where x, y are the nominal positions; x_p, y_p represent the offsets of the tool tip; $\delta_u(v)$ is the translational error along the u-direction under motion in the v-direction; ϵ_u is the rotation of the u-axis; and α represent the out-of-squareness error. Therefore, a volumetric error model can be derived with respect to the table system

$$\overrightarrow{OP} = \overrightarrow{OO_1} + R_1^{-1}\overrightarrow{O_1O_2} + R_1^{-1}R_2^{-1}\overrightarrow{O_2P}. \tag{5.5}$$

Substituting Equations (5.1)–(5.4) into Equation (5.5) and noting that $\epsilon_u\epsilon_v \approx 0, \epsilon_u\delta_u(v) \approx 0, \epsilon_u\alpha \approx 0$ since $\epsilon_u, \delta_u(v), \alpha$ are very small, the geometrical error to be compensated along the x and y-directions are respectively

$$\Delta x = \delta_x(x) + \delta_x(y) - y_p(\epsilon_x + \epsilon_y) + x_p, \tag{5.6}$$

$$\Delta y = \delta_y(x) + \delta_y(y) + x\epsilon_y - x\alpha + x_p(\epsilon_y + \epsilon_x) + y_p. \tag{5.7}$$

It should be noted that the error sources are all calibrated using only appropriate combinations of linear displacement measurements.

5.8 Look-up Table for Geometrical Errors

The geometrical errors to be compensated are usually stored in the form of a look-up table. The look-up table is built based on points collected and calibrated in the operational working space of the machine. It captures the overall position errors in a matrix where each element of the matrix is assigned to a calibrated point within the workspace. When linear interpolation is used between the points for which errors are recorded, the data for only the six adjacent points (for 3D calibration) are recovered and interpolated.

A 1D error compensation is illustrated in Figure 5.17. Assume an axis is calibrated with a laser measurement system at equally spaced points according to encoder feedback. Denote $x_1, x_2, ..., x_n$ as the encoder measurements and $e_1, e_2, ..., e_n$ as the corresponding positioning errors derived from the laser measurement system. For a certain point x between x_{i-1} and x_i, the associated error e can be estimated *via* a linear interpolation process as

$$e = (x - x_1)\frac{e_2 - e_1}{x_2 - x_1} + e_1 \tag{5.8}$$

Many servo motion controllers will allow for geometrical error compensation *via* look-up tables. For example, the Programmable Multi-axis Controller (PMAC) from Delta Tau Data Systems, Inc. is a family of high-performance servo motion controllers which is capable of performing the look-up table compensation. PMAC has sufficient capacity to store up to eight of these compensation tables.

For a 2D compensation under PMAC, the amount of servo compensation for either motor will depend on the position of both motors (see Figure 5.18).

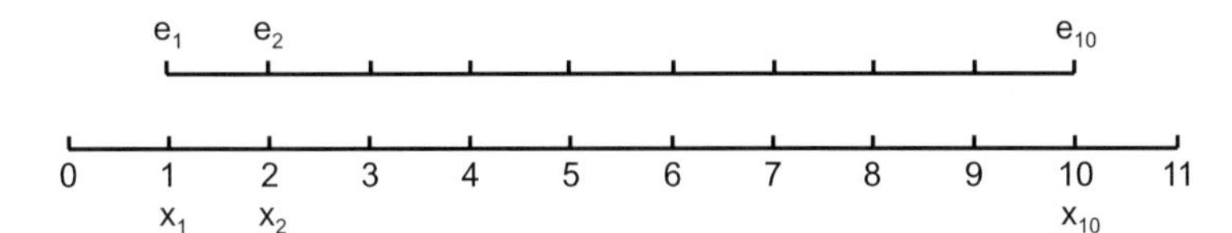

Fig. 5.17. 1D error compensation table

Each motor has one associated look-up table. Unless otherwise specified, the position information from the motors (source data) is used to extract the appropriate correction entry (target data) from the table. A 2D compensation table is thus associated with two source motors and one target motor. The compensation is performed from within the servo loop during every servo cycle. Typically this is between two entries in the table, so PMAC linearly interpolates between these two entries to obtain the correction for the current servo cycle.

During operation, PMAC computes the compensation for a given location in the plane of the two source motors as the weighted average of the four specified compensation values surrounding that location.

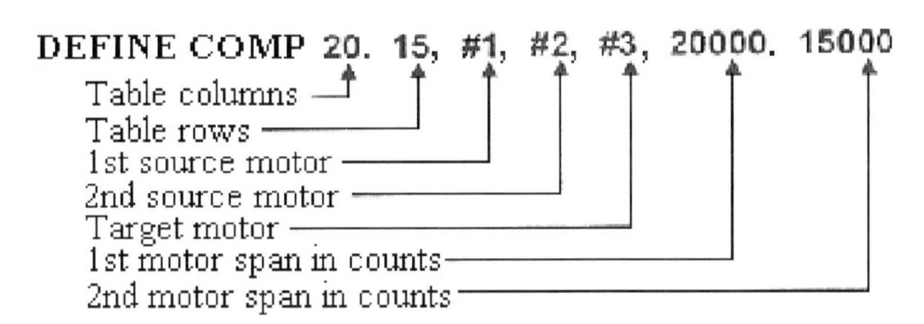

Fig. 5.18. PMAC compensation table

There are several disadvantages associated with the look-up table which clearly become significant with increasing precision requirements. First, the look-up table has extensive memory requirements. When the number of data points calibrated in a 3D workspace increases by a factor of N, the number of table entries increases by the order of N^3. This difficulty is thus especially significant for high precision machines, where a huge amount of calibration effort is necessary in order to compensate errors to within an acceptably precise threshold. Second, for the look-up table, the errors associated with intermediate points of the recorded data are compensated by using linear interpolation. This assumes the error to vary linearly between the calibrated points, and neighbouring points are not utilised to improve the interpolation. Linear interpolation may suffice if the calibration is done at very fine intervals compared to the precision requirements. However, this will in turn imply tremendous memory requirements which may be beyond the capacity of a typical look-up table. Third, the look-up table does not have a structure which is amenable to direct expansion when considering other factors affecting positioning accuracy, such as thermal and other environmental effects. When these factors are to be considered for a more precise compensation, additional tables are usually set up according to the schedules of the environmental parameters. Finally, for continuous on-line error compensation, a search through the look-up table will be necessary at every sampling interval. This is tedious when the table is large in size, especially when the calibration does not occur at regular intervals.

5.9 Parametric Model for Geometrical Errors

Since each error component varies with displacement in a non-linear manner, it is more naturally inclined to represent the non-linear profile using a non-linear function compared to using a look-up table. The Radial Basis Functions (RBF) are general tools for modelling non-linear functions since they can approximate any non-linear function to any desired level of accuracy. The RBF has desirable features which are useful when compared to the deficiencies of the look-up table. First, when appropriately tuned, the RBF can reduce the multitude of data points to a more manageable number of RBF parameters. Second, the RBF essentially uses a non-linear interpolation for intermediate points which are not calibrated, resulting in a smoother error modelling. Third, the RBF can be recursively refined based on additional points calibrated, and it may be expanded easily to include other factors to be considered in the error compensation, such as the thermal effects *etc.* Finally, the RBF is a parametric model. The output is directly computed based on the input, and no search for the correct entries is necessary. In this section, the RBF will be employed to model each error component.

5.9.1 Error Modelling with Radial Basis Functions

The main property of the RBF, used here for estimation purposes, is the function approximation property. RBF networks are a kind of feedforward networks. They form mappings from an input vector χ to an output vector Υ.

Let $f(\chi)$ be a smooth function from R to R. Then, given a compact $S \in R$ and a positive number ϵ_M, there exists an RBF network such that

$$f(\chi) = \sum_{i=0}^{m} w_i \phi_i(||\chi - c_i||^2/\sigma_i^2) + \epsilon, \tag{5.9}$$

where w_i is the representative value vector and $\phi_i(\alpha) = exp(-\alpha)$ is the radial basis function with $||\epsilon|| < \epsilon_M$ for all $\chi \in S$.

It has been shown that, under mild assumptions, RBF is capable of universal approximations, *i.e.*, approximating any continuous function over a compact set to any degree of accuracy. Therefore, RBF will be used to approximate the non-linear functions associated with the various error components. To obtain the function weights w_i, some weight tuning algorithms should be adopted. A commonly used weight tuning algorithm is the gradient search algorithm based on a backpropagated error, where the RBF is trained off-line to match specified exemplary pairs (χ_d, Υ_d), with χ_d being the ideal RBF input that yields the desired RBF output Υ_d. The discrete-time version of the backpropagation algorithm for the RBF is given by

$$w_i(t+1) = w_i(t) - \eta_w E\phi_i(||\chi - c_i||^2/\sigma_i^2), \tag{5.10}$$

$$c_i(t+1) = c_i(t) - \eta_c E w_i(\chi_d - c_i)\phi_i(||\chi - c_i||^2/\sigma_i^2), \tag{5.11}$$

where η_w, η_c are positive design parameters governing the speed of convergence of the algorithm. σ_i is chosen as a constant. The backpropagated error E is selected as the desired RBF output minus the actual RBF output $E = \Upsilon_d - \Upsilon$. A terminating condition is usually formulated in terms of this error to end the iterative weights tuning process. Thus, the optimum weighting values $\mathcal{W}^* = \{w_i^*, i = 0, 1, 2, ..., m\}$ can be obtained. It is usually a tradeoff in terms of the quality of fit and the iteration time. Since the tuning process is done off-line, more emphasis may be given to deriving a better fit at the expense of incurring a longer tuning time.

5.9.2 Parameter Error Approximations

In this section, the application of RBF to model individual geometrical error components and the adequacy of the resultant RBF-based models will be illustrated. The error data sets to be used for training the RBF are collected using a Hewlett Packard (HP) HP5529 laser interferometer. Figure 5.19 shows the experimental set-up.

Calibration is done at 1 mm intervals along the 100 mm travel for both the X- and Y-axes. Hence, 100 points are collected for each error component for each axis. Figure 5.20 shows the linear errors along the X-direction collected from five cycles of complete bi-directional travel of the X-carriage. The average value from the five cycles is computed to minimise the effects of any random influence arising.

Linear Errors

Linear errors may arise from various sources, including geometrical deficiencies along the guideway and measurement offsets/errors. For the XY table under study, the largest error source is probably due to the nonlinearities

Fig. 5.19. Experimental set-up

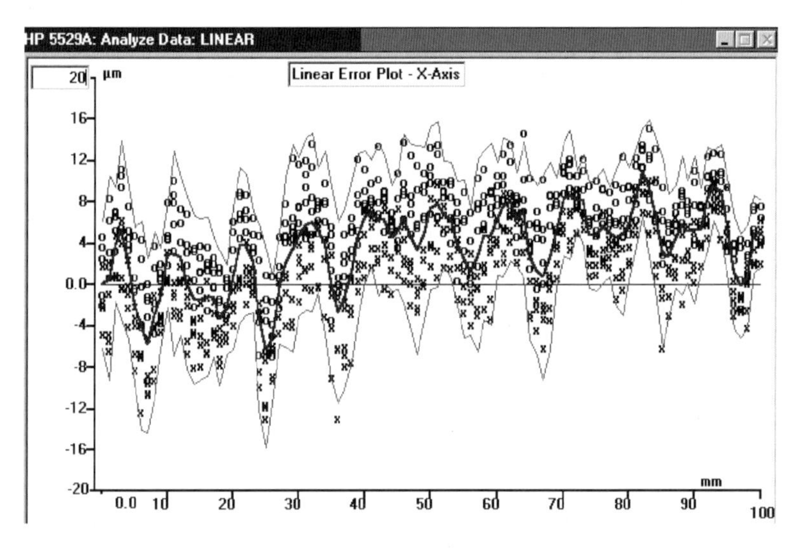

Fig. 5.20. Raw data set for linear x errors

in motion arising from the screw thread and associated backlash errors. Figure 5.21 shows the motion transfer mechanism from the screw thread to the moving carriage. The air gaps present in the mechanical interface can cause the actual displacement to vary rather significantly. This probably also explains the differences in linear error measurements in the forward and reverse directions.

For the modelling of linear errors, m is chosen as $m = 80$ for the RBF. The terminating condition for the gradient weights tuning algorithm is defined as $e_{ms} < 0.01$, where e_{ms} is the mean squared error e_{ms}. The spread of the RBF

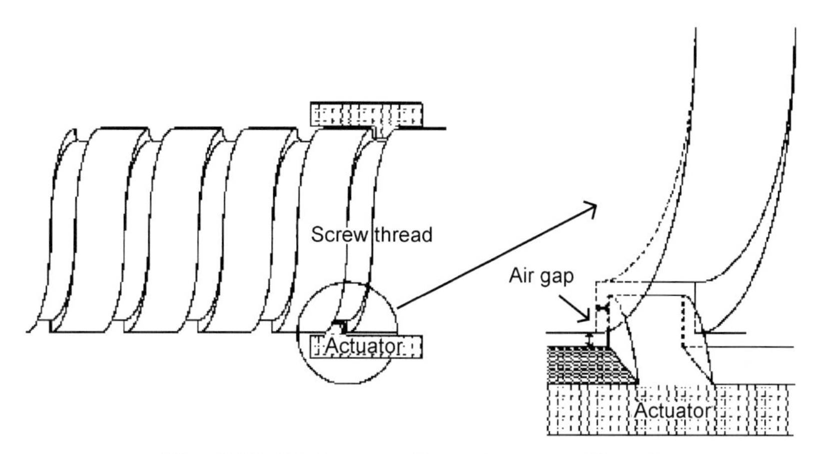

Fig. 5.21. Motion coupling using screw thread

is chosen as $\sigma = 2$. After 10^5 iterations of the parameter self-learning, the algorithm converges according to the terminating condition. The weights are then available to commission the RBF. A model of the linear error is then available as

$$\delta_x(x) = f_{lin,x}(x; \mathcal{W}^*_{linx}), \tag{5.12}$$

where x_i is the input nominal distance along the X-axis, $f_{linx}(\cdot)$ represents the RBF network, and $\mathcal{W}^*_{linx}$ is a set of the weighting values of the trained RBF. Figure 5.22 shows that the RBF network output closely follows the linear error measurements. Similarly, it follows that

$$\delta_y(y) = f_{lin,y}(y; \mathcal{W}^*_{liny}). \tag{5.13}$$

Figure 5.23 compares the output of the RBF network with the linear error measurements for the Y-axis.

Usually, the direction of motion is not addressed in the calibration processes. The average of the measurements taken from machine in the forward and reverse runs are simply used to construct the compensator, which may be a look-up table or in this case, an RBF. It may be noted that for a machine with more severe directional asymmetry in geometrical properties, it can be worthwhile to construct two RBF approximations for each error component, corresponding to the direction of motion. Figure 5.24 shows an example of asymmetrical linear errors.

Straightness Errors

Straightness errors arise mainly from the guideway. The straightness error measurements are derived from the perpendicular deviations from a reference

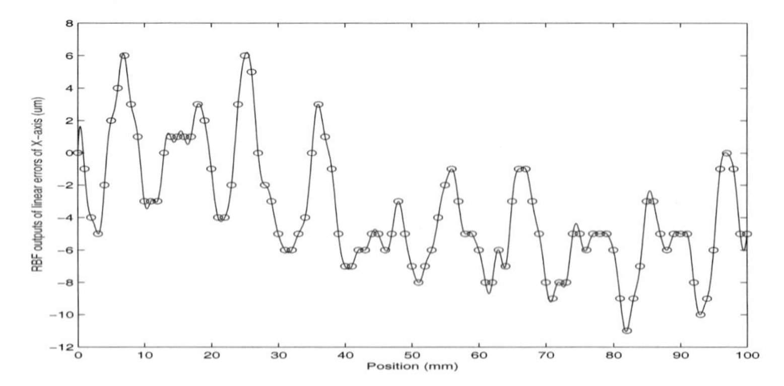

Fig. 5.22. RBF approximation of the linear errors (X-axis): *solid line* is RBF approximation and *circles* represent the measured data

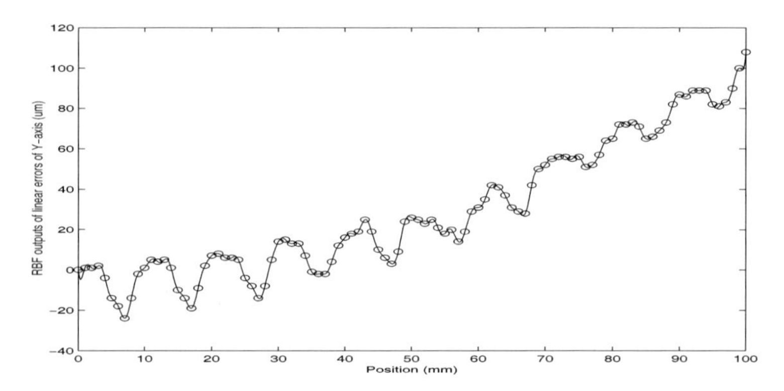

Fig. 5.23. RBF approximation of linear errors (Y-axis): *solid line* is RBF approximation and *circles* represent the measured data

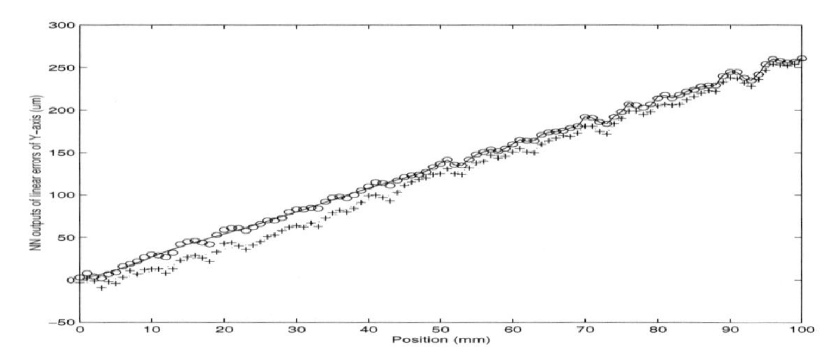

Fig. 5.24. Asymmetrical linear errors: *solid line* is RBF approximation and *circle* represents the measured data (forward run); *dottedline* is RBF approximation and *+line* represents the measured data (reverse run)

straight line. For the XY table there are two straightness error components to be determined: straightness of the X-axis which is concerned with deviation along the Y-axis, and straightness of the Y-axis which is concerned with deviation along the X-axis.

To model the straightness errors along the axes, the size of the RBF and the terminating condition are set to be the same as that used for the linear error RBF. After a total of about 10^6 iterations of the tuning algorithm, the weights converge in terms of the terminating condition specified. The corresponding RBF-based models for straightness are respectively

$$\delta_y(x) = f_{str,x}(x; \mathcal{W}^*_{strx}), \tag{5.14}$$

$$\delta_x(y) = f_{str,y}(y; \mathcal{W}^*_{stry}). \tag{5.15}$$

The excellence of the RBF-based models in approximating the straightness errors is illustrated in Figure 5.25 and Figure 5.26 respectively.

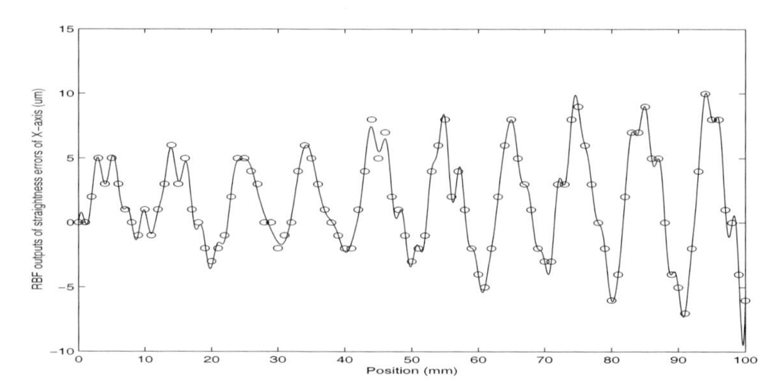

Fig. 5.25. RBF approximation of straightness errors (X-axis): *solid line* is RBF approximation and the *circles* represent the measured data

Angular Errors

Non-uniformity and distortion of the guideway also contribute to angular errors. According to Equations (5.6)-(5.7), only the yaw errors (Y-axis) need to be measured for the XY table with zero tool offsets. Pitch and roll errors are not relevant here since the XY table is a 2D motion system. Yaw error measurements are thus made along the travel path of the Y-axis to test for rotation about the axis perpendicular to the XY plane. The same design parameters as before are adopted. The weights tuning process converges after 10^6 iterations.

The yaw errors can be expressed as

$$\epsilon_y = f_{yaw,y}(y; \mathcal{W}^*_{yawy}). \tag{5.16}$$

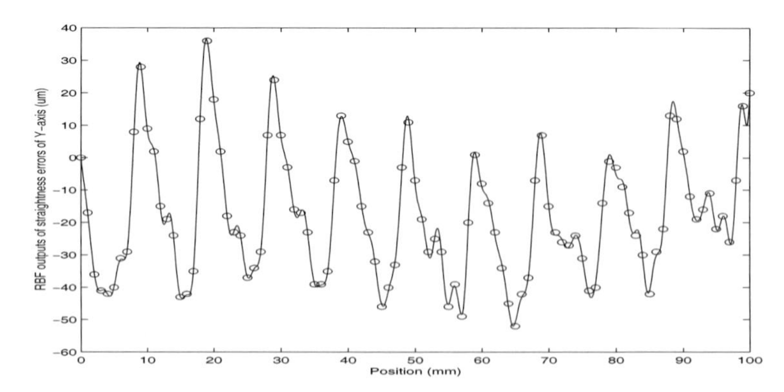

Fig. 5.26. RBF approximation of straightness errors (Y-axis): *solid line* is RBF approximation and the *circles* represent measured data.

The outputs of the RBF-based model follow the actual error measurements very closely as shown in Figure 5.27.

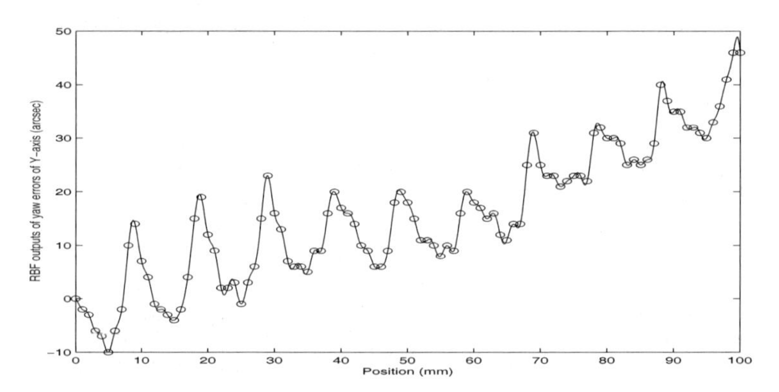

Fig. 5.27. RBF approximations of yaw errors (Y-axis): *solid line* is RBF approximation and the *circles* represent measured data

Squareness Error

Squareness between two axes characterises how far from a 90^0 orientation the two nominal axes are positioned relative to each other. It arises mainly during the assembly phase, where it is difficult to fix precisely a right angle between the X- and Y-axes. The squareness measurement can be accomplished by performing two straightness measurements, one of which is made based on a 90^0 reference. In this experiment, the Y-axis is chosen as the reference line for the squareness measurement, *i.e.*, the straightness of X is measured with respect to Y. Since the squareness measurement only yields a single constant of 141 arcsec, the RBF is not needed, in this case, for squareness error $\alpha = 141$ arcsec.

5.9.3 Experiments

An XY table is used as the testbed for the study. The tool attached to the table may be moved in either the X or Y-direction. The X and Y travel together span a 100×100 (mm) 2D space. The digital encoder resolution is 2.5 μm after a fourfold electronic interpolation, which also corresponds to the minimum step size. The motor uses screw threads for translating a rotation into a linear motion. Highly non-linear displacement errors are thus expected, in which case, linear interpolation may not be adequate and a non-linear error model will be necessary if high-precision requirements are to be satisfied. Figure 5.28 shows a picture of the XY table used. It is the primary objective of this section

Fig. 5.28. XY table testbed

to introduce an improved calibration method to reduce the positioning errors of the XY table arising from the geometrical errors.

Calibration of the XY Table

Error modelling typically begins with a calibration of the errors at selected points within the operational space of the machine. For a 3D working space, the resultant geometrical errors in positioning may be decomposed into 21 underlying components (Satori *et al.* 1995). For the XY table with zero tool offsets, the error sources reduce to six components, including two linear errors, two straightness errors, one angular error, and the orthogonality error between the X- and Y-axes. These errors may be measured accurately using an independent metrology system such as a laser interferometer which can typically measure linear displacement to an accuracy of 1 nm and angular displacement to an accuracy of 0.002 arcsec. These errors are subsequently cumulated using the overall error model to yield the overall positional error.

Assessment of Error Compensation

With the respective geometrical error models in terms of the RBF, error compensation can thus be carried out. The error compensations, based on the RBF approximations, are

$$\Delta x = f_{lin,x}(x; \mathcal{W}^*_{linx}) + f_{str,y}(y; \mathcal{W}^*_{stry}) \tag{5.17}$$

$$\Delta y = f_{lin,y}(y; \mathcal{W}^*_{liny}) + f_{str,x}(x; \mathcal{W}^*_{strx}) + f_{yaw,y}(y; \mathcal{W}^*_{yawy})x - \alpha x \tag{5.18}$$

Note that $x_p = y_p = 0$, since there is no probe used in this experiment. The error compensation with the RBF approximations is implemented using MATLAB®. To assess the performance of the error compensation, the two-axes are servo-controlled so that the carriage translated along two diagonals of the working area (see Figure 5.29). This provides a fair basis to gauge the adequacy of the RBF-based models, and complies with the recommended method by the British Standard (1989).

The linear errors are measured across the diagonals using an HP laser interferometer system. Ten points are measured of the linear displacement along the diagonal with and without the error compensation. Figure 5.30 shows the linear errors along the diagonal motion of the XY-table as the X- and Y-axes are servo-controlled from one corner A to the opposite corner D in the positive direction while Figure 5.31 shows the linear errors along the diagonal motion from one corner B to the opposite corner C in the positive direction.

For comparison, the same positions are calibrated without the error compensation to predict the same diagonals shown in Figures 5.32-5.33. It is clear that the diagonal errors have been reduced from about 160 μm to less than 55 μm after compensation with the RBF compensation method.

5.9.4 Error Modelling with Multi-layer Neural Networks

Geometrical error may also be modelled using multi-layer neural networks (NN). The approximation accuracy using multi-layer NN is generally better than that achieved using the linear-two-layer RBF. Moreover, the neuron number used in the multi-layer NN can possibly be less than the number of weights used in the linear-in-the parameter RBF for approximating the same system to a similar degree of accuracy. The main principles of error modelling with multi-layer NN will be highlighted in this subsection.

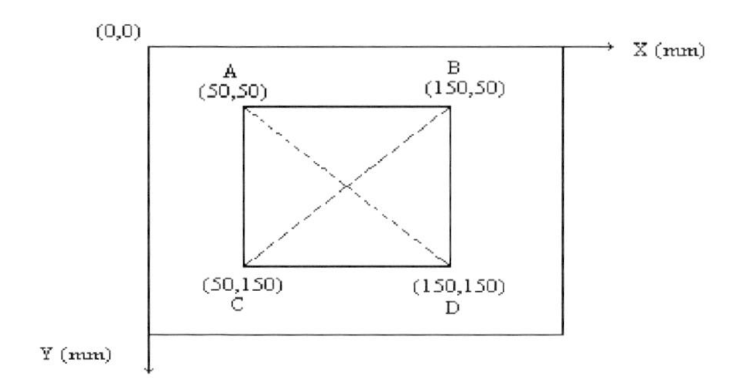

Fig. 5.29. Diagonal performance tests

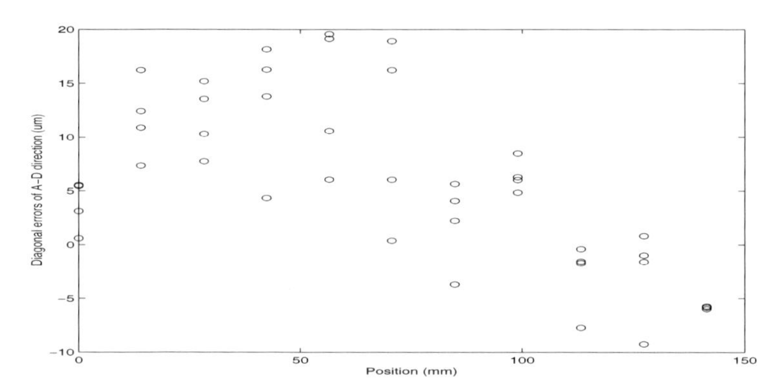

Fig. 5.30. Diagonal errors of A–D direction after compensation

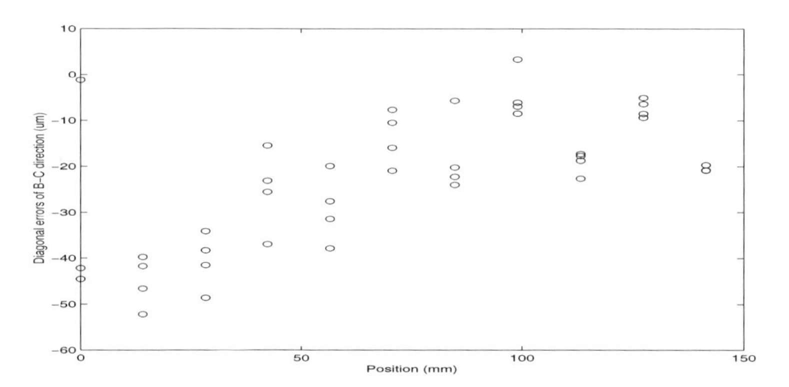

Fig. 5.31. Diagonal errors of B–C direction after compensation

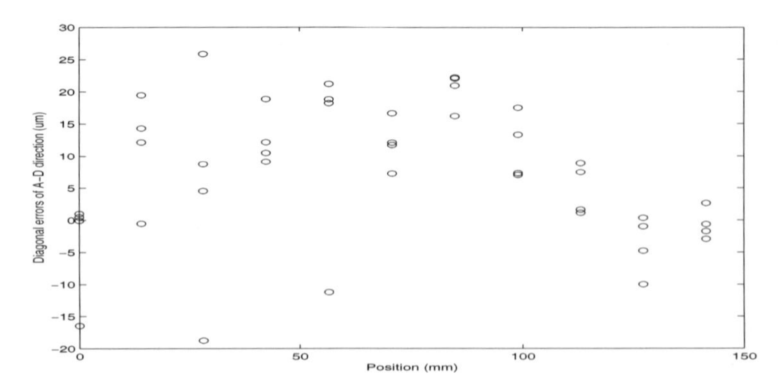

Fig. 5.32. Diagonal errors of A–D direction before compensation

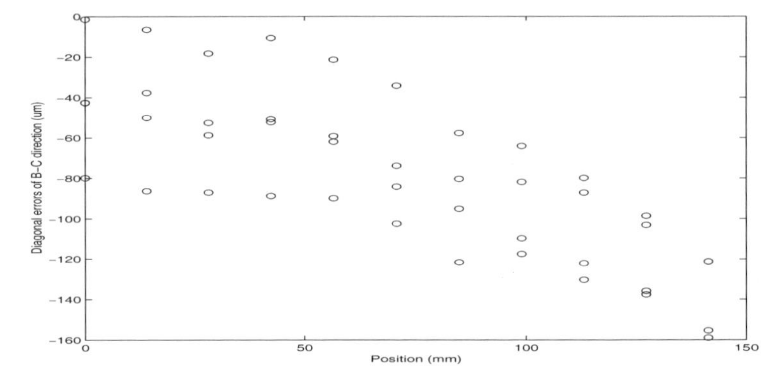

Fig. 5.33. Diagonal errors of B–C direction before compensation

The basic structure of a three-layer NN is shown in Figure 5.34. A three-layer NN can be used to approximate any continuous function to the desired accuracy. Given $X \in R^N$, a three-layer NN has a net output given by

$$X_{2k} = f(X; \mathcal{W}) = \sum_{h=1}^{N_1} \sigma \left[\sum_{j=1}^{N} \left[W_{1jk} \sigma \left[\sum_{i=1}^{N_0} W_{ij} X_i + \theta_{0j} \right] + \theta_{1k} \right] \right] \quad (5.19)$$
$$+\theta_2, k = 1, \ldots N_2,$$

with $\sigma(.)$ being the activation function, W_{ij} the first-to-second layer interconnection weights, and W_{1jk} the second-to-third layer interconnection weights. θ_{1j}, θ_{2k}, are threshold offsets. It is usually desirable to adapt the weights and thresholds of the NN off-line or on-line to achieve the required approximation performance of the net, *i.e.*, the NN should exhibit a "learning behavior".

To obtain the NN weights W, appropriate weight tuning algorithms should be adopted. A commonly used weight tuning algorithm is the gradient search algorithm based on a backpropagated error, where the NN is trained off-line to match specified exemplary pairs (χ_d, Υ_d), with χ_d being the ideal NN input that yields the desired NN output Υ_d. The learning procedure aims at driving the total error to near zero *via* suitable adjustments of the learning parameters. This essentially constitutes a minimisation problem that the gradient search techniques will attempt to solve. With the appropriate choice of hidden nodes, the net can usually be driven close to the desirable accuracy. This can usually be done by specifying a large number of hidden nodes in the network structure or by starting with a small number of hidden units and increasing the number until it becomes possible to drive the approximation error to within a small threshold.

For the application of geometrical error modelling, the NN can be designed as a single-input and single-output (SISO) function. The input is connected directly to a neuron node. The input will correspond to the nominal measurements from the respective encoders. The NN output will attempt to follow closely the laser measurement, with non-linear interpolation where necessary. The gradient search algorithm based on the backpropagation error for the NN is summarised as follows

A. Compute the output of the INPUT layer, $\bar{X}_1$

$$\bar{X}_1 = \frac{1}{1 + exp(-\bar{O}_1 + \bar{\theta}_{01})},$$

where $\bar{O}_1 = W_1 X_1$, and X_1 is the input (or input sample) of the NN.

B. Compute the output of the HIDDEN layer, X_{1j}

$$X_{1j} = \frac{1}{1 + exp(-O_{1j} - \theta_{1j})}, \tag{5.20}$$

where $O_{1j} = W_{1j}\bar{X}_1, j = 1, 2, ..., N$.

C. Compute the output of OUTPUT layer, X_{21}

$$X_{21} = \sum_{j=1}^{N} W_{1j1} X_{1j} + \theta_2, \tag{5.21}$$

where X_{21} is the output of the NN.

D. Update the weights from HIDDEN to OUTPUT layer, W_{1j1} according to

$$W_{1j1}^{t+1} = W_{1j1}^{t} + \eta_1 \delta_1 X_{1j}, \tag{5.22}$$

where $\delta_1 = -(X_{21}^d - X_{21})$ with X_{21}^d being the desired output (or output sample), and X_{21} being the NN output.

E. Update the weights from INPUT to HIDDEN layer, W_{1j}

$$W_{1j}^{t+1} = W_{1j}^{t} + \eta_2 \delta_{2j} \bar{X}_1, \tag{5.23}$$

where $\delta_{2j} = [\delta_1 W_{1j1}] X_{1j}(1 - X_{1j})$.

F. Update the weights for INPUT layer, W_1,

$$W_1 = W_1 + \eta_3 \delta_3 X_1, \tag{5.24}$$

where $\delta_3 = [\sum_{j=1}^{N} \delta_{2j} W_{1j}] \bar{X}_1 (1 - \bar{X}_1)$.

G. Update the thresholds, $\theta_2, \theta_{1j}, \theta_{01}$

$$\theta_2^{t+1} = \theta_2^t + \eta_{1\theta}\delta_1, \theta_{1j}^{t+1} = \theta_{1j}^t + \eta_{2\theta}\delta_{2j}, \theta_{01}^{t+1} = \theta_{01}^t + \eta_{3\theta}\delta_3, \tag{5.25}$$

where $\eta_1, \eta_{2j}, \eta_3, \eta_{1\theta}, \eta_{2\theta}$, and $\eta_{3\theta} > 0$ are gain factors.

A terminating condition for the training process is usually formulated, that is

$$\mathcal{E} = \frac{1}{2} \sum_{l=1}^{M} (X_{21}^{dl} - X_{21}^{l})^2, \tag{5.26}$$

where l represents the sample number. The iterative weights tuning process is terminated when the errors converge within a specified threshold. Thus, the optimum weights W can be obtained. The training is usually a trade-off in terms of the quality of fit and the iteration time. Since the training process can be done off-line, more emphasis may be given to deriving a better fit at the expense of incurring a longer tuning time.

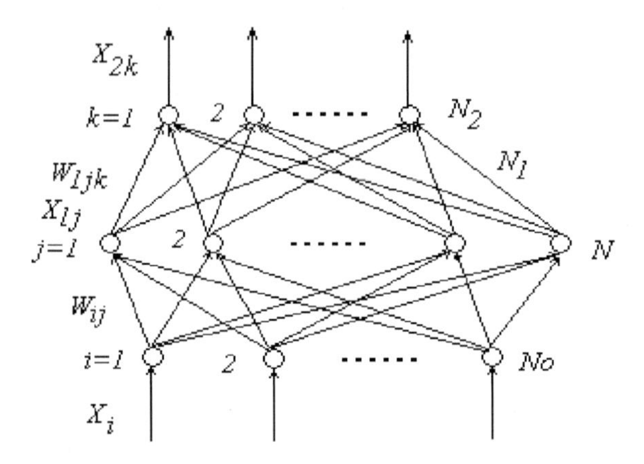

Fig. 5.34. Basic structure of a multi-layer NN

In this way, the NN can be applied to model the individual geometrical error components, similar to the case when RBF is used.

5.10 Compensation of Machines with Random Errors

The effectiveness of any soft error compensation scheme relies very much on the reliability of the geometrical error information obtained from the prescribed calibration methodology. By error compensation, it is the systematic component of the geometrical error which is relevant. Errors of a random nature cannot be compensated. Unfortunately, the error measurements obtained during the machine calibration will inevitably contain both systematic and random components. It is essential to separate the two components effectively so as to compensate correctly for the systematic error. Most of the existing error compensation methods may not adequately address the influence of random errors on the compensation of the systematic errors. The geometrical error to be compensated is usually based on the mean value of the error samples calibrated. This mean value is essentially assumed to be the equivalent systematic error component which can be compensated, since the random ones are assumed to be filtered after the averaging process. Unfortunately, this is not true in general and compensation based on the mean value may lead to a grossly inadequate compensation, especially when there exists significant random errors during the calibration process. Invalid samples or outliers arising due to short and momentary disturbances or noise during the calibration exercise can distort the mean to deviate far from the actual systematic error present. Consequently, using the error mean can lead to insufficient or excessive compensation. Thus, a weighted approach taking into account the probabilities associated with different magnitudes of the random error component in the error sample can extract a more accurate and reliable data set of systematic errors for compensation. Such a statistical approach requires a large set of samples to be calibrated, in order to have a sufficiently large data density to operate on. This can be easily facilitated by modern automated and DSP-based data logging and analysis systems.

In this section, the use of a statistical approach is explored to reduce the adverse influence of random errors, as an inevitable part of the calibrated data set, on the compensation of systematic errors. The basic idea is to deduce and isolate the most likely systematic geometrical error components from a data set which is infiltrated with random ones by appropriately analysing the probability of the magnitude of random errors. The approach is simple and directly amenable to practical applications. It consists of three main steps. First, from the geometrical error information collected, the error band is split into small consecutive classes. Second, the probability of the random error falling into each class is calculated based on the density of data clusters falling within the class. These probabilities are used in a statistical analysis to deduce the most probable systematic error from the data set. Finally, compensation is made based on this statistically deduced error. The statistical approach has the advantage that the influence of random errors, including those arising due to intermittent faults, noise and disturbances, can be largely removed and isolated from adversely affecting the compensation effort. This advantage becomes significant when the precision requirements of the machine get correspondingly higher. Experimental results are provided on the linear error compensation of a single-axis piezo-ceramic motion system. Consistent with expectations, they show that the compensation error obtained statistically according to the method is quite different from the error mean of conventional methods. Enhanced compensation can be achieved as a result.

5.10.1 Probabilistic Methodology

The random error $\delta(x)$ is defined as the deviation of the actual error measurement from the mean computed over the entire data set. Mathematically, this can be expressed as

$$\delta(x) = \Delta y(x) - \Delta \bar{y}(x), \tag{5.27}$$

where Δy is the actual error measurement, and $\Delta \bar{y}$ is the aforementioned mean computed from the data set, *i.e.*,

$$\Delta \bar{y} = \frac{1}{n} \sum_{k=0}^{n} \Delta y^{(k)}. \tag{5.28}$$

n is the total number of measurements available. In general, the expectation $E[\delta] \neq 0$.

It is well known that only systematic errors can be compensated. Random errors which can possibly occur during the calibration process should be minimised ideally by a proper system design, since they cannot be compensated. The main difficulty, which is being addressed in this section, is not to attempt to compensate for random errors, but to reduce the influence of the random errors from adversely affecting the compensation of the systematic errors, since both the systematic and the random components co-exist in the same error measurement. The challenge is how to distillate the two components effectively so that compensation can be based purely on the systematic error only.

Most approaches use $\Delta \bar{y}$ to make the final compensation. In these approaches, $\Delta \bar{y}$ is assumed to be free of the influence of random errors and thus it is taken to be equivalent to the systematic component which can be compensated. Unfortunately, these methods will yield a grossly inadequate compensation when a significant amount of random errors is present in the error measurements. Invalid samples or outliers arising due to short and momentary disturbances or noise during the calibration exercise can distort the mean to deviate far from the actual systematic error present. Consequently, using just the sample mean can lead to insufficient or excessive compensation. Thus, a weighted approach to filter the influence of random errors using a statistical analysis may be used to extract a more accurate and reliable data set of systematic errors for machine compensation.

To this end, the error band is split into sub-classes. Consider a one-dimensional space that consist of $2k+1$ error classes s_j, $j = 0, \pm1, \pm2, ..., \pm k$. The random error δ defined in Equation (5.27) is assumed to fall into the s_j class if

$$(2j-1)L < \delta \leq (2j+1)L,\ j = 0, \pm1, \pm2, ..., \pm k, \tag{5.29}$$

where L is the class size. While it is desirable for the class size to be as small as possible to maximise the resolution of errors across the classes, a small L will correspondingly requires a large data set. Thus, the actual class size used should depend on the resources available in the acquisition of the raw data set.

The probability of a random error with a magnitude falling into each of these classes can be computed from the number of error samples within the class. If the class s_j contains n_j random error samples, then

$$P(s_j) = \frac{n_j}{n}. \tag{5.30}$$

This definition of probability clearly satisfies the axioms of the axiomatic theory when $n \to \infty$.

Ideally, if $\Delta\bar{y}$ is an adequate representation of the systematic component of the geometrical error, most random errors are expected to be contained in the set s_0, *i.e.*, the random error class containing $\delta = 0$ should be registered with the highest probability. Otherwise, any bias phenomenon in the random error should be used to offset the error mean for a more accurate re-construction of the systematic error component. This mean offset $\bar{\delta}$ can be obtained from the probabilities associated with the various error classes $(P(s_j),\ j = 0, \pm1, \pm2, ..., \pm k\ )$ as

$$\bar{\delta} = \sum_j \tilde{\delta}_j P(s_j), \tag{5.31}$$

where $\tilde{\delta}_j = 2jL$ refers to the center of the s_j class, *i.e.*, $[(2j-1)L, (2j+1)L]$. The "most likely" systematic error, in a statistical sense, can thus be computed to be $\Delta\bar{y} + \bar{\delta}$.

Finally, the error compensation can be based on this statistically deduced systematic error.

5.10.2 Experiments

In this section, experimental results are provided to illustrate the effectiveness of the statistical method when applied to linear error compensation of a linear piezo-ceramic motion system. Figure 5.35 shows the experimental set-up.

Fig. 5.35. Experimental set-up

The piezo-ceramic motor used is a single axis servo motor manufactured by Nanomotion Ltd, which has an optical encoder with an effective resolution of 0.1 μm. The main sources of random errors arising in this application include the varying contact friction conditions, the stick-slip effect, clearances in shift mechanisms, deformations of bearing elements, transients from the sensor electronics, changes in ambient temperature, and even minor structural vibrations.

dSPACE control development and rapid prototyping system, in particular the all-in-one DS1102 board, is used for both calibration and compensation purposes. The error compensation is carried out in the form of a look-up table embedded into a Simulink® control block.

Calibration is carried out using a laser interferometer at specific points located at 5 mm intervals along the 50 mm travel. Hence, the linear errors at 10 points along the linear piezo-motor are measured. Figure 5.36 shows the raw data set of linear errors collected from 40 complete bi-directional travels, *i.e.*, $n = 40$. The average values at each of these points from the 40 samples are computed and tabulated in Table 7.1.

The combined error data band is split into sub-classes with L=2 μm and the probability of the random error component falling within each class computed accordingly. Figures 5.37–5.39 show a clear bias phenomenon in the random error from zero. The probability of random error falling in class s_0 is lower than that in class s_1 in eight out of the ten calibration positions.

Table 5.1. Mean errors

SP	5	10	15	20	25	30	35	40	45	50
ME	1.1	2.2	3.4	4.9	6.0	7.8	8.9	10.7	12.7	14.0

SP: Sample Point (mm), ME: Mean Error $\Delta\bar{y}$ (μm)

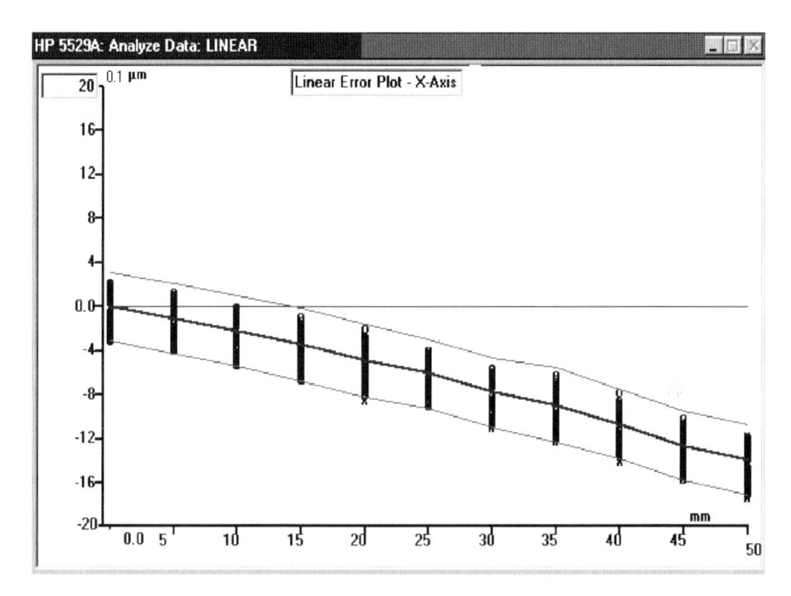

Fig. 5.36. Raw data set for linear errors

This indicates that the sample mean does not adequately reflect the actual systematic error present.

Compensation is then made based on these mean values. Re-calibration is carried out after compensation at the same positions, and the experimental results are shown in Figures 5.40–5.42. As expected, from these figures, it is evident that, after compensation, the "likely" random error class continues to exhibit a strong bias from class s_0.

Using the statistical method, the mean offsets ($\bar{\delta}$) are computed and tabulated in Table 7.2. These values are used to adjust the mean values in Table 5.1 for compensation. Calibration is again carried out at the same ten positions after the error compensation. The random errors arising after compensation are shown in Figures 5.43–5.45. Comparing these with Figures 5.40–5.42, the bias phenomenon has been significantly reduced. The statistical method has achieved a more accurate compensation of systematic errors with respect to the influence of random errors.

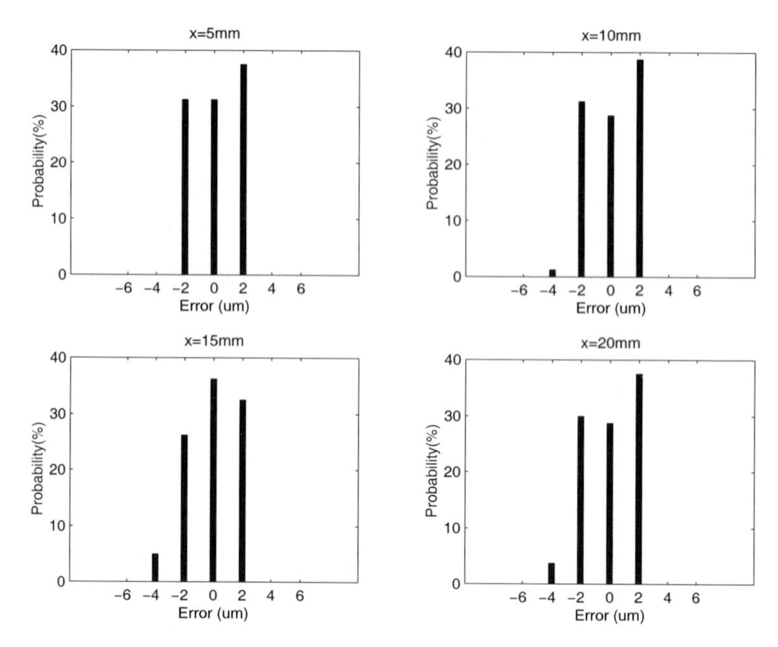

Fig. 5.37. Error probability analysis (5mm–20mm)

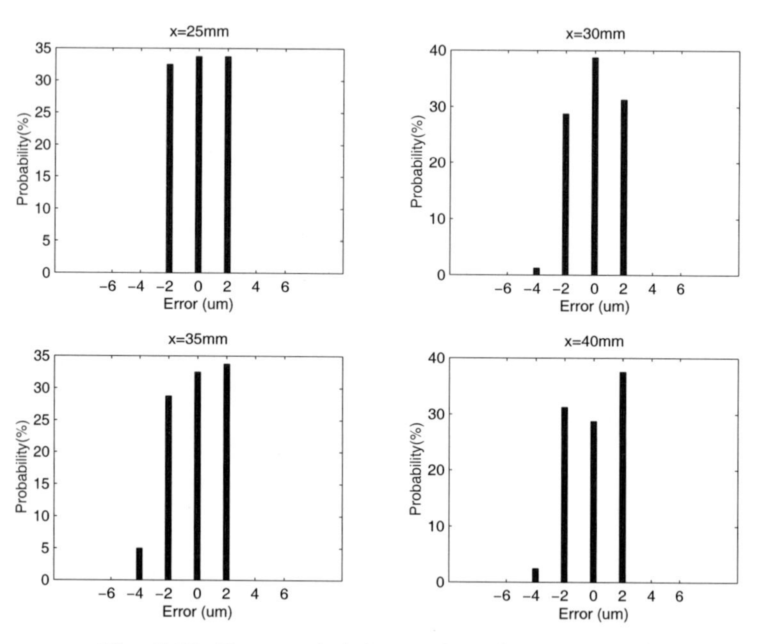

Fig. 5.38. Error probability analysis (25mm–40mm)

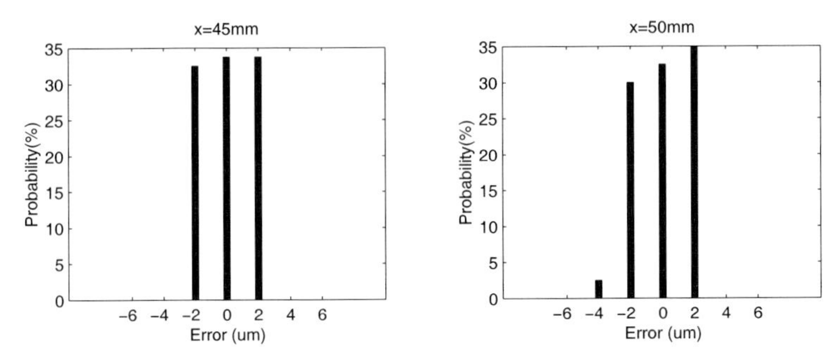

Fig. 5.39. Error probability analysis (45mm–50mm)

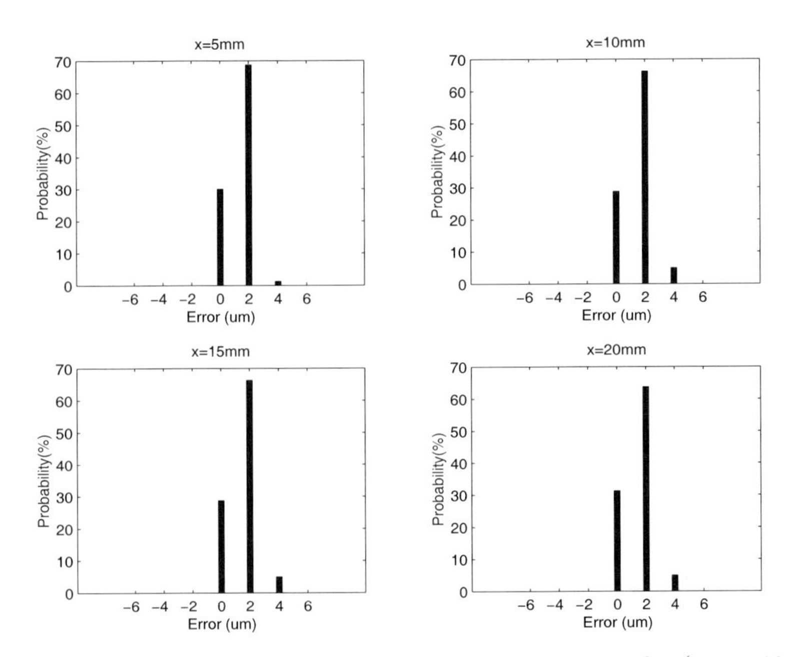

Fig. 5.40. Probability analysis after compensation with mean value (5 mm–20mm)

Table 5.2. Mean offsets

SP	5	10	15	20	25	30	35	40	45	50
MOV	2	2	0	2	1	0	2	2	1	2

SP: Sample Point (mm), MOV: Mean Offset Value $\bar{\delta}$ (μm)

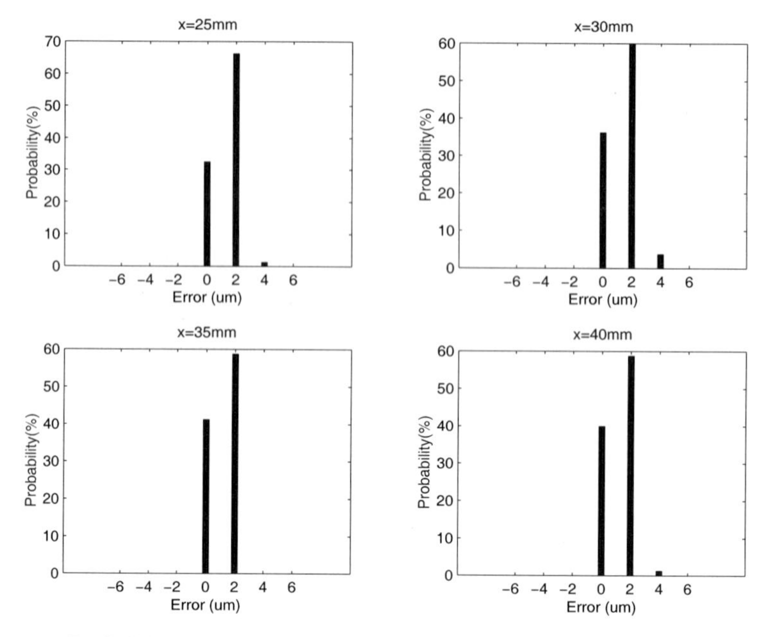

Fig. 5.41. Probability analysis after compensation with mean value (25mm–40mm)

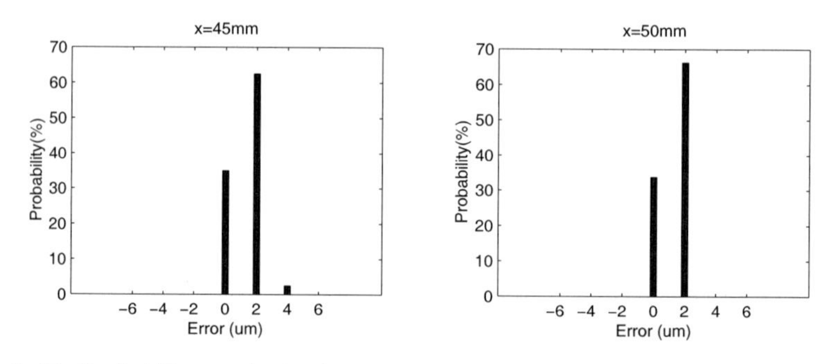

Fig. 5.42. Probability analysis after compensation with mean value (45mm–50mm)

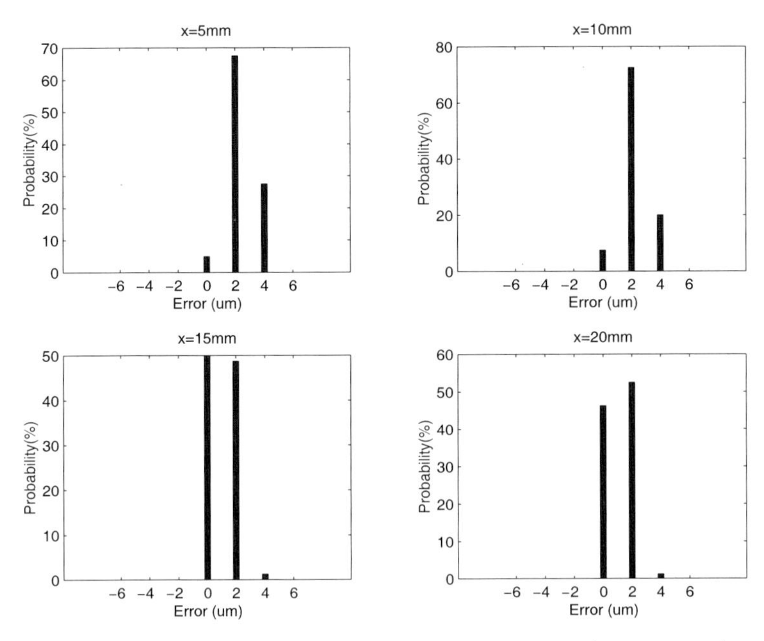

Fig. 5.43. Probability analysis after compensation (5mm–20mm)

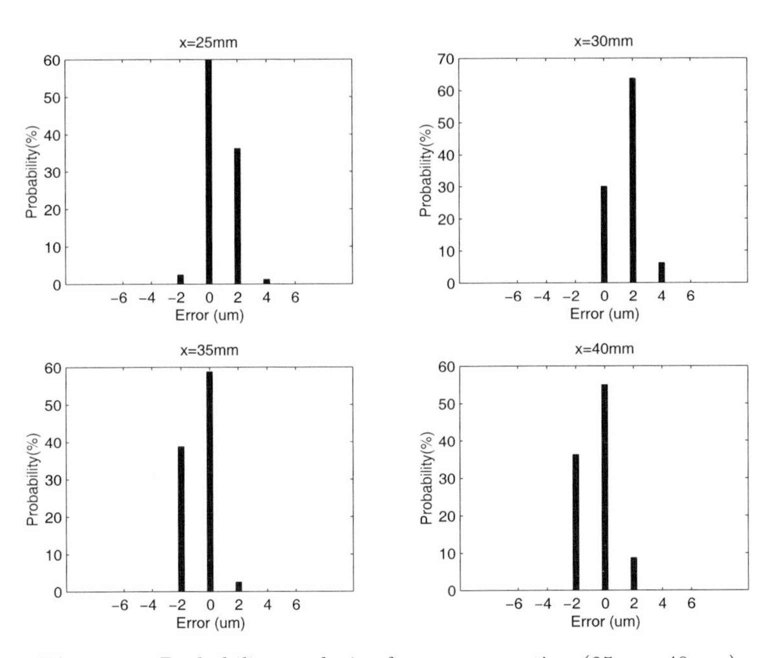

Fig. 5.44. Probability analysis after compensation (25mm–40mm)

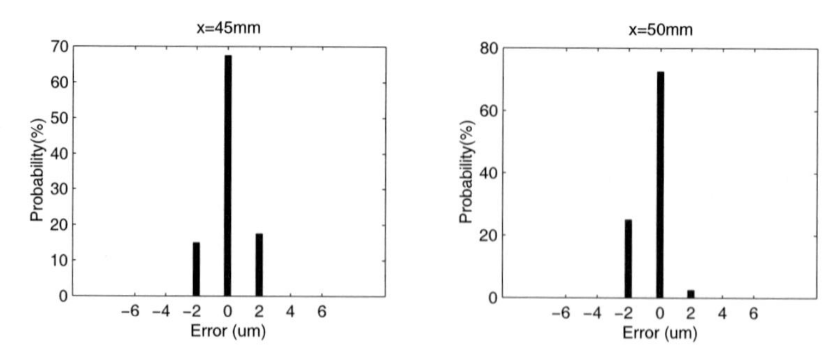

Fig. 5.45. Probability analysis after compensation (45mm–50mm)

6

Electronic Interpolation Errors

High-precision and resolution motion control relies critically on the precision and resolution of positional measurements achievable from the encoders. These factors are in turn limited by the technology behind the manufacturing of encoders. To date, the scale grating on linear optical encoders can be manufactured to less than 4 μm in pitch, but clearly, further reduction in pitch is greatly constrained by physical considerations in the manufacturing processes. This implies an optical resolution of 1 μm is currently achievable after a fourfold interpolation. Analog interpolation using soft techniques will provide an interesting possibility to improve further on the encoder resolution, by processing the analog encoder signals online to yield the small intermediate positions.

The error sources associated with position information obtained this way can be classified into pitch and interpolation errors. Pitch errors are due to scale manufacturing tolerances and mounting distortion. They can be compensated *via* the same procedures which are carried out for general geometrical error compensation. Interpolation errors are associated with the accuracy of subdivision within a pitch, affecting any calibration performed. Ideal signals from encoders are a pair of sinusoids with a quadrature phase difference between them. Interpolation operates on the relative difference in amplitude and phase of these paired sinusoids. Therefore, interpolation errors will occur if the pair-periodic signals deviate from the ideal waveforms on which the interpolation computations are based. These deviations must be corrected before interpolation, using digital signal processing techniques, to reduce the interpolation errors.

This chapter addresses the nature of electronic interpolation errors, and various approaches to address the calibration and compensation of these errors.

6.1 Heydemann Interpolation Method

The signals from analog encoders usually deviate from the ideal quadrature sinusoidal signals. The technology to compensate the mean value errors, phase and amplitude errors for two quadrature sinusoidal signals was first introduced by Heydemann (1981). He used least squares fitting to compute these error components efficiently and made correction for the two non-ideal sinusoidal signals. Using this method, Birch (1990) was able to calculate optical fringe fractions to nanometric accuracy. By making use of the amplitude variation with angle, Birch divided one period of sinusoidal signal into N equiangular segments to increase the effective electrical angle resolution. The principle of Heydemann interpolation method will be described in this section.

Denote the ideal phase quadrature signals as u_1 and u_2 respectively. These are identical sinusoidal signals displaced by a phase of $\pi/2$ with respect to each other, given by

$$\begin{aligned} u_1 &= A\cos\delta, \\ u_2 &= A\sin\delta, \end{aligned}$$

where δ denotes the instantaneous phase. If these signals are used to produce a Lissajous figure, a rotating vector will be obtained which will describe a circle of radius A. One revolution of the vector is equivalent to an optical phase change of 2π. The instantaneous phase can thus be directly obtained from

$$\delta = \tan^{-1}\frac{u_2}{u_1}.$$

δ varies from $-\pi/2$ to $\pi/2$, whereas, for ease of fringe fractioning, a phase range of 0 to 2π is desired. This can be achieved by an appropriate polar adjustment:

$$\theta = \delta + \sigma,$$

where

$$\begin{aligned} \sigma &= 0, \quad u_1, u_2 > 0, \\ &= \pi, \quad u_1 < 0, \\ &= 2\pi, \quad u_1 > 0, u_2 < 0. \end{aligned}$$

From the viewpoint of the encoder, θ corresponds to an incremental translation Δx given by

$$\Delta x = \frac{\theta}{2\pi} x_p,$$

where x_p is the pitch period of the scale. Therefore, the resolution with which δ may be identified determines the maximum interpolation achievable with the particular optical encoder. Commercial interpolation of 4096 times has been available using 12-bit A/D converters. Theoretically, arbitrarily high interpolation can be achieved using sufficiently long A/D registers.

In reality, however, the encoder signals will deviate from the ideal. First, the actual signals may have dissimilar mean values, m_1 and m_2 respectively, due mainly to the bias influence of electronic signal processing. Second, their amplitudes may not be the same, due to variable gains in the associated detection systems. Their ratio is denoted as $G = \frac{A_1}{A_2}$, where ideally, $G = 1$. Third, there may be an additional phase shift ϵ on top of the $\pi/2$ phase shift. Finally, there is still the presence of waveform distortion, noise as well as any drifting influence due to varying enviromental factors which imply the non-ideal factors may also be slowly time-varying.

The actual equations describing the encoder signals are thus

$$\begin{aligned}\tilde{u}_1 &= u_1 + m_1,\\ \tilde{u}_2 &= \frac{A\sin(\delta - \epsilon)}{G} + m_2 = \frac{u_2 \cos\epsilon - u_1 \sin\epsilon}{G} + m_2.\end{aligned}$$

Combining the equations yields

$$A^2 = (\tilde{u}_1 - m_1)^2 + \frac{[(\tilde{u}_2 - m_2)G + (\tilde{u}_1 - m_1)\sin\epsilon]^2}{cos^2\epsilon}.$$

Direct simplification yields

$$k_1\tilde{u}_1^2 + k_2\tilde{u}_2^2 + k_3\tilde{u}_1\tilde{u}_2 + k_4\tilde{u}_1 + k_5\tilde{u}_2 = 1, \tag{6.1}$$

where

$$\begin{aligned}k_1 &= [A^2\cos^2\epsilon - m_1^2 - G^2 m_2^2 - 2Gm_1 m_2 \sin\epsilon]^{-1},\\ k_2 &= k_1 G^2,\\ k_3 &= 2k_1 G \sin\epsilon,\\ k_4 &= -2k_1[m_1 + Gm_2 \sin\epsilon],\\ k_5 &= -2k_1 G[Gm_2 + m_1 \sin\epsilon].\end{aligned}$$

Equation (6.1) is in the linearly parameterised form which is suitable for using a least squares fitting routine to derive the estimates of k_1 to k_5, from which the offset parameters may be derived as follow:

$$\begin{aligned}\epsilon &= \sin^{-1}\left[\frac{k_3}{\sqrt{4k_1k_2}}\right],\\ G &= \sqrt{\frac{k_2}{k_1}},\\ m_1 &= \frac{2k_2k_4 - k_3k_5}{k_3^2 - 4k_1k_2},\\ m_2 &= \frac{2k_1k_5 - k_3k_4}{k_3^2 - 4k_1k_2},\\ A &= \frac{\sqrt{4k_2(1 + k_1m_1^2 + k_2m_2^2 + k_3m_1m_2}}{4k_1k_2 - k_3^2}.\end{aligned}$$

The estimates derived in this way are unbiased and their effects to noise (variance) can be reduced using a larger data set.

Consequently, the corrected signals $\bar{u}_1$ and $\bar{u}_2$ may be obtained as

$$\tilde{u}_1 = (\bar{u}_1 - m_1)/A,$$
$$\tilde{u}_2 = \frac{(\bar{u}_1 - m_1)\sin\epsilon + G(\bar{u}_2 - m_2)}{A\cos\epsilon}.$$

The corrected phase may be obtained as

$$\bar{\delta} = \tan^{-1}\frac{\bar{u}_2}{\bar{u}_1}.$$

$\bar{\theta}$ can be calculated accordingly, and the relative displacement is derived as

$$\Delta\bar{x} = \frac{\bar{\theta}}{2\pi}x_p,$$

where x_p denotes the grating pitch. The uncorrected displacement is

$$\Delta\tilde{x} = \frac{\tilde{\theta}}{2\pi}x_p.$$

Therefore, the interpolation error to be compensated is given by

$$e_\Delta(\tilde{x}) = \Delta\bar{x} - \Delta\tilde{x} = \frac{x_p}{2\pi}(\bar{\theta} - \tilde{\theta}).$$

This error can be stored for subsequent online correction purposes.

6.1.1 Interpolation Bounds

In order that the translation Δx can be measured to a resolution of R, it follows that θ must be derived to a resolution of $\theta_r \leq 0.5\pi R$. Therefore, if R=0.01 μm, the phase must be derived to a precise resolution of $\theta_r = 0.0157$ radians. This subsequently poses requirements on the precision with which the signals u_1 and u_2 are to be measured which may be directly derived. Table 6.1 provides sufficient measurement precision requirements for a Heidenhein linear encoder LIP401 with a typical signal period of 4 μm with 1 Vpp analog output signals.

The signal measurement precision also determines the maximum noise amplitude allowable for the interpolation desired. Conversely, given the noise amplitude, the achievable encoder interpolation is fixed irregardless of the measurement precision.

Table 6.1. Recommended signal measurement resolution

Interpolation	Position resolution R, (μm)	Signal resolution, (V)	A/D word length, (no. of bits)
40	0.1	0.05	5
400	0.01	0.005	8
4000	0.001	0.0005	12

6.1.2 Calibration and Compensation

Signals $\tilde{u}_1$ and $\tilde{u}_2$ can be obtained over the entire length of the scale by translating the encoder head along the scale at low velocity. The offset parameters may be obtained from the signals at specific indices of the encoder. The interpolation errors e_Δ are calculated accordingly.

Compensation of the interpolation errors involves a three-steps process. First, the errors are calibrated over the entire length of the optical scales. Second, the errors are modelled, usually in a non-parametric form (*e.g.*, a look-up table), or alternatively using a parametric model. The error compensation is then done on-line based on the error model.

6.2 Enhanced Interpolation Method

The Heydermann interpolation approach generally requires explicit high-precision analog-to-digital converters in the control system, and a high speed DSP to compute the electrical angle to the required resolution. Therefore, they are inapplicable to the typical servo controller with only digital incremental encoder interface. Furthermore, it is cumbersome to integrate sinusoidal correction with interpolation since the correction parameters must be calibrated off-line. As a result, most servo controllers, which are able to offer interpolation, have assumed perfect quadrature sinusoids. Consequently, specifications relating to resolution may be achievable, but the accompanying accuracy cannot be guaranteed. Current effort for sinusoids correction also does not consider error in the form of waveform distortion, *i.e.*, the actual signal may be periodic, but it may not be perfectly sinusoidal. These errors are certainly significant when sub-micron resolution and accuracy is required.

This section presents a new method to carry out both correction and interpolation, independent of the servo controller. As a result, the method is applicable to most servo controllers, including those with only digital incremental encoder interfaces. The basic idea is to derive high order sinusoids based on existing quadrature sinusoids from the encoder. These high order signals may in turn be converted to a series of high frequency binary pulses, which are readily decoded by standard servo controllers. A look-up table is used to implement the idea with little computational requirements. Sinusoidal corrections, including mean and phase offsets, amplitude difference and waveform distortion, can be directly reflected in the look-up table. The table can be updated adaptively and online to reflect any subsequent changes or drift in the encoder signals. Simulation and experiment results are provided to highlight the principles and applicability of the enhanced method.

6.2.1 Principle of Enhanced Interpolation Method

The basic idea of the enhanced interpolation method is to derive high order sinusoids based on the fundamental one. From these, binary pulses can be generated which can be readily decoded by standard servo controllers for position information. As an example, given the values of $\sin(\alpha)$ and $\cos(\alpha)$, $\sin(2\alpha)$ and $\cos(2\alpha)$ can be obtained from the trigonometry relations:

$$\begin{aligned} \sin(2\alpha) &= 2\sin(\alpha)\cos(\alpha), \\ \cos(2\alpha) &= 1 - 2\sin^2(\alpha). \end{aligned} \tag{6.2}$$

In general, assuming $\sin(\alpha)$ and $\cos(\alpha)$ are known with sufficient precision, $\sin(n\alpha)$ and $\cos(n\alpha)$ ($n \in$Z$>$ 1) can be derived from the following general equations:

$$\begin{aligned} \sin(n\alpha) &= n\cos^{n-1}(\alpha) - C_n^3\cos^{n-3}(\alpha)\sin^3(\alpha) \\ &\quad + C_n^5\cos^{n-5}(\alpha)\sin^5(\alpha) - \dots, \\ \cos(n\alpha) &= \cos^n(\alpha) - C_n^2\cos^{n-2}(\alpha)\sin^2(\alpha) \\ &\quad + C_n^4\cos^{n-4}(\alpha)\sin^4(\alpha) - \dots. \end{aligned} \tag{6.3}$$

Using an electronic comparator to detect zero crossings, quadrature binary pulses may in turn be obtained from $\sin(n\alpha)$ and $\cos(n\alpha)$. These pulses are more readily decoded using most standard servo controllers or CNC systems for position information. A further four times interpolation can be obtained from these signals. The method eliminates the need for precision analog-to-digital signal acquisition and processing units within the control system for interpolation purposes, since interpolation has been done independent of the controller.

A look-up table will serve as the inferencing engine to provide the signal interpolation (Section 6.2.2). Errors in the originating encoder signals can then be directly reflected in the entries of the look-up table without any separate correction mechanisms. These errors will include waveform distortion error, apart from the usual mean and phase offsets.

6.2.2 Construction of a Look-up Table

Before interpolation, it is important to correct the errors in the originating encoder signals according to Section 6.1, While $\sin(n\alpha)$ and $\cos(n\alpha)$ can be computed from Equation (6.3), it is too inefficient to be viable when the encoder signals are to be processed at high speed, especially when n is large. A look-up table can be designed instead for this purpose. The table can output directly the values of $\sin(n\alpha)$ and $\cos(n\alpha)$, given the inputs $\tilde{u}_1 = \sin(n\alpha)$ and $\tilde{u}_2 = \cos(n\alpha)$.

Look-up Table Based on $\tilde{u}_1$ Only

To simplify the inferencing procedure, the values of $\sin(n\alpha)$ and $\cos(n\alpha)$ can be pre-computed and recorded corresponding to pre-determined samples of either $\tilde{u}_1$ and $\tilde{u}_2$, and the sign of the other (for illustration, $\tilde{u}_1$ and the sign of $\tilde{u}_2$ will be used for this purpose). To simplify the addressing of the table, these samples are obtained at equal intervals over the entire amplitude range from –1 to 1 (instead of over the entire range of electrical angle over one period).

As an example, consider $n = 16$ and $s = 1024$. The look-up table is accordingly set up as in Table 6.2 for one period.

Table 6.2. Look-up table based on $\tilde{u}_1$ only

Index	$1,\ldots,s-1,s$	...	$3s+1,\ldots,4s-1,4s$
Range	$0 \sim \pi/2$	...	$3\pi/2 \sim 2\pi$
$\tilde{u}_1$	$1/s,\ldots,(s-1)/s,1$	...	$-1,\ldots,-(s-1)/s,-1/s$
$\sin(16\alpha)$	$0.016,\ldots,-0.649,0.000$	...	$0.649,\ldots,-0.016,0.0$
$\cos(16\alpha)$	$0.999,\ldots,0.760,1.000$	...	$0.760,\ldots,0.999,1.0$

Given the real-time value of $\tilde{u}_1$ and sign of $\tilde{u}_2$, the associated table entry can be directly located since the sample interval is fixed and known. Table 6.3 serves as the search table to locate the relevant entries efficiently. Indices N_s, N_c, N_0 are first defined as

$$\begin{aligned} N_s &= round(\mathrm{s}\times\tilde{u}_1), \\ N_c &= \tilde{u}_2, \\ N_s &= s \end{aligned} \tag{6.4}$$

One potential problem with this tabulation method arises due to the large non-linear variation of the amplitude of $\tilde{u}_1$ with the electrical angle α. Using pre-recorded samples of $\tilde{u}_1$, equally spaced in amplitude, will mean a varying interval of the corresponding angle as shown in Figure 6.1.

This angle resolution is poor near the vicinity of $\tilde{u}_1 = \sin(\alpha) \approx 1$. Thus, to have sufficient information pre-recorded from this part of the signal, s must

Table 6.3. Index table

Condition	n_{index}	Range of α	$\tilde{u}_1$ or $\tilde{u}_2$ used
$N_s > 0$ and $N_c > 0$	N_s	$0 \sim \pi/2$	$\tilde{u}_1$
$N_s > 0$ and $N_c < 0$	$2N_0 - N_s$	$\pi/2 \sim \pi$	$\tilde{u}_1$
$N_s \leq 0$ and $N_c > 0$	$3N_0 + N_s$	$3\pi/2 \sim 2\pi$	$\tilde{u}_1$
$N_s \leq 0$ and $N_c < 0$	$3N_0 - N_s$	$\pi \sim 2\pi/3$	$\tilde{u}_1$

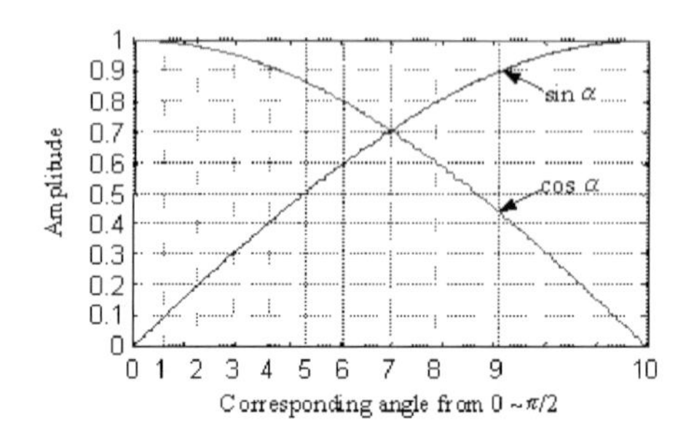

Fig. 6.1. Variation of amplitude against angle

be very large which will correspondingly imply a large look-up table. Figure 6.2 shows the interpolation result, when $s = 5000$ and $n = 64$. The waveforms of $\sin(64\alpha)$ and $\cos(64\alpha)$ are distorted around $\sin(\alpha) \approx 1$.

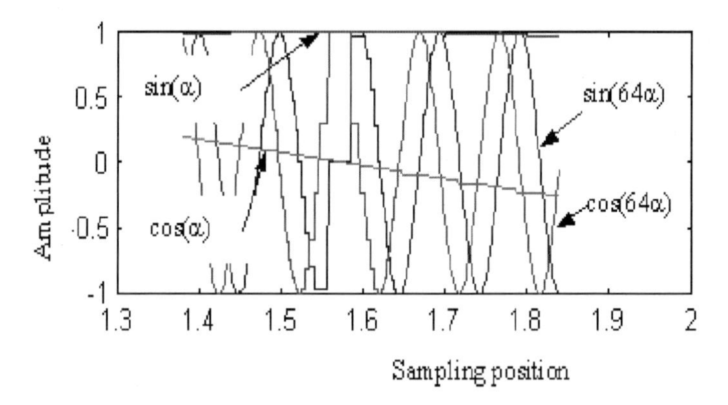

Fig. 6.2. Interpolation based on $\tilde{u}_1$

Look-up Table Based on Both $\tilde{u}_1$ and $\tilde{u}_2$

To overcome this difficulty, amplitudes of both $\tilde{u}_1 = \sin(n\alpha)$ and $\tilde{u}_2 = \cos(n\alpha)$ may be pre-recorded, since for the region around $\sin(\alpha) \approx 1$, $\tilde{u}_2 = \cos(\alpha)$ has a more even relationship between the amplitude and phase angle. Therefore,

$\tilde{u}_2$ can be used more effectively for the inferencing procedure instead in these areas. To this end, one approach is that for $|\tilde{u}_1| < 0.707$, $\tilde{u}_1$ will be used as the basis to search for the table entry. Otherwise, the amplitude of $\tilde{u}_2$ is used instead. Essentially, this means the look-up table now consists more portions corresponding to various parts of $\tilde{u}_1$ and $\tilde{u}_2$. The look-up table for $n = 16$ is given in Table 6.4.

Table 6.4. Look-up table based on both $\tilde{u}_1$ and $\tilde{u}_2$

Index	$1, \ldots, s-1, s$	$s+1, s+2, 2s$	$\ldots$	$7s+1, \ldots, 8s-1, 8s$
Range	$0 \sim \pi/4$	$\pi/4 \sim \pi/2$	$\ldots$	$7\pi/4 \sim 2\pi$
$\sin(16\alpha)$	$0.999, \ldots, 1.000$	$0.999, \ldots, 0.0$	$\ldots$	$0.000, \ldots, 0.000$
$\cos(16\alpha)$	$0.016, \ldots, 0.000$	$0.023, \ldots, 1.000$	$\ldots$	$1.000, \ldots, 1.000$

To facilitate the efficient and quick access to the appropriate part of the look-up table, an index table (similar to Table 6.3) is useful. To this end, indices N_s, N_c, N_0 are defined as

$$\begin{aligned} N_s &= round\left(\frac{s}{\sin(\pi/4)} \times \tilde{u}_1\right), \\ N_c &= round\left(\frac{s}{\sin(\pi/4)} \times \tilde{u}_2\right), \\ N_s &= round\left(s \times \sin(\pi/4)\right) \end{aligned} \tag{6.5}$$

Based on these indices, the index table (Table 6.5) yields the actual points where the appropriate $\sin(n\alpha)$ and $\cos(n\alpha)$ can be directly located (n_{index}), corresponding to the various parts of $\tilde{u}_1$ and $\tilde{u}_2$ respectively.

Table 6.5. Index table

Condition	n_{index}	Range of α	$\tilde{u}_1$ or $\tilde{u}_2$ used
$N_s > N_0$ and $N_c > 0$	$2N_0 - N_c$	$\pi/4 \sim \pi/2$	$\tilde{u}_2$
$N_s > N_0$ and $N_c < 0$	$2N_0 + N_c$	$\pi/2 \sim 3\pi/4$	$\tilde{u}_2$
$N_s \leq -N_0$ and $N_c > 0$	$6N_0 - N_c$	$5\pi/4 \sim 3\pi/2$	$\tilde{u}_2$
$N_s \leq -N_0$ and $N_c < 0$	$6N_0 + N_c$	$3\pi/2 \sim 7\pi/4$	$\tilde{u}_2$
$N_c > N_0$ and $N_s > 0$	N_s	$0 \sim \pi/4$	$\tilde{u}_1$
$N_c > N_0$ and $N_s < 0$	$8N_0 + N_s$	$7\pi/4 \sim 2\pi$	$\tilde{u}_1$
$N_c \leq -N_0$ and $N_s > 0$	$4N_0 - N_s$	$3\pi/4 \sim \pi$	$\tilde{u}_1$
$N_c \leq -N_0$ and $N_s < 0$	$4N_0 + N_s$	$\pi \sim 5\pi/4$	$\tilde{u}_1$

Figure 6.3 shows the interpolation results when $s = 707$ and $n = 64$. There is no waveform distortion even though $s = 707$, which is smaller than that used in Figure 6.1 ($s = 5000$).

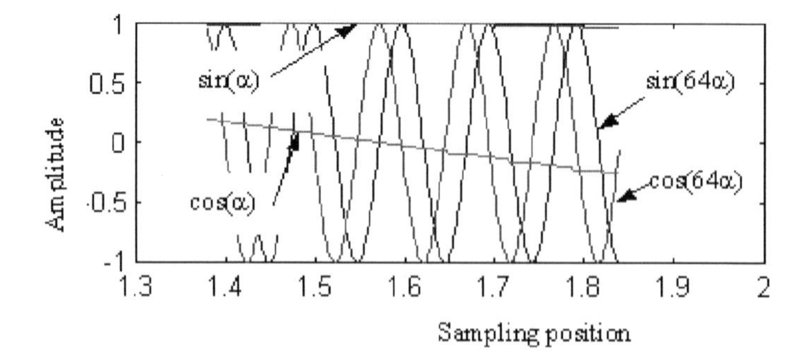

Fig. 6.3. Interpolation based on both $\tilde{u}_1$ and $\tilde{u}_1$

Maximum Interpolation

The maximum interpolation n^* achievable is limited by the minimum number of samples to be recorded in one period of the raw encoder sinusoid signal, and the minimum number of samples required to appear over one period of the high order sinusoid to be generated according to the following equation:

$$n^* = \frac{4s_1}{s_2}, \tag{6.6}$$

where s_1 is the minimum number of samples recorded in one period of $\sin(\alpha)$ and s_2 is the minimum number of samples to appear over one period of $\sin(n\alpha)$. For example, if $s_2 = 6$ and an interpolation of 1024 is required (*i.e.*, $n^* = 1024$), then $s_1 = 1536$, *i.e.*, at least 1536 samples need to be acquired over one period of the raw signal.

Waveform Distortion

The Heydermann method has assumed that the signals from encoder are periodic sinusoidal signal, with no waveform distortion. In practice, the waveform of the actual encoder signals deviate from the ideal sinusoidal waveform. Therefore, corrections based on the ideal sinusoidal waveform assumption may yield inaccurate position information which may not be acceptable for applications with high-precision requirements. It is more reasonable to assume that the encoder signal is periodic and reproducible in waveform which is not necessarily sinusoidal. In this case, since the non-sinusoidal waveforms are available, an error mapping method can be used to map them into sinusoidal ones. The idea is depicted in Figure 6.4. The look-up tables of Section 6.2.2 continue to be applicable

Conversion to Binary Pulses

In order for the encoder signals to be received by a general-purpose incremental encoder interface, the quadrature sinusoidal signals must be converted to

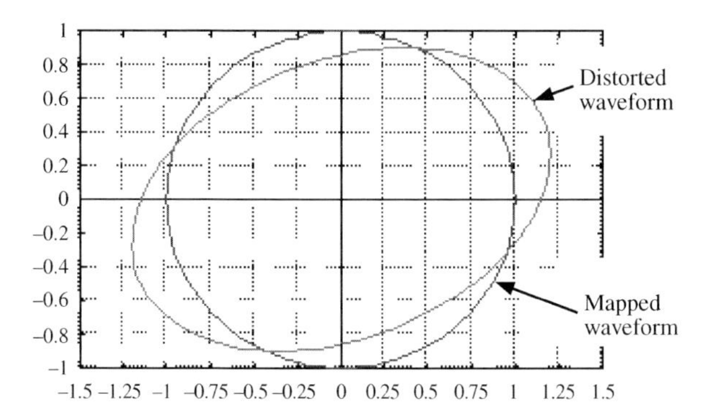

Fig. 6.4. Waveform error mapping

a series of binary pulses. An analog comparator may be used to transform the high order sinusoids into pulses. As shown in Figure 6.5, the comparator will simply switch the pulse signals when the associated sinusoidal signal crosses zero. The rest of the analog information will not be used.

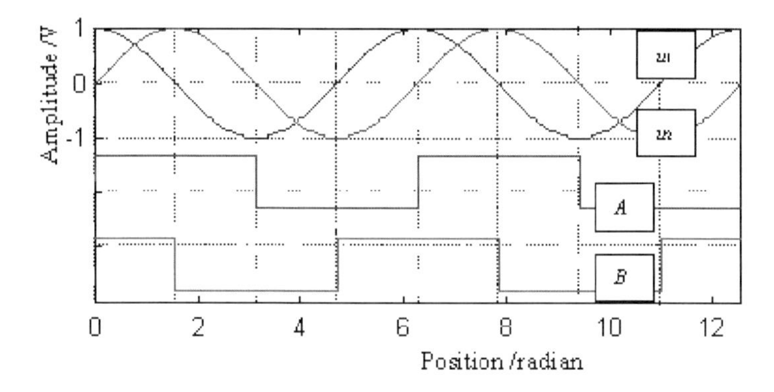

Fig. 6.5. Quadrature sinusoidal signal decoding

Alternatively, this transformation can be more efficiently done within the look-up table. The $\sin(n\alpha)$ and $\cos(n\alpha)$ entries in the table can be directly converted into binary values (A and B respectively) according to the following equations:

$$A = \begin{cases} 1 & \text{when } \sin(n\alpha) \geq \delta, \\ -1 & \text{when } \sin(n\alpha) \leq -\delta. \end{cases} \tag{6.7}$$

$$B = \begin{cases} 1 & \text{when } \cos(n\alpha) \geq \delta, \\ -1 & \text{when } \cos(n\alpha) \leq -\delta. \end{cases} \tag{6.8}$$

Thus, A and B can be generated which are quadrature square curves directly from Table 6.4 . δ can be 0 or a small value set according to the threshold of measurement noise.

Direct Conversion to Digital Position

The pulse information in Table 6.4 can be easily converted into digital position values, which can be directly used for control purposes without further computations. This is especially true of the aforementioned interpolation method integrated into a general digital controller. Alternatively, the encoder card can be made PC-bus based and the general motion controller can acquire the digital position value directly from the register or shared memory. In this case, the D/A converters for the encoder card are not required

6.2.3 Experiments

A dSPACE controller with a high-speed A/D card is first used to acquire the raw quadrature sinusoidal signals from a Heidenhein linear encoder LIP481 for the pre-interpolation signal conditioning. The compensation parameters are: $m_1 = -0.0126, m_2 = 1.4483e - 004, A_1 = 0.1331, A_2 = 0.1221$. Interpolation is subsequently carried out based on the enhanced method. Figure 6.6 shows the interpolation result with $n = 4$. Figure 6.7 shows the results with $n = 16$.

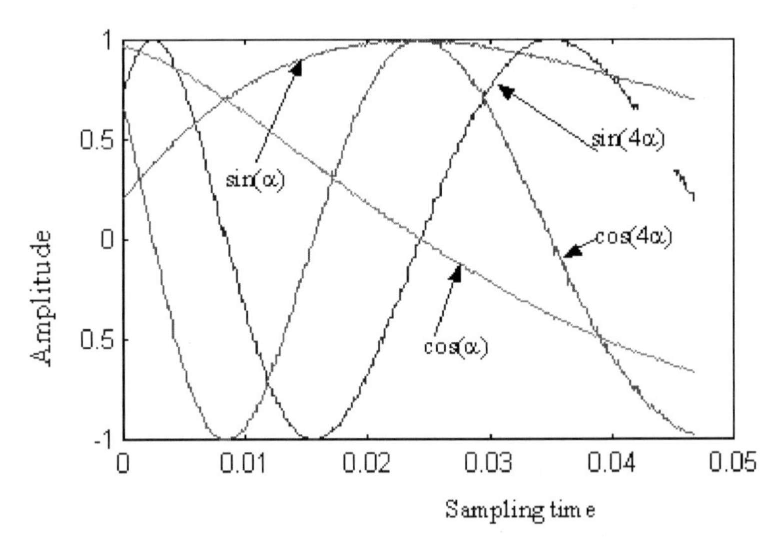

Fig. 6.6. Interpolation ($n = 4$)

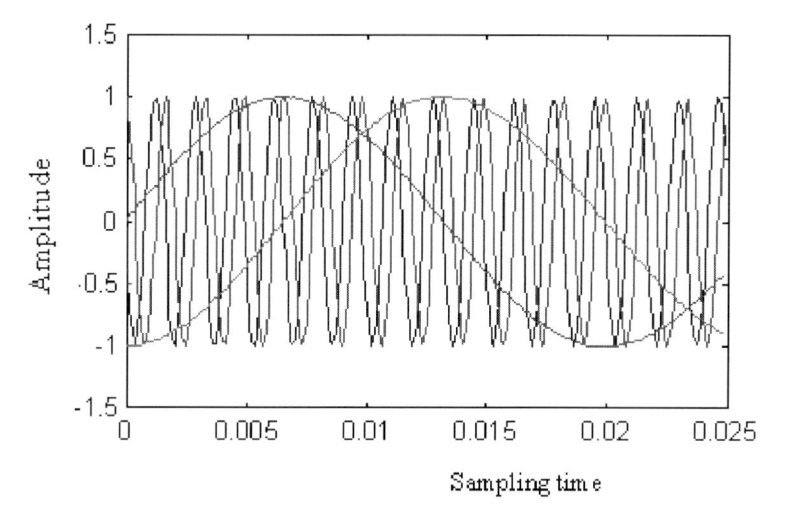

Fig. 6.7. Interpolation ($n = 16$)

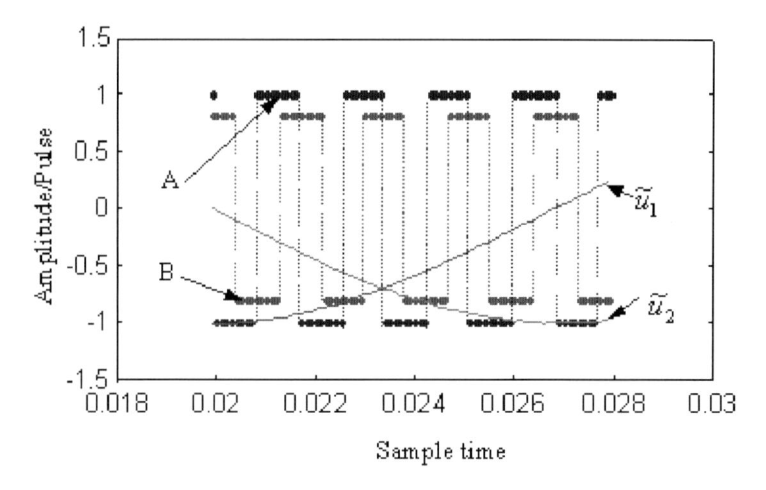

Fig. 6.8. Interpolation and conversion to quadrature pulses ($n = 16$)

Figure 6.8 shows the results with $n = 16$ where in addition, the look-up table entries are converted to binary values to yield binary pulses directly. To allow the pulses (with similar amplitudes) to be shown more clearly in Figure 6.8, the amplitude of B is deliberately set to 0.8. Figure 6.9 shows the results with $n = 32$.

To illustrate the situation with non-sinusoidal encoder signals and the correction using mapping more clearly, triangular waveforms are simulated

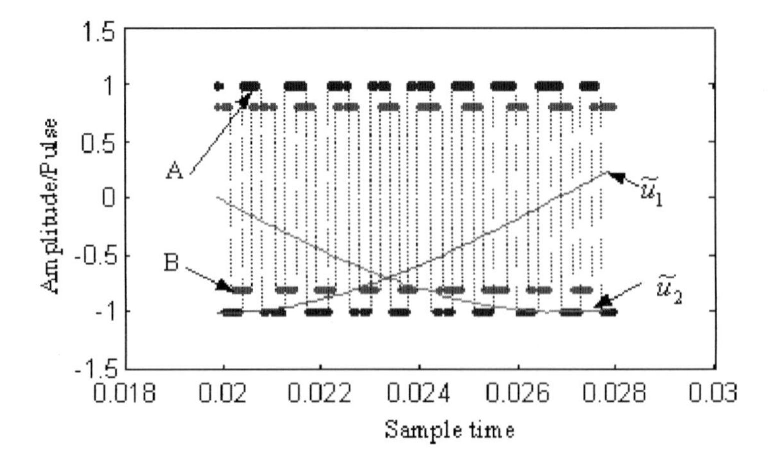

Fig. 6.9. Interpolation and conversion to quadrature pulses ($n = 32$)

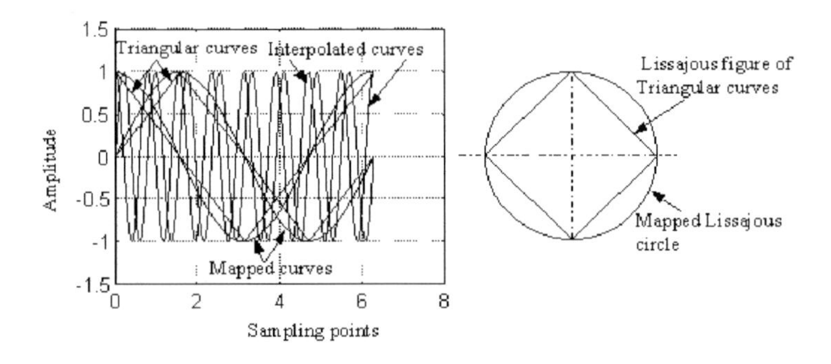

Fig. 6.10. Mapped interpolation ($n = 8$)

and mapped to sinusoidal ones. The interpolation results for $n = 8$ and their Lissajous figures are shown in Figure 6.10.

6.3 Parametric Model for Interpolation

Interpolation with a look-up table can enhance the encoder resolution. However, this approach incurs similar disadvantages as the look-up table approach for geometrical compensation highlighted in Chapter 5. The radial basis function (RBF) neural network can approximate any smooth nonlinear function arbitrarily well. This is especially true for functions where only the input/output pairs are available and the explicit relationships are unknown. The effective interpolation of the available sinusoidal signals can be seen as a generalisation process for the available data. The training of the RBF network entails finding a surface in the multi-dimensional space that best fits the available data. One main challenge, to be addressed in this section, is to realise an adequate fit with the simplest RBF structure possible by minimizing the redundancy present in the data mapping process.

A two-stage RBF network is used in the implementation of the present approach. The first RBF stage is concerned mainly with the correction of incoming non-ideal encoder signals, including the compensation of mean, phase offsets, amplitude deviations and waveform distortion. This RBF can be updated adaptively online to reflect any subsequent change or drift in the characteristics of the encoder signals. The second RBF stage serves to derive high-order sinusoids from the corrected signals from the first stage, based on how a series of high-frequency binary pulses can be converted which, in turn, can be readily decoded by standard servo controllers. Factors affecting the limit and accuracy of interpolation will be discussed in the section. Experimental results are provided to highlight the principles and practical applicability of the developed method.

6.3.1 Principles of Interpolation Approach

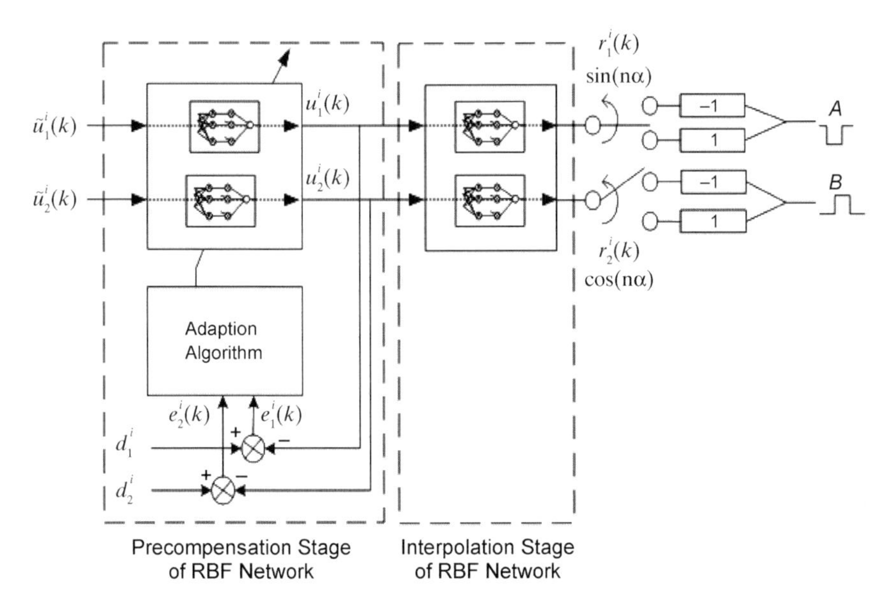

Fig. 6.11. Overall configuration of the two-stage RBF network

The overall configuration of the two-stage RBF network (for the purpose of encoder interpolation) is shown in Figure 6.11. It consists of two stages: the precompensation stage and the interpolation stage. The inputs to the precompensation stage are the quadrature signals direct from the encoders $\bar{u}_1$ and

$\bar{u}_2$. The outputs of the precompensation stage are fed as inputs to the interpolation stage. The outputs from the interpolation stage are the higher order sinusoids, $sin(n\alpha)$ and $cos(n\alpha)$ (where n refers to the order of interpolation).

Each of the two stages uses the configuration of a two-layered RBF network. The precompensation stage corrects the errors in the raw encoder signals. An adaptation algorithm is used to refine the correction process, since the error characteristics in the raw signals exhibit a tendency to drift from time to time. An online batch updating process is used to update the first RBF (precompensation stage) whenever a new batch of M time samples of the signals becomes available. The updating process is based on the modified Recursive Least Squares algorithm (Ljung 1997). The interpolation stage is used to derive the high order sinusoids based on the corrected signals forthcoming from the precompensation stage.

For both stages, the objective may be described as follows.

Given a set of N different points in a p dimensional input space, $\{i.e., x^k \in \Re^p, k = 1, 2, .., N\}$ and a corresponding set of W points in a q dimensional output space, $\{i.e., d^k \in \Re^q, k = 1, 2, .., W\}$, it is necessary to find a mapping function $\Im : \Re^p \rightarrow \Re^q$ that fulfils the relationship, such that $\Im(x^k) = d^k, \quad k = 1, 2, ..., N$. For the precompensation stage, the mapping function will map the raw encoder signals ($\bar{u}_1$, $\bar{u}_2$) to the corrected ones (u_1, u_2) which in turn become the inputs to the interpolation stage. For the interpolation stage, the mapping function will fulfil the mapping from u_1, u_2 to the higher-order sinusoids r_1, r_2.

The following notations to appear in subsequent developments are first defined, corresponding to the k-th frame (batch) of data for $i = 1, 2, ..., M; \quad j = 1, 2$:

$$e_j{}^i(k) = d_j{}^i - u_j{}^i(k),$$

$$E_j(k) = \frac{\sum_{i=1}^{M} [e_j{}^i]^2(k)}{M},$$

$$u_j{}^i(k) = \sum_{r=1}^{N} w_{r_j}(k) \phi_{r_j}^i(\bar{u}_j{}^i(k)),$$

$$\phi_{r_j}^i(\bar{u}_j{}^i(k)) = exp[-\frac{\| u_j{}^i(k) - c_j(k) \|^2}{2\sigma_j(k)^2}],$$

$$W_j(k) = [w_{1_j}(k) \quad w_{2_j}(k) \quad .... \quad w_{N_j}(k)]^T,$$

$$\Phi_j^i(k) = [\phi_{1_j}^i(\bar{u}_j{}^i(k)) \quad \phi_{2_j}^i(\bar{u}_j{}^i(k)) \quad .... \quad \phi_{N_j}^i(\bar{u}_j{}^i(k))]^T.$$

6.3.2 Precompensation Stage

Commonly encountered errors in the encoder signals include mean, phase offsets, amplitude deviation and waveform distortion. To reduce interpolation errors, it is necessary to correct these errors prior to interpolation. Figure 6.11

shows how these error components can be corrected in an adaptive manner in the precompensation stage of the RBF network. As mentioned earlier, an adaptive approach is useful for this purpose, since the error characteristics in the raw encoder signals can drift with time.

Ideally, the quadrature encoder signals (denoted by u_1 and u_2 respectively) are identical sinusoidal signals displaced by a phase of $\pi/2$ with respect to each other, described by

$$u_1 = A \cos \alpha, \ u_2 = A \sin \alpha.$$

α denotes the instantaneous phase and A denotes the amplitude of the signals. If there is no waveform distortion, the general equations relating the ideal and practical encoder signals can be obtained according to Heydermann (1981),

$$\bar{u}_1 = u_1 + m_1, \ \bar{u}_2 = \frac{A_1 \cos(\alpha - \varepsilon)}{G} + m_2,$$

where m_l and m_2 are the mean values of the signals and ε is the phase shift. $\bar{u}_1$ and $\bar{u}_2$ are values obtained from the encoder. $G = \frac{A_1}{A_2}$ and A_1, A_2 are the actual amplitudes of the encoder signals. The offset parameters m_1, m_2, ε and G can be estimated using a least squares estimation method operating on the raw signals.

Using a two-layered RBF network, this correction can be easily accomplished as a mapping from raw signals to ideal signals. In addition, unlike the Heydermann's method, waveform distortion can be addressed directly in the mapping function. To enable the precompensation stage of the RBF network to fine tune its parameters adaptively in concert with possible variation in the error characteristics, an adaptation algorithm is necessary. The adaptation algorithm used here is a modified version of the Recursive Least Squares algorithm (Ljung 1997). The parameters of the RBF network are updated in the Lyapunov sense so that the error in Equation (6.9) can converge to zero asymptotically. The following algorithm is used to update the parameters of the RBF network (in the algorithm, only the tuning of the weights of the RBF network is illustrated; tuning of the other parameters of the RBF network, *e.g.*, the centres, are similar):

$$W_j(k) = W_j(k-1) + \delta_j^i(k) \lambda_j^i(k),$$

where

$$\delta_j^i(k) = \frac{\Phi_j^i(k)}{\| \Phi_j(k) \|^2} \left[1 - \rho \frac{e_j^i(k-1)}{| \lambda_j^i(k) |} \right], \tag{6.9}$$

and

$$0 \leq \rho < 1,$$
$$\lambda_j^i(k) = d_j^i - W_j^T(k-1) \Phi_j^i(k),$$

for $i = 1, 2, ..., M; j = 1, 2$.

The convergent properties of the algorithm will be shown. Suppose the Lyapunov energy function is chosen as $V(k) = E_j(k)$. Therefore

$$
\begin{aligned}
\Delta V(k) &= V(k) - V(k\text{-}1) \\
&= E_j(k) - E_j(k-1) \\
&= \frac{\sum_{i=1}^{M}[e_j{}^i]^2(k)}{M} - E_j(k-1) \\
&= \frac{1}{M}\sum_{i=1}^{M}(d_j{}^i - u_j{}^i(k))^2 - E_j(k-1) \\
&= \frac{1}{M}\sum_{i=1}^{M}(d_j{}^i - W_j{}^T(k)\Phi_j{}^i(k))^2 - E_j(k-1) \\
&= \frac{1}{M}\sum_{i=1}^{M}(d_j{}^i - (W_j^T(k-1) + \lambda_j^{i^T}(k)\delta_j^{i^T}(k))\Phi_j{}^i(k))^2 \\
&\quad - E_j(k-1) \\
&= \frac{1}{M}\sum_{i=1}^{M}(d_j{}^i - W_j^T(k-1)\Phi_j{}^i(k) - \lambda_j^{i^T}(k)\delta_j^{i^T}(k)\Phi_j{}^i(k))^2 \\
&\quad - E_j(k-1) \\
&= \frac{1}{M}\sum_{i=1}^{M}(\lambda_j^{i^T}(k) - \lambda_j^{i^T}(k)\delta_j^{i^T}(k)\Phi_j{}^i(k))^2 - E_j(k-1) \\
&= \frac{1}{M}\sum_{i=1}^{M}(\lambda_j^{i^T}(k)(1 - \delta_j^{i^T}(k)\Phi_j{}^i(k)))^2 - E_j(k-1).
\end{aligned}
$$

Substituting Equation (6.9) into Equation (6.10),

$$
\begin{aligned}
\Delta V(k) &= \frac{1}{M}\sum_{i=1}^{M}(\rho\frac{e_j^i(k-1)}{\mid \lambda_j^i(k) \mid})^2 - E_j(k-1) \\
&= -\frac{1}{M}[1-\rho^2]E_j(k-1) < 0.
\end{aligned}
$$

Therefore, following the Lyapunov theory on stability, the approximation error at Equation (6.9) is stable and will converge to within a hypersphere centred at origin with radius τ (where τ is a small value). Figure 6.12 shows the fundamental encoder signals before and after the precompensation stage of the RBF network. The various error components underlying in the raw fundamental signals would have been corrected for after this stage. The subsequent interpolation stage will only deal with ideal sinusoidal signals.

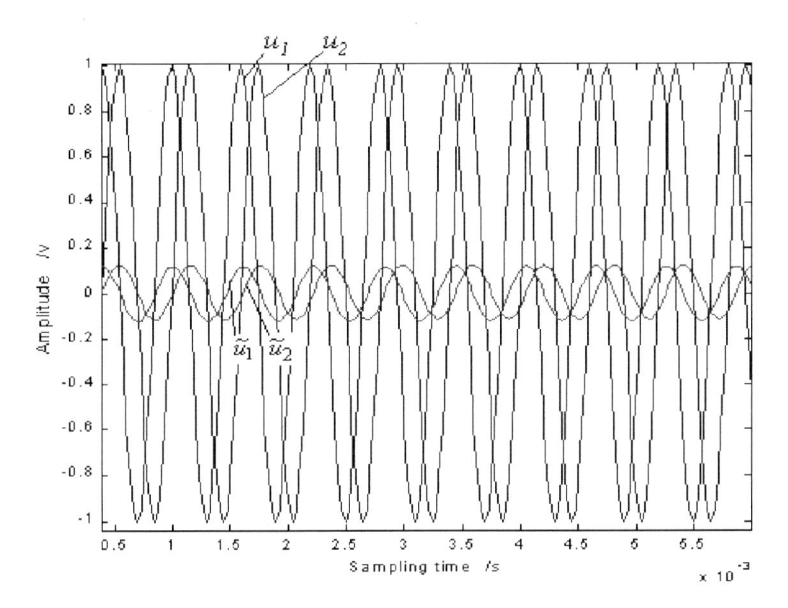

Fig. 6.12. Encoder signals before and after the precompensation stage

6.3.3 Interpolation Stage

The inputs to the interpolation stage are u_1 and u_2 from the precompensation stage. The outputs from the interpolation stage are the instantaneous values of the higher order sinusoids, *i.e.*, r_1 and r_2 (the output is dependent on the order of interpolation n). The RBF network is used to fulfil this mapping. The network is trained off-line, where the weights (ω_is) and the centres (c_is) of the RBF network are the free parameters to be tuned. As the order of interpolation increases, the memory requirements of the network also increases accordingly, since the mapping function $\Im(.)$ will become more complicated. Thus, more computing units (ϕ_is) (and subsequently more weights w_is to be tuned) are needed to implement the interpolation.

To reduce the memory requirements of this single-stage RBF network, it is useful to minimise the level of redundancy within the RBF network. To this end, it is noted that there is a strong degree of symmetry in a pure sinusoid. By considering only a quarter of the full sinusoid, the mapping function $\Im(.)$ between the absolute value of the inputs ($|u_1|$ and $|u_2|$) and the absolute value of the higher-order sinusoid outputs (r_1, r_2) can be fully represented. The sign of the higher order sinusoids (r_1 and r_2) can be subsequently restored by inferring the signs of u_1 and u_2, according to Table 6.6.

There are many different techniques available to tune the parameters of the RBF network. They include the 'Fixed-centres-selected-at-random', 'Self-organised-selection-of-centers' and 'Supervised-selection-of-centres'. The reader is referred to Haykin (1994) for more detailed discussions of the avail-

Table 6.6. Schedule table

$u_1{}^i(k)$ condition	$u_2{}^i(k)$ condition	Range	Output
$u_1{}^i(k) \geq 0$	$\rightarrow u_2{}^i(k) \geq 0$	$0 \sim \pi/2$	$\rightarrow r_1{}^i(k)$ $\rightarrow r_2{}^i(k)$
	$\rightarrow u_2{}^i(k) < 0$	$\pi/2 \sim \pi$	$\rightarrow r_1{}^i(k)$ $\rightarrow -r_2{}^i(k)$
$u_1{}^i(k) < 0$	$\rightarrow u_2{}^i(k) \geq 0$	$3\pi/2 \sim 2\pi$	$\rightarrow -r_1{}^i(k)$ $\rightarrow r_2{}^i(k)$
	$\rightarrow u_2{}^i(k) < 0$	$\pi \sim 3\pi/2$	$\rightarrow -r_1{}^i(k)$ $\rightarrow -r_2{}^i(k)$

able tuning techniques. In this section, the 'Supervised-selection-of-centres' (Haykin 1994) will be used, where the parameters to be tuned undergo a supervised batch learning process, using error-correction learning (*i.e.*, gradient descent procedure). The main objective of the supervised learning process is to minimize the value of the cost function $\xi^i = \frac{1}{2}\sum_{k=1}^{K} e_k^{i^2}$, where K is the number of data used to tune the parameters, and e_k^i is the error signal (between the training data and the output values of the RBF network), defined as

$$e_k^i = d_k^i - \Im(x_k^i), \tag{6.10}$$

where d_k^i and x_k^i $(k = 1, .., K)$ are the desired output and the training data respectively, and $\Im(.)$ is the function that is being modelled by the RBF network. The desired output values d_k^1 and d_k^2 are obtained from the amplitudes of the ideal mathematical functions of sine and cosine, respectively.

6.3.4 Experimental Study

In this section, real experimental results will be presented to illustrate the performance of the online adaptive correction and interpolation approach. In the experiment, raw data is acquired from a linear encoder (model: Heidenhein LIP481A) attached to the slide of a linear motor. These raw signals are then fed to a dSPACE controller with a high-speed A/D card, on which the RBF-based algorithms are implemented. The raw signals are accordingly precompensated and interpolated to higher order sinusoids. Figures 6.13 and 6.14 show the interpolation results with $n=$ 64 and 4096 respectively. Figure 6.15 shows the interpolated encoder signals converted to pulses with $n=$ 4096. The pulses are scaled to different amplitudes for easy observation.

There are advantages associated with this approach when compared to the look-up table approach, in the use of storage memory and execution speed. Under the developed approach, it is only necessary to reserve memory space for storing the parameters (*i.e.*, the weights and centres) of the RBF networks. The number of data points used to train the RBF networks for the

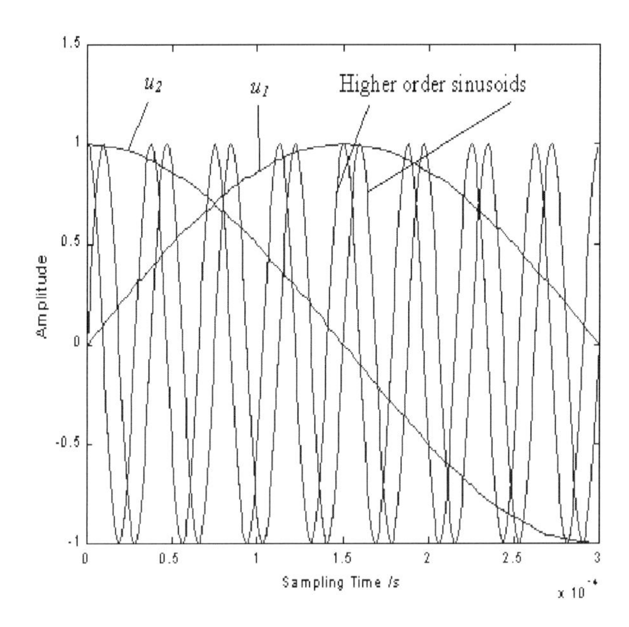

Fig. 6.13. Encoder signals before and after interpolation, with $n = 64$

precompensation stage and the interpolation stage is the number of weights required in the RBF network. In the experimental study, only seven points are required to map a complete sine or cosine function (Figure 6.16) when the redundancy present is eliminated. For $n = 16$, 308 data points are needed for the RBF approach, while 12,288 data points are needed for a look-up table method. With a lesser demand on memory storage space, the execution speed of the RBF approach is also much increased.

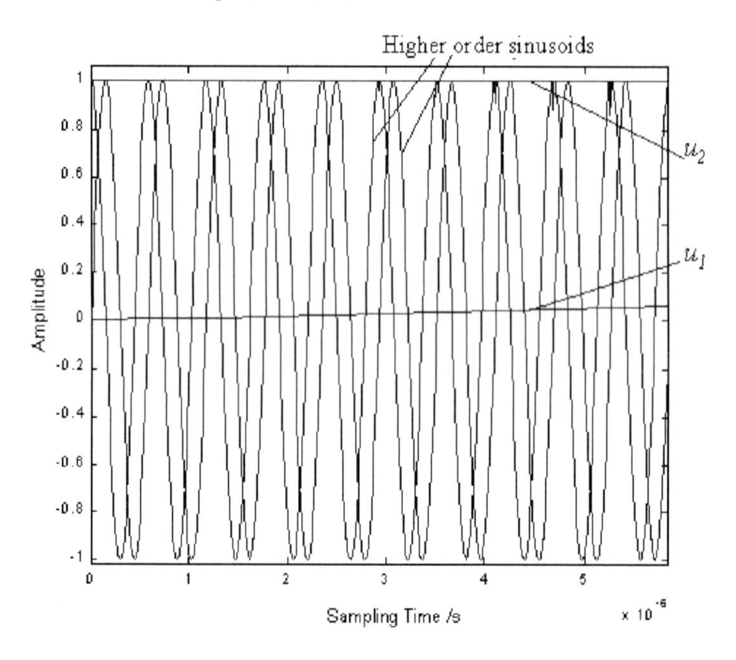

Fig. 6.14. Encoder signals before and after interpolation, with $n = 4096$

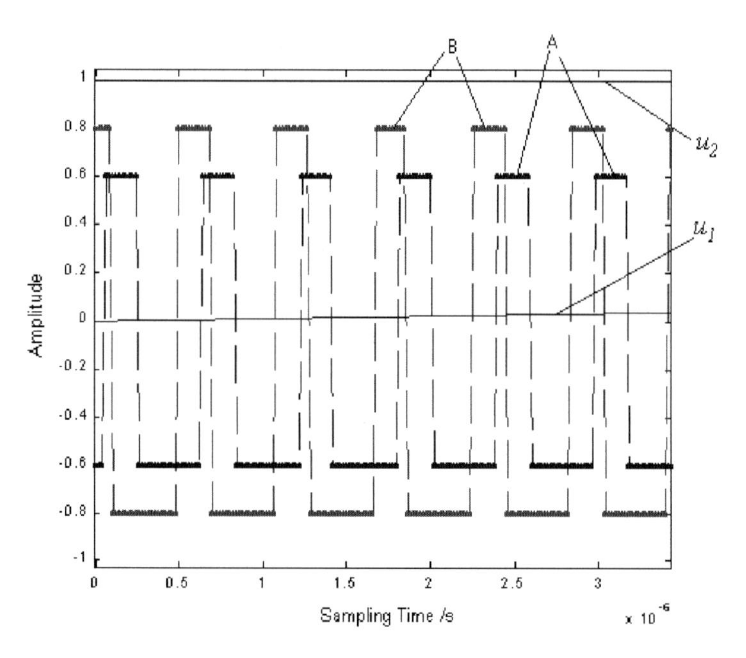

Fig. 6.15. Encoder signals converted to pulses, with $n = 4096$

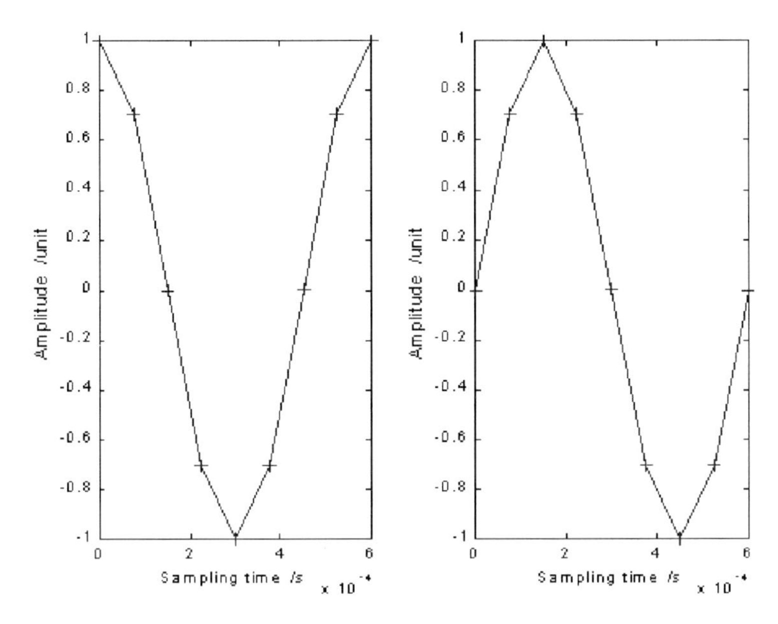

Fig. 6.16. Number of data points required to model the sine and cosine function for the RBF approach

7

Vibration Monitoring and Control

Mechanical vibration in machines and equipment can occur due to many factors, such as unbalanced inertia, bearing failures on turbines, motors, generators, pumps, drives, turbofans, *etc.*, poor kinematic design resulting in a non-rigid support structure, component failure and/or operations outside prescribed load ratings. The machine vibration signal can typically be characterised as a narrow-band interference signal anywhere in the range from 1 Hz to 500 kHz. To prevent equipment damage from the severe shaking that occurs when machines malfunction or vibrate at resonant frequencies, a real-time monitoring or control device is very useful. When the machine is used to perform highly precise positioning functions, undue vibrations can lead to poor repeatibility properties, impeding any systematic error compensation effort. This results directly in a loss of precision and accuracy achievable.

This chapter provides three possible approaches to deal with mechanical vibrations. The first approach will focus on a proper mechanical design, based on the determinacy of machine structure, to reduce mechanical vibration to a minimum. While the system design approach is certainly a first and key step to minimise vibration in mechanical systems, a parallel monitoring and suppression mechanism is necessary to cope with additional and usually unpredictable sources of vibration seeping in during the course of operations. The second approach utilises an adaptive notch filter to identify the resonant frequencies and subsequently to terminate any signal transmission at these frequencies *via* a narrow-bandstop filter. The adaptive notch filter can be directly incorporated into the control system. The third approach uses a real-time analyser to detect excessive vibration. This solution can be implemented independent of the control system, and as such can be applied to existing equipment without modification. A vibration signature is derived from the vibration signal acquired using an accelerometer attached to the machine running under normal conditions. Subsequently, a pattern recognition template is used to compare the real-time vibration signal against the signature. An alarm can be activated when the difference deviates beyond an acceptable threshold. Rectification actions can be invoked before damage is inflicted on the machine.

7.1 Mechanical Design to Minimise Vibration

In the development of high-speed and high-precision motion systems, the notion of determinism is a key consideration (Evan 1989) which implies that a physical system complies with the law of cause and effect, and this behavior allows the physical system to be modeled mathematically. The governing equations describing the model can then be used to predict the behavior of the system and thus allow for the compensation of possible errors to meet the demand of a tight error budget. A mechatronic approach, in which the structural design and the control design are to be seamlessly integrated, is one of the possible approaches for machine design. This approach has been adopted by many scientists and engineers, and the benefits are clearly evident in the end products such as the wafer scanner and stepper.

In this section, the key issues to address in a sound mechanical design to keep mechanical vibration to a minimum will be highlighted. The issue of mechanical design represents a very large area in precision motion systems. In this section, only key pointers will be highlighted in general, to enable designers to design "rigid" structure during the initial phase even before the physical modeling stage. Design, being an iterative process, always requires the designer to re-visit the drawing board frequently until an optimum design is achieved. The section will give qualitative ideas with abundant figures to illustrate key ideas, rather than using a purely quantitative approach, the reason being that during the initial phase of a design, intensive quantification is normally not necessary for decision making. Iterative and optimization which are normally mathematically intensive should be addressed during the next stage of the design process.

7.1.1 Stability and Static Determinacy of Machine Structures

A structure is a supporting framework which houses all the sub-assemblies that make up the machine, or it can be a collection of many smaller structures, or even a single component. The reaction forces of the high speed moving parts will excite the structural dynamics resulting in mechanical vibrations. These vibrations can be attenuated by reducing either the excitation or the response of the structure to that excitation (Beards 1983). The first factor can be overcome by relocating the source within the structure or by isolating it from the structure so that the generated vibration is not transmitted to the structure *via* the supports. As for the second factor, changing the mass, the stiffness, or the damping can alter the structural response. In order to understand the dynamic responses of the structure, the real structure can be transformed into a physical model, which is usually a simplified model representative of the real structure, for example; a real machine can be modeled as a number of coupled spring-mass systems. With a physical model of the real system derived, it can be translated to a mathematical model which can be solved *via* software or by hand, thereby allowing engineers from different disciplines to communicate and refine their portion of the design.

Every designer producing structures for machines needs to answer a very important question. Is the designed structure rigid and stable? A structure is rigid if its shape cannot be changed without deforming the members in the structure (Fleming 1997) and a structure is stable if rigid body translation or rotation cannot occur. A good way to tell whether a structure is stable or not is the degree of indeterminacy. A structure is considered statically determinate if all the support reactions and internal forces in the members can be determined solely by the equations of static equilibrium. Otherwise, the structure is considered statically indeterminate. Statically indeterminate structures arise due to the presence of extra supports, members, reaction forces or reaction moments. For a structure to be statically determinate, it must first be constructed correctly, then supported correctly.

7.1.2 Two-dimensional Structures

Most machine structures, in practice, are three-dimensional. However, it is useful to look at a two-dimensional problem first before extending the problem to a three-dimensional one. Generally, machine structures are stationary. Therefore the sum of the forces and moments acting on it must be zero, which is in accordance with Newton's second law. In mathematical form,

$$\sum F_x = 0, \tag{7.1}$$

$$\sum F_y = 0, \tag{7.2}$$

$$\sum M_z = 0, \tag{7.3}$$

where F_x and F_y are the forces in the x- and y-axis, respectively, while M_z is the moment about the z-axis where the z-axis is pointing out of the page. For a plane structure, we will make use of the sign conventions as depicted in Figure 7.1.

Since the static determinacy of a structure is a twofold issue, it is possible to proceed without first considering the support. Each structural configuration can be tested to verify whether the plane structure satisfies the equation

$$2j = m + 3, \tag{7.4}$$

where j denotes the number of joints and m denotes the number of members; then, there are three possible cases, namely,

1. If $2j = m + 3$, then the structure is statically determinate
2. If $2j > m + 3$, then the structure is unstable
3. If $2j < m + 3$, then the structure is statically indeterminate

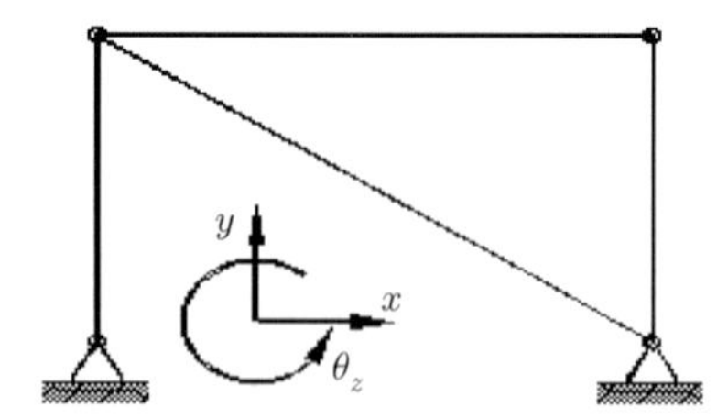

Fig. 7.1. Sign convention for plane structure

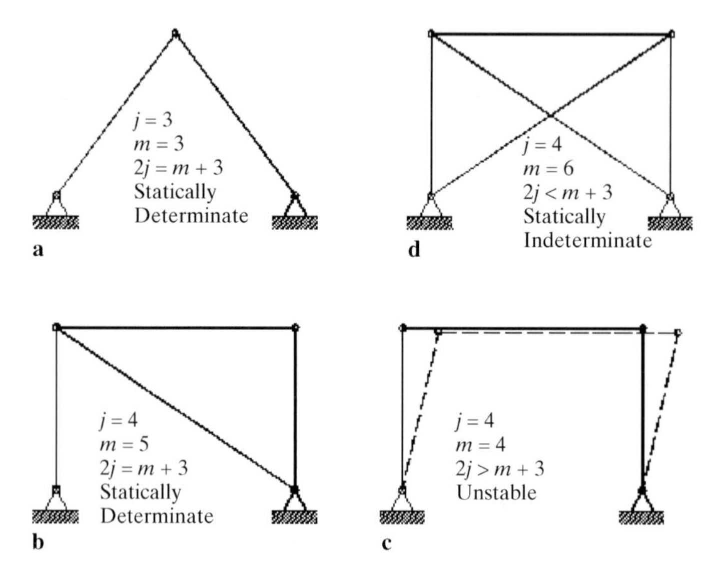

Fig. 7.2. Plane structures: **a,b** statically determinate structure; **c** unstable structure and **d** statically indeterminate structure

The three conditions are depicted in the Figure 7.2.

When a structure is unstable due to member deficiency, it appears that the structure becomes a four-bar mechanism, and with one degree of freedom as shown in Figure 7.2c. This degree of freedom is undesirable, since it is the structure which is being designed and not the mechanism! If, however, there are too many members present, the structure becomes statically indeterminate. Under such a condition, it will be difficult to assemble the fifth bar of the structure shown in Figure 7.2d if the dimension of the fifth bar is not exact. Assembly is probably possible with brute force and internal stresses will be "built-into" the structure even without any external loading. When a structure is statically determinate, it will be stress-free when it is not loaded externally other than by its own weight. In the event of thermal expansion of its member, due to a rise in temperature, statically determinate structures allow expansion of their members, without inducing any stress resulting from an over-constrained condition due to the redundant members.

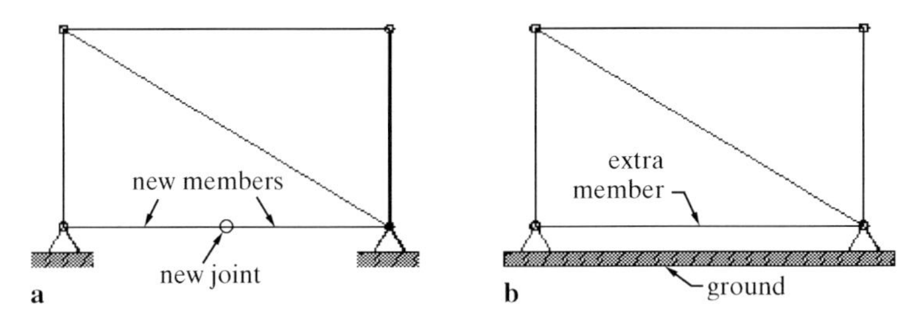

Fig. 7.3. a Unstable structure. **b** Extra member

The triangle is the basic shape for a plane structure as shown in Figure 7.2a. Statically determinate plane structure can be expanded from this basic structure simply by linking two new members to two different existing joints for every new joint added, as shown in Figure 7.2b. However, the axis of the two new members must not form a line; in other words, the three joints must not be in the same line as shown in Figure 7.3a. It should also be remembered that the ground constitutes one member as well, and all joints are pin-joints, as shown in Figure 7.3b.

The second part of structure design lies in its supports. From this aspect, the whole structure can be treated as a rigid body. For a plane-structure, it has three degrees of freedom, *i.e.*, the plane-structure is capable of motion in the x and y directions, and rotation about the z-axis. Therefore, three members are needed providing three reactive forces to constrain exactly the plane-structure in the plane. Figure 7.4a–c shows some possible support for plane structure, while Figure 7.4d shows an unstable support scenario.

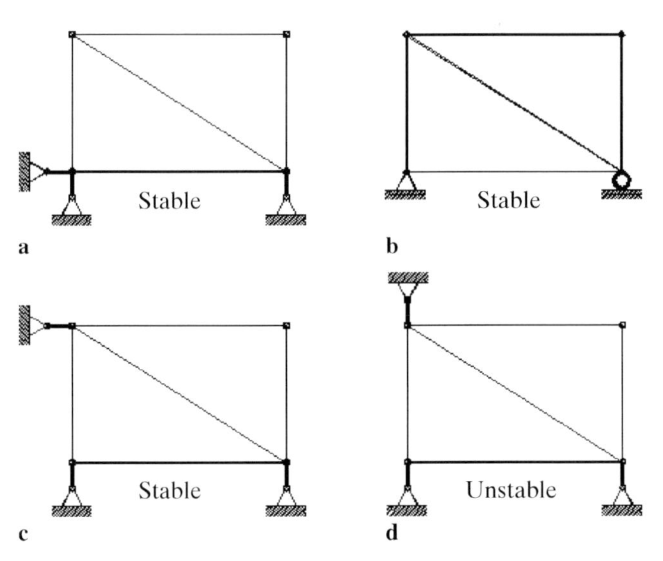

Fig. 7.4. a-c Stable and exactly constraint supports. **d** Unstable support

It should be highlighted that the condition of having two support members at the same location can be replaced by a single pin-joint, as shown in Figure 7.5. It is apparent that in both cases, they constitute two reaction forces, and do not constrain the angular motion about the z-axis present in the plane-structure.

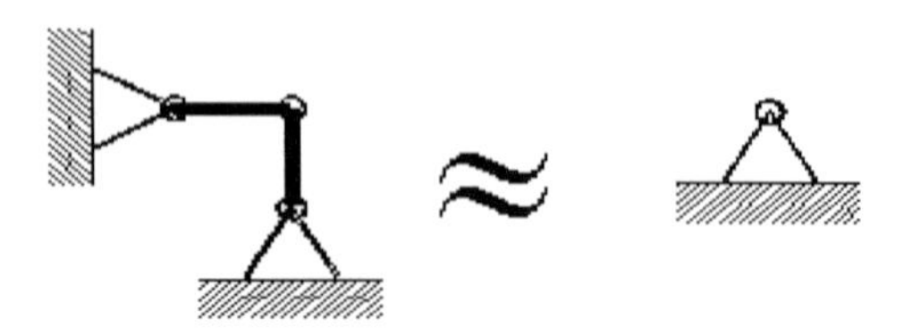

Fig. 7.5. Equivalent of a two-member support

The correct number of members in a structure as well as the correct number of supports must be in place for a stable and statically determinate structure. At this juncture, the issue of where the loads are to be applied onto the structure must be addressed. To this end, it is necessary to examine the members that make up the structure. The stiffness of a bar member is affected by the way the load is applied with respect to its axial axis, its cross sectional geometry (*e.g.*, the diameter, for a round bar) and its modulus of elasticity, E of its material. In most cases, the bar is either loaded in tension, compression or bending, as shown in Table 7.1 corresponding to three configurations (see Figure 7.6).

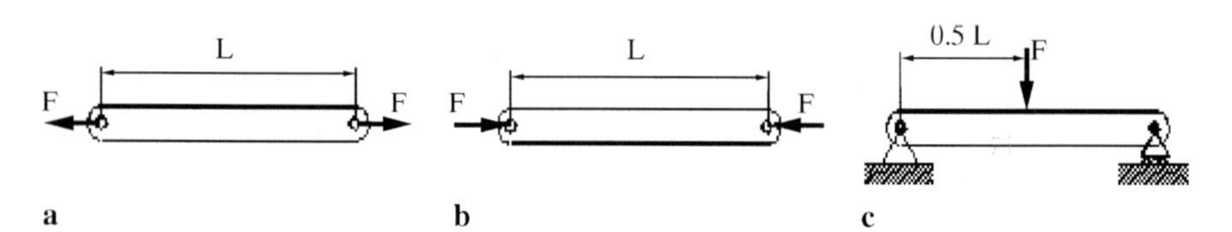

Fig. 7.6. a–c Three configurations of a bar

It is apparent by examining Table 7.1, that the stiffness of a bar is much better in axial loading as compared to bending loading. For a value of d=0.05 m, and L=1.2 m, the ratio of k_t/k_b is 192. That is, a bar is 192 times stiffer when loaded axially as compared to bending. Therefore, when designing a rigid and stiff structure, the members must be loaded in tension or compression, never in bending. At times, re-designing the way an external load is applied onto a structure can greatly improve the stiffness of the structure. Various configurations are shown in Figure 7.7, while the comparison of stiffness is

Table 7.1. Comparison of stiffness for axial loading verse bending loading

Figure 7.6	Loading	Stiffness	Normalized stiffness
(a)	Tension	$k_t = 0.25\pi Ed^2/L$	1
(b)	Compression	$k_c = 0.25\pi Ed^2/L$	1
(c)	Bending	$kb = 0.75\pi Ed^4/L^3$	$3(d/L)^2$

Units: stiffness(N/m), E [N/m2],d[m], L[m] and L>>d.

shown in Table 7.2. As a general rule to observe, the loading point should be located at the joints.

Table 7.2. Comparison of stiffness for various loading configurations

Figure 7.7	Stiffness	Normalized stiffness	Compare
(a)	$k_t = 0.25\pi Ed^2/L$	1	1
(b)	$k_b = 0.75\pi Ed^4/L^3$	$3(d/L)^2$	1/192
(c)	$k_\wedge = 0.5\pi Ed^2 sin^2\beta/L$	$sin^2\beta$	1/2
(d)	$k_{cl} = 0.047\pi Ed^4/L^3$	$0.1875(d/L)^2$	1/3072
(e)	$k_> = 0.25\pi Ed^2/L$	1	1

7.1.3 Three-dimensional Structures

Next, space-structures or three-dimensional structures will be considered. These are structures that are of interest in most applications. In a very general sense, space-structures can be perceived as a combination of many plane-structures, arranged in a manner that all the planes are not coplanar. Therefore, for a space-structure to be rigid, every plane-structure that makes up the space-structure must be rigid in its own right. This is one reason to have a good understanding of plane structural rigidity.

Since machine structures are stationary, the sum of the forces and moments acting on it must be zero; which is in accordance with Newton's second law. Mathematically, this implies

$$\sum F = 0, \tag{7.5}$$

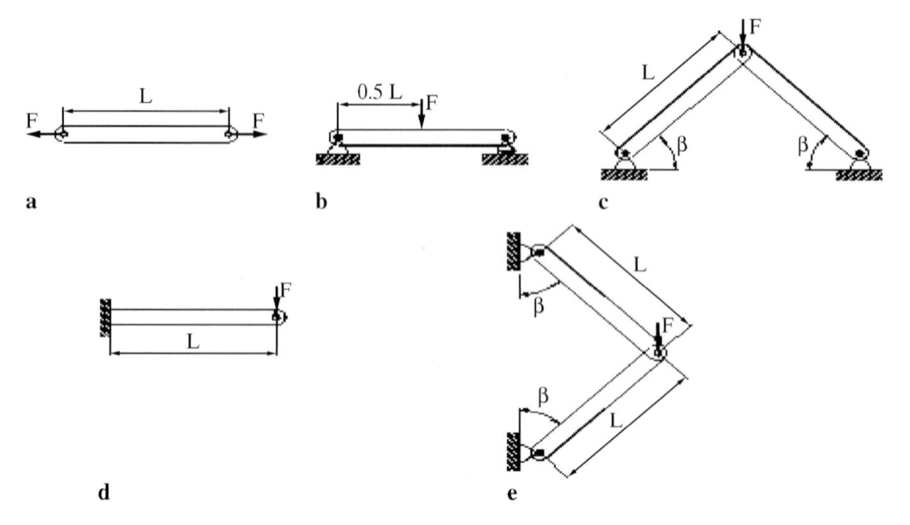

Fig. 7.7. a–e Various configurations

$$\sum M = 0, \tag{7.6}$$

where F and M are three-dimensional force and moment vectors, respectively. The sign conventions as depicted in Figure 7.1 will be used.

As before, each structural configuration can be tested to verify if the plane-structure satisfies the equation

$$3j = m + 6, \tag{7.7}$$

where j denotes the number of joints and m denotes the number of members; then, there are three possible cases, namely

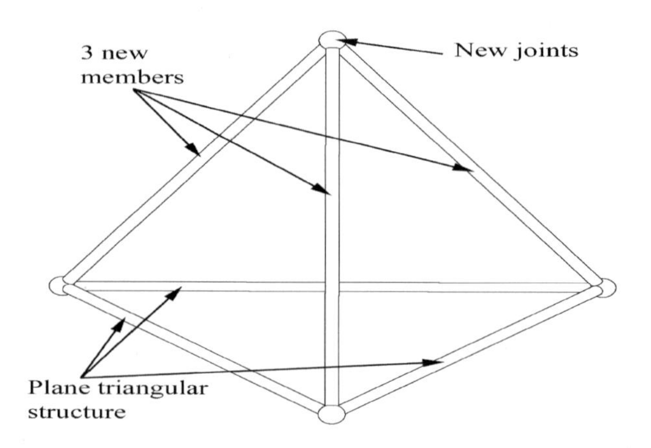

Fig. 7.8. Basic space-structure - the tetrahedron-structure

1. If $3j = m + 6$, then the structure is statically determinate
2. If $3j > m + 6$, then the structure is unstable
3. If $3j < m + 6$, then the structure is statically indeterminate

In the plane-structure, the triangle is the basic shape, which is rigid and statically determinate. In a space-structure, the basic form for rigidity and statically determinant is the tetrahedron, which is depicted in Figure 7.8. Adding a new non-coplanar joint to the three existing joints of a triangular plane-structure derives the tetrahedron-structure. This new joint is connected to the existing joints with three new members. By following this procedure, rigid and statically determinate space-structure can be derived.

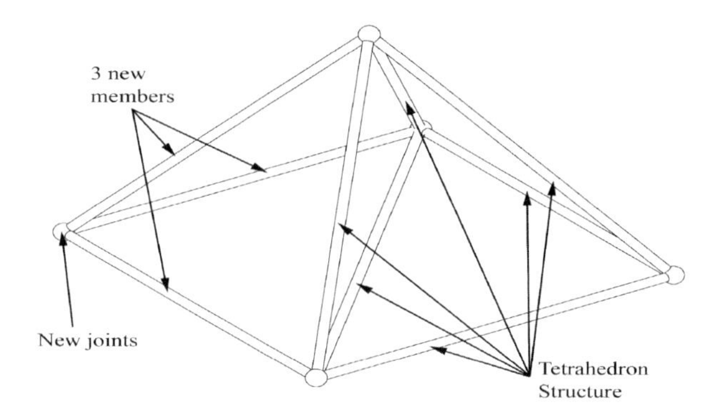

Fig. 7.9. Pyramid-structure derived from tetrahedron-structure

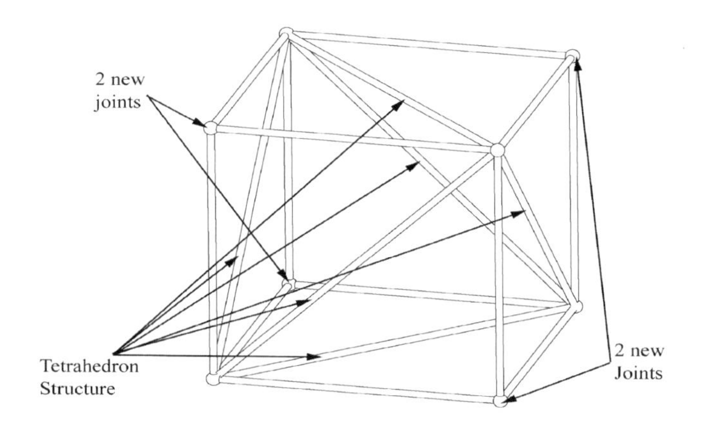

Fig. 7.10. Box-structure derived from tetrahedron-structure

Other space-structures are shown in Figures 7.9 and 7.10. It is also noteworthy that the members are connected with ball-joints.

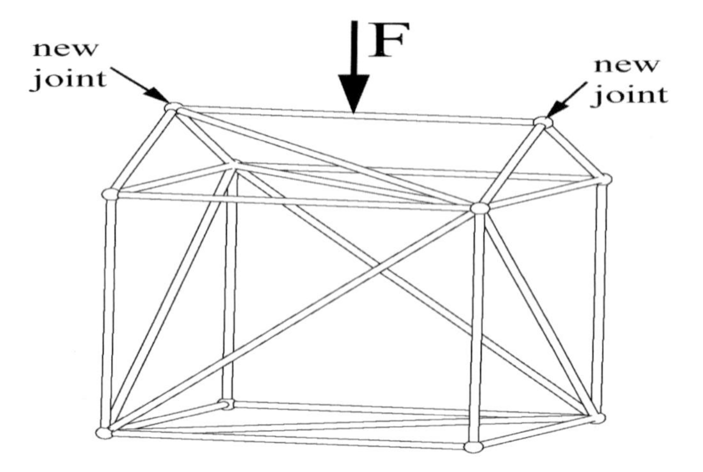

Fig. 7.11. Gantry space-structure

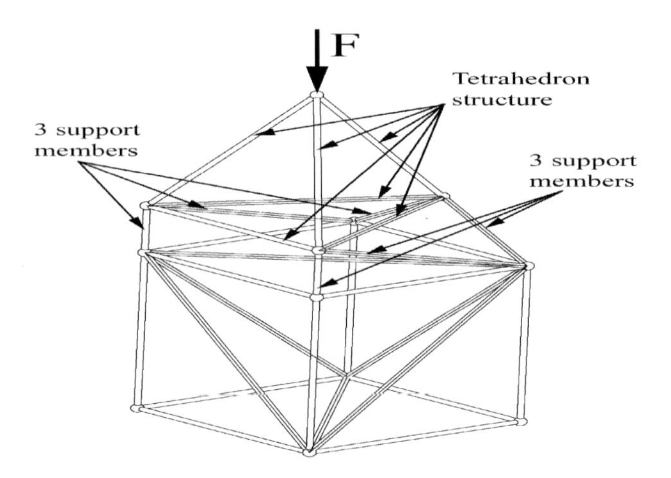

Fig. 7.12. Coupling of Tetrahedron-structure to box-structure by six members

Thus far, the approach to obtain the tetrahedron space-structure from the triangle plane-structure, the pyramid from the tetrahedron, and the box from the tetrahedron has been illustrated. The next aspect of the design is to combine some of these structures. The structures can be treated as being coupled together as rigid bodies, and a rigid body in space has six degrees of freedom, *i.e.*, the structure is capable of translations in the x, y and z directions, and rotation about the x, y and z axes. Therefore, six members are needed providing six reactive forces to exactly constrain the structure in space. Figure 7.11 shows a typical gantry configuration, which is used extensively in many coordinate-measuring machines (CMM). However, one of the members is bearing a bending load, which has been shown earlier to be very detrimental

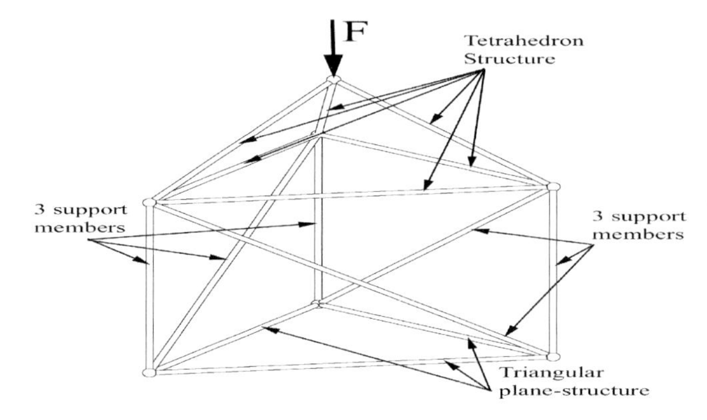

Fig. 7.13. Coupling of triangular plane-structure to tetrahedron space-structure with six members

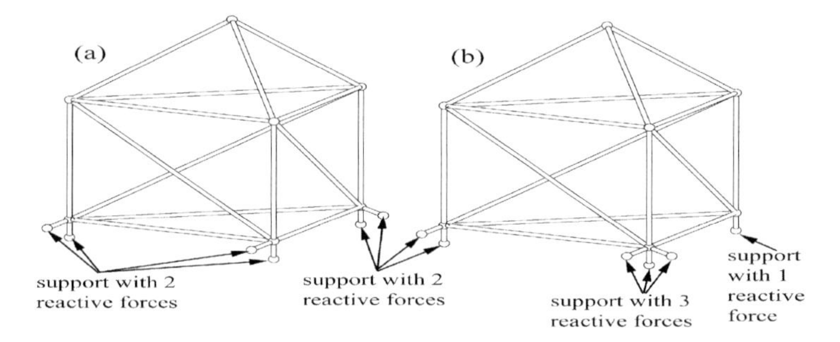

Fig. 7.14. a,b Examples for applying six constraints to a rigid body: **a** three set of twin reactive forces; **b** 3-2-1 reactive forces

to the stiffness of the structure. There are alternative structure configurations as shown in Figures 7.12 and 7.13, although some redesign maybe needed if such a configuration is utilized.

If the ground is perceived as another rigid body in which the space-structure is to be coupled, then the design of the supports for a space-structure is similar to those of coupling two space-structures together, *i.e.*, six reactive forces are needed to exactly constrain the space-structure. Some ways to arrange the six supporting members constraining a space-structure are suggested in Figure 7.14.

Examples of physical supports offering one, two or three reactive forces are shown in Figure 7.15.

This method of design, known as kinematical design, requires the use of point contact at the interfaces. Unfortunately, this method has some disadvantages, namely:

- Load carrying limitation
- Stiffness may be too low for application

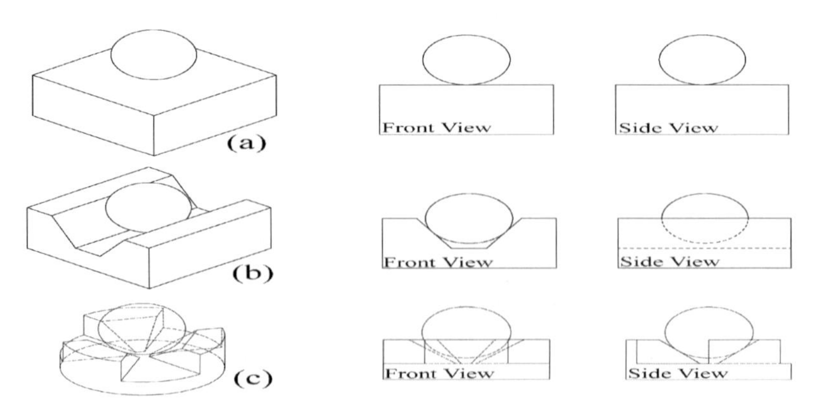

Fig. 7.15. a–c Examples of support with: **a** one reactive force; **b** two reactive forces; **c** three reactive forces

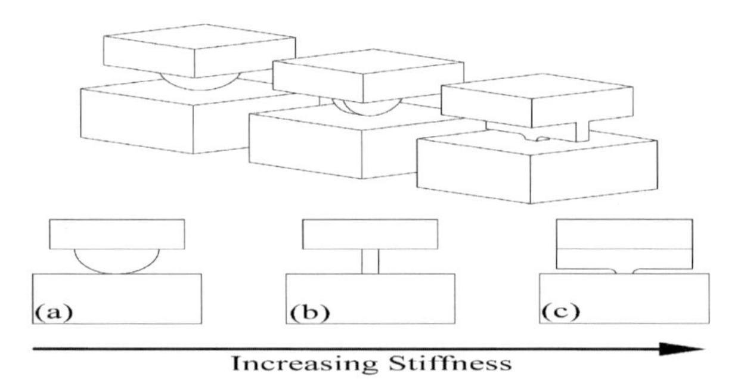

Fig. 7.16. Kinematical versus Semi-kinematical design. (a) Ideal condition - point contact (b) Line contact; and (c) area contact.

- Low damping

There are, however, ways to overcome the disadvantages which are *via* the semi-kinematical approach. This approach is a modification of the kinematical approach, and it targets to overcome the limitations of pure kinematical design. The direct way is to replace all point contact with a small area, as shown in Figure 7.16. Doing so decreases the contact stress, but increases the stiffness and load carrying capacity. However, the area contact should be kept to a reasonably small area.

This section has only illustrated some fundamental concepts in designing rigid and statically determinate machine structures. Interested readers may refer to (Blanding, 1999) for more details on designing machine using the exact constraints principles.

7.2 Adaptive Notch Filter

The task of eliminating/suppressing undesirable narrow-band frequencies can be efficiently accomplished using a notch filter (also known as a band-stop filter). Ideally, the filter highly attenuates a particular frequency component and leaves the others relatively unaffected. Thus, an ideal notch filter has a unity gain at all frequencies and a zero gain at the null frequencies. A single-notch filter is effective in removing a single frequency or a narrow-band interference; a multiple-notch filter is useful for the removal of multiple narrow-bands, necessary in applications requiring harmonics cancellation.

Digital notch filters are widely used to retrieve sinusoids from noisy signals, eliminate sinusoidal disturbances, track and enhance time-varying narrow-band signals in wide-band noise. They have found extensive applications in the areas of radar, signal processing, communications, biomedical engineering, and control/instrumentation systems.

To create a null in the frequency response of a filter at a normalised frequency β_0, a pair of complex-conjugate zeros can be introduced to the unit circle at angles $\pm\beta_0$ respectively. The zeros are defined as

$$z_{1,2} = e^{\pm j\beta_0} = \cos\beta_0 \pm j\sin\beta_0, \tag{7.8}$$

where the normalised null frequency β_0 is defined as

$$\beta_0 = 2\pi\frac{f_0}{f_s}. \tag{7.9}$$

f_s is the sampling frequency in Hz (or rad) and f_0 is the notch frequency in Hz (or rad). This yields a *Finite Impulse Response* (FIR) filter given by

$$H(z) = 1 - 2\cos\beta_0 z^{-1} + z^{-2}. \tag{7.10}$$

A FIR notch filter has a relatively large notch bandwidth, which means that the frequency components at the neighbourhood of the desired null frequency are also severely attenuated as a consequence. The frequency response can be improved by introducing a pair of complex-conjugate poles. The poles are placed inside the circle with a radius of α at angles $\pm\beta_0$. The poles are thus defined as

$$p_{1,2} = \alpha e^{\pm j\beta_0} = \alpha(\cos\beta_0 \pm j\sin\beta_0), \tag{7.11}$$

where $\alpha \leq 1$ for filter stability, and $(1-\alpha)$ is the distance between the poles and the zeros.

The poles introduce a resonance in the vicinity of the null, thus reducing the bandwidth of the notch. The transfer function of the filter is given by

$$H(z) = \frac{(z-z_1)(z-z_2)}{(z-p_1)(z-p_2)}. \tag{7.12}$$

Substituting the expression for z_i and p_i, and dividing through by z^2, the resulting filter has the following transfer function:

$$H(z) = \frac{a_0 + a_1 z^{-1} + a_2 z^{-2}}{1 + b_1 z^{-1} + b_2 z^{-2}}, \tag{7.13}$$

$$= \frac{1 - 2\cos\beta_0 z^{-1} + z^{-2}}{1 - 2\alpha\cos\beta_0 z^{-1} + \alpha^2 z^{-2}}. \tag{7.14}$$

Digitally, the filtered signal y is thus obtained from the raw signal u *via* the recursive formula in the time domain as follows:

$$y(n) = a_0 u(n) + a_1 u(n-1) + a_2 u(n-2) - b_1 u(n-1) - b_2 u(n-2), \tag{7.15}$$

where the coefficients a_i and b_i can be inferred from Equations (7.13) and (7.14), replacing z with the time-shift operator.

The bandwidth and the Q-factor of the notch filter are respectively given by:

$$BW = \frac{2\sqrt{2}(1-\alpha^2)}{[16 - 2\alpha(1+\alpha)^2]^{\frac{1}{2}}}, \tag{7.16}$$

$$Q = \omega_0 \frac{[16 - 2\alpha(1+\alpha)^2]^{\frac{1}{2}}}{2\sqrt{2}(1-\alpha^2)}. \tag{7.17}$$

$H(z)$ has its zeros on the unit circle. This implies a zero transmission gain at the normalised null frequency β_0. It is interesting to note that the filter structure at Equation (7.14) allows independent tuning of the null frequecny and the 3-dB attenuation bandwidth by adjusting β_0 and α respectively. The performance of the notch filter depends on the choice of the constant α, which controls the bandwidth BW. The bandwidth, which is a function of the distance of the poles and zeros $(1-\alpha)$, narrows when α approaches unity. Clearly, when α is close to 1, say $\alpha = 0.995$, the corresponding transfer function behaves virtually like an ideal notch filter.

Complete narrow-band disturbance suppression requires an exact adjustment of the filter parameters to align the notches with the resonant frequencies. If the true frequency of the narrow-band interference to be rejected is stable and known *a priori*, a notch filter with fixed null frequency and fixed bandwidth can be used. However, if no information is available *a priori* or when the resonant frequencies drift with time, the fixed notch may not coincide exactly with the desired null frequency if the bandwidth is too narrow (*i.e.*, $\alpha \approx 1$). In this case, a tunable/adaptive notch filter is highly recommended. In Ahlstrom and Tompkins (1985) and Glover (1987), it is proposed to adapt the null bandwidth of the filter to accommodate the drift in frequency; In Bertran and Montoro (1998), it is suggested that an active compensator is used to suppress the vibration signals. Kwan and Martin (1989) adapt the null frequency β_0, while keeping the poles radii α constant. In other words, the parameters a_1 and b_1 of Equation (7.13) are adjusted such that the notch centers at the unwanted frequency while retaining the null bandwidth of the notch filter.

7.2.1 Fast Fourier Transform

The *Discrete Fourier Transform* (DFT) is a tool which link the discrete-time to the discrete-frequency domain. It is a popular off-line approach widely used to obtain information of the frequency distribution required for the filter design. However, the direct computation of DFT is prohibitively expensive in terms of required computation effort. Fortunately, FFT is mathematically equaivalent to DFT, but it is a more efficient alternative for implementation purposes and can be used when the number of samples n is a power of two. For vibration signals where the concerned frequencies drift with time, FFT can be continuously applied to the latest n samples to update the signal spectrum. Based on the updated spectrum, the filter characteristics can be continuously adjusted for notch alignment.

7.2.2 Simulation

Computer simulation is carried out to explore the application of the adaptive notch filter in suppressing undesirable frequency transmission in the control system for a precision positioning system based on permanent magnet linear motors (PMLM). In the simulation, a sinusoidal trajectory is to be closely followed and an undesirable vibration signal is simulated which drifts from a frequency of 500 Hz in the first cycle to a frequency of 505 Hz in the second cycle of the trajectory. Figure 7.17 shows the tracking performance of the precision machine without a notch filter. Figure 7.18 shows the performance when a fixed notch filter is used and Figure 7.19 shows the performance with an adaptive notch filter. It is evident that a time-invariant narrow-band vibration signal can be effectively eliminated using just a fixed notch filter. However, when the vibration frequencies drift, an adaptive notch filter is able to detect the drift and align the notch to remove the undesirable frequencies with only a short transient period.

7.2.3 Experiments

A notch filter is implemented in the control system for a linear drive tubular linear motor (LD3810) equipped with a Renishaw optical encoder with an effective resolution of 1 μm is used as the testbed.

Figure 7.20 shows the performance of the PMLM when no filter is used. Figure 7.21 shows the improvement in the control performance when the notch filter is incorporated into the control system.

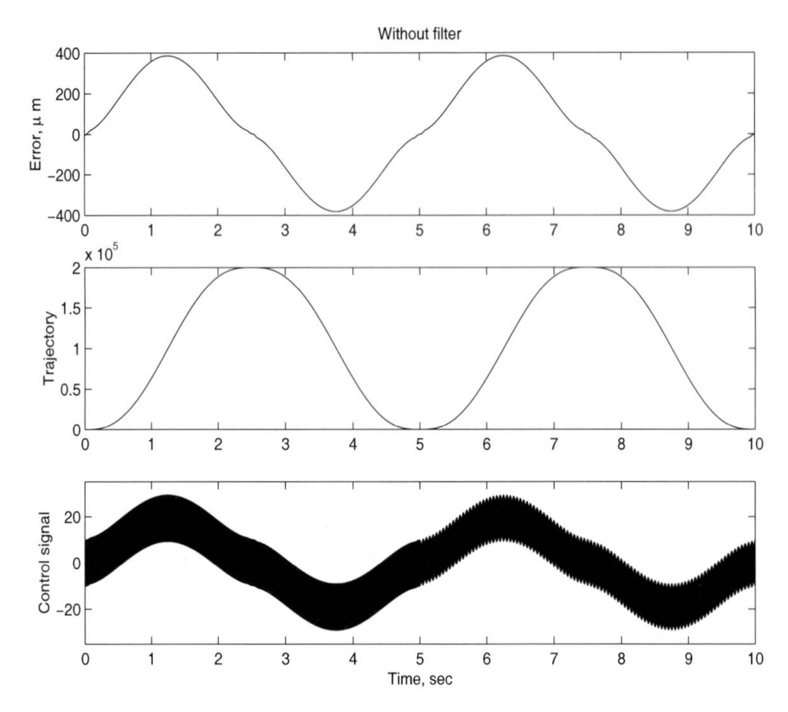

Fig. 7.17. Control performance without a notch filter: error (μm) (*top*); desired trajectory (μm)(*middle*); control signal (V)(*bottom*)

7.3 Real-time Vibration Analyser

Another approach towards real-time monitoring and analysis of machine vibration will be described in this section. The main idea behind this approach is to construct a vibration signature based on pattern recognition of "acceptable" or "healthy" vibration patterns. The vibration analyser can operate in three modes: learning, monitoring or diagnostic mode. The learning mode, to be initiated first, will yield a set of vibration signatures based on which the monitoring and diagnostic modes can operate. In the monitoring mode, with the machine under normal closed-loop control, the analyser only uses the naturally occuring vibration signal to deduce the condition of the machine. No deliberate and additional signal is input to the machine. More than one criterion can be used in the evaluation of the condition of the machine, in which case, a fusion approach can be used to provide one output (machine condition) based on the multiple inputs. In the diagnostic mode, explicit signals are input to the machine and the output signal (vibration) is logged for analysis with respect to the associated vibration signature. In what follows, the details of the various components/functions of the analyser will be described systematically.

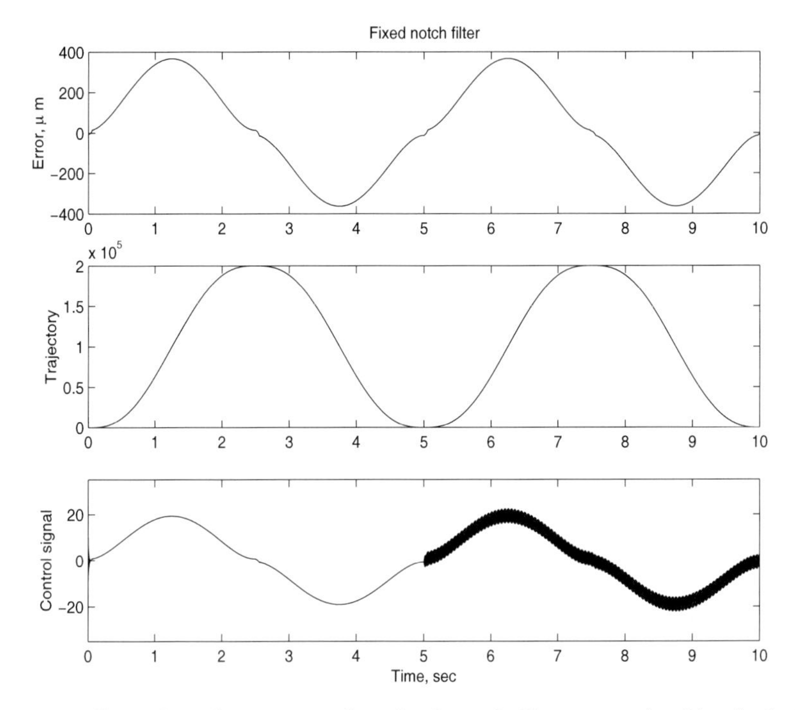

Fig. 7.18. Control performance with a fixed notch filter: error (μm)(*top*); desired trajectory (μm)(*middle*); control signal (V)(*bottom*)

The block diagram of the real-time vibration analyser is shown in Figure 7.22. It consists of an accelerometer which is mounted on the machine to be monitored. The accelerometer measures a multi-frequency vibration signal and transmits it to an intelligent DSP module. This module can be a standalone device, or one integrated to the *Personal Computer* (PC). The vibration analysis algorithm is downloaded to this DSP module. With this algorithm in operation, it can establish whether the condition of the machine is within a pre-determined acceptable threshold. If the condition is determined to be poor, the DSP module will trigger an operator alarm to enable corrective action.

The construction of the real-time vibration analyser is inexpensive and requires only commercially available, low-cost components. New micromachined accelerometers have dramatically reduced both the cost and the required signal conditioning circuitry. The cost of these new surface micromachined sensors is much lower than the piezoelectric ones which used to be the industry standard for condition monitoring. In addition, the typical sensitivity of micromachined sensors will stay within a 1% tolerance over the industrial temperature range which eliminates the need for temperature compensation and recalibration. The installation can be hassle free, as the accelerometer is

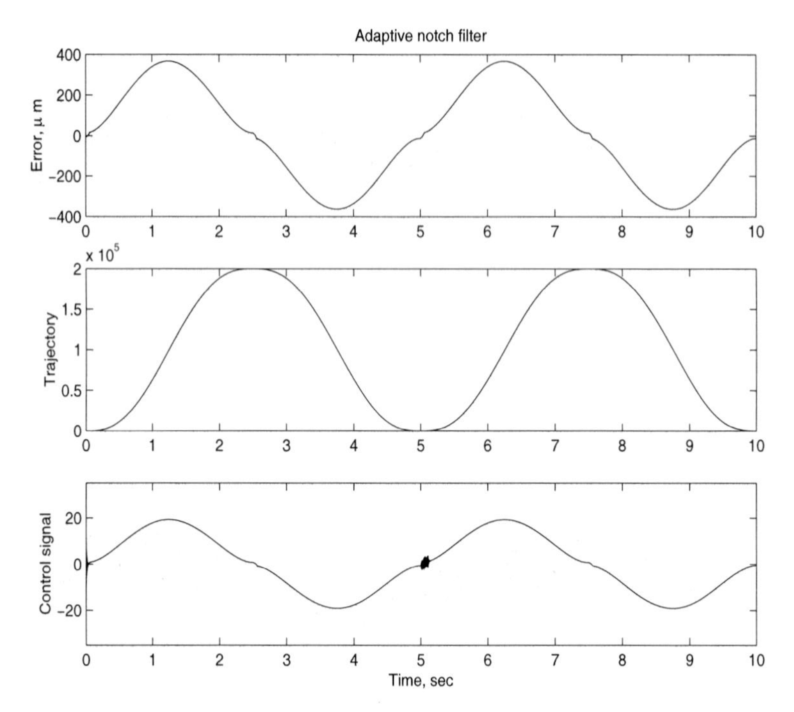

Fig. 7.19. Control performance with an adaptive notch filter: error (μm)(*top*); desired trajectory (μm)(*middle*); control signal (V)(*bottom*)

able to gather vibration signals, independent of the machine's own control. Thus, there is no need to disrupt any productive operation of the machine. In the prototype reported here, the DSP emulator board (TMS320C24x model) from Texas Instruments is used for the standalone DSP module. This C24x series emulator board is built around the F240 DSP controller, operating at 20MIPS with an instruction cycle time of 50ns. It is optimised for digital motor control and conversion applications. Other key components supported on this DSP module are analog-to-digital convertors (ADCs), dual access RAM (DARAM), on-chip flash memory and RS-232 compatible serial port. The vibration analysis algorithm (to be described) will be downloaded to the DSP board after satisfactory evaluation and tests on the PC. This DSP module and the accelerometer constitute the only hardware requirements of the real-time vibration analyser (Figure 7.22).

7.3.1 Learning Mode

In the learning mode, the vibration signals, with the machine operating under normal conditions, are acquired by the accelerometer and stored in the DSP module. A suitable vibration signature (Ramirez 1985) is then extracted from the vibration signals. There are many types of vibration signatures which can

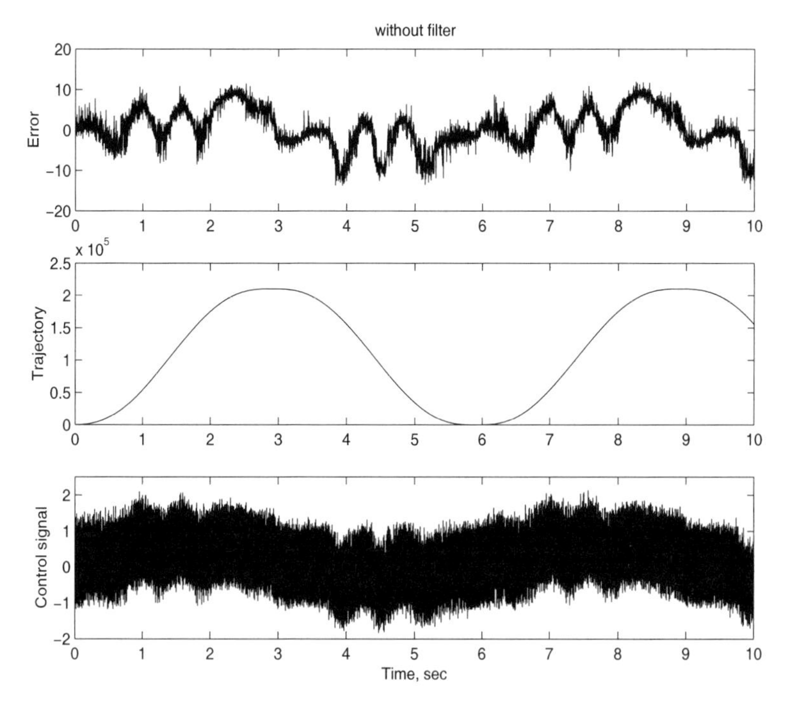

Fig. 7.20. Control performance without notch filter: error (μm)(*top*); desired trajectory (μm)(*middle*); control signal (V)(*bottom*)

be adequate for the purpose of machine monitoring. For example, one form of vibration signature may be based on the amplitude of the vibration; another form may be based on a time series analysis of the vibration; yet another form may be based on the spectrum of the vibration which can be efficiently obtained using the FFT algorithm. Here, the spectrum of the vibration signal will be used as a significant component of the signature. Whichever form, the vibration signatures are also dependent on the type of input signals driving the machine. For example, a square wave input will produce a vibration spectrum which can be quite different from that resulting from an input of a chirp signal (*i.e.*, repeating sine wave of increasing frequency) or a pure sinusoid. Thus, a particular input signal will produce a unique spectrum based on which a unique vibration signature can be derived. Multiple vibration signatures corresponding to the natural vibrations of the machine (useful for the monitoring mode), or corresponding to different input signals (useful for the diagnostic mode) can thus be captured for the subsequent diagnosis and monitoring of the machine.

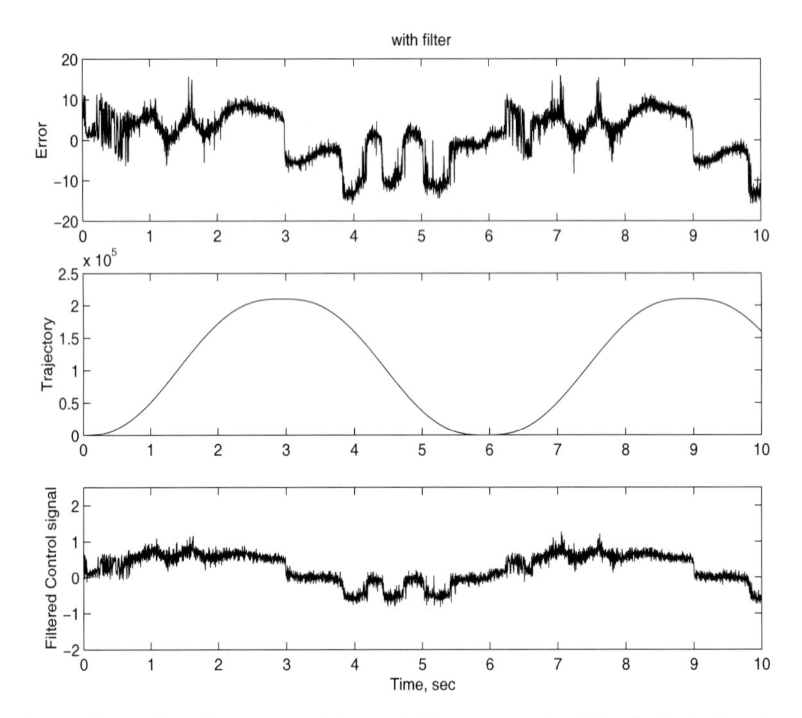

Fig. 7.21. Control performance with notch filter: error (μm)(*top*); desired trajectory (μm)(*middle*); control signal (V)(*bottom*)

7.3.2 Monitoring Mode

In the monitoring mode, the vibration signals are sampled periodically from the machine to monitor its condition. No deliberate or additional input signal is required, so the machine is not disrupted from its operation. The updated spectra are analysed against the relevant vibration signatures. The analysis and comparison may be in terms of the shift in frequency or amplitude of the spectrum, or a combination of both. For example, one evaluation criterion (EV) may be based on the mean-square (ms) value of the error (Ramirez 1985) between the current real-time vibration spectrum and the vibration signature:

$$\mathtt{EV}_1 = \frac{\sum_{q=1}^{N}(S_q - S_q^*)^2}{N}, \tag{7.18}$$

where S_q is the discretised current real-time vibration spectrum, S_q^* is the corresponding vibration signature, q is the index for the data points and N is the total number of data points. Another EV may be formulated based on the difference in the amplitude of the current time series vibration pattern and its corresponding vibration signature:

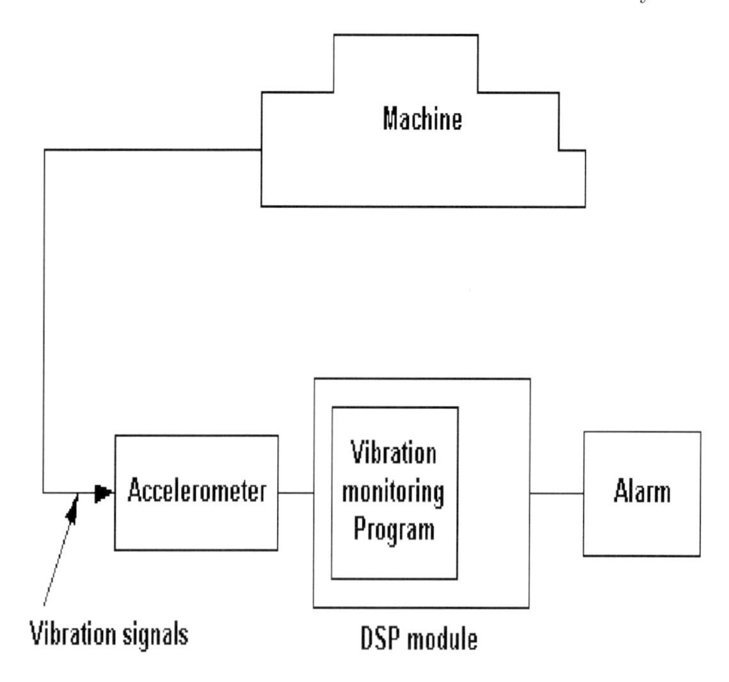

Fig. 7.22. Schematic diagram of the real-time vibration analyser

$$\mathtt{EV}_2 = \max(T_q) - \max(T_q^*), \tag{7.19}$$

where $\max(T_q)$ represents the highest amplitude of the current time series vibration pattern T_q and $\max(T_q^*)$ is the highest amplitude of its corresponding vibration signature.

More than one evaluation criterion may be used in the determination of the machine condition. In this case, a fusion technique is necessary. The key idea of the fusion technique is to associate the machine with a HEALTH attribute which is computed from multiple evaluation criteria. These criteria are expected to influence, to varying degrees, the HEALTH of the machine. The HEALTH attribute is thus an appropriate function $\Im$ of the various criteria ($\mathtt{EV_i}$s), *i.e.*,

$$\mathtt{HEALTH} = \Im(\mathtt{EV}_1, \mathtt{EV}_2, ..., \mathtt{EV}_\mathrm{n}), \tag{7.20}$$

where n refers to the number of criteria being evaluated.

A fuzzy weighted approach may be used to realise the $\Im$ function as follows:

The HEALTH attribute is treated as a fuzzy variable (*i.e.*, HEALTH $\in [0, 1]$). HEALTH $= 0$ will reflect absolute machine failure while HEALTH$= 1$ reflects a perfectly normal machine condition. This attribute may be computed from a fuzzy operation on a combination of the evaluation criteria ($\mathtt{EV_i}$s) obtained *via* an analysis of the vibration signals against their signatures. The final decision on the condition of the machine will be derived from the HEALTH attribute.

Interested readers may refer to Zadeh (1973) for a comprehensive review on fuzzy logic, as an alternative branch of mathematics.

A Takagi and Sugeno (1985) type of fuzzy inference is used here. Consider the following p rules governing the computation of the attribute:

$$\textbf{IF}\ \mathtt{EV}_1^{\mathtt{i}}\ \textbf{IS}\ \mathtt{F}_1^{\mathtt{i}} \otimes ... \otimes \mathtt{EV}_{\mathtt{n}}^{\mathtt{i}}\ \textbf{IS}\ \mathtt{F}_{\mathtt{n}}^{\mathtt{i}}\ \textbf{THEN}\ u^i = \alpha^i, \quad i = 1...p. \tag{7.21}$$

$u^i \in (0, 1]$ is a crisp variable output representing the extent to which the i-th evaluation rule affects the final outcome. Thus, α_i represents the weight of the i-th rule, with $\sum_i \alpha^i = 1$. $\mathtt{F}_{\mathtt{j}}^{\mathtt{i}}$ represents the fuzzy sets in which the input linguistic variables ($\mathtt{EV}_{\mathtt{i}}$s) are evaluated. $\otimes$ is a fuzzy operator which combines the antecedents into premises.

The value of the attribute is then evaluated as a weighted average of the u^is:

$$\mathtt{HEALTH} = \frac{\sum_{i=1}^{p} \omega^i u^i}{\sum_{i=1}^{p} \omega^i}, \tag{7.22}$$

where the weight ω^i implies the overall truth value of the premise of rule i for the input and it is calculated as

$$\omega^i = \Pi_{j=1}^{n} \mu_{\mathtt{F}_{\mathtt{j}}^{\mathtt{i}}}(\mathtt{EV}_{\mathtt{j}}^{\mathtt{i}}). \tag{7.23}$$

$\mu_{\mathtt{F}_{\mathtt{j}}^{\mathtt{i}}}(\mathtt{EV}_{\mathtt{j}}^{\mathtt{i}})$ is the membership function for the fuzzy set $\mathtt{F}_{\mathtt{j}}^{\mathtt{i}}$ related to the input linguistic variable $\mathtt{EV}_{\mathtt{j}}^{\mathtt{i}}$ (for the i-th rule). For example, in this application, the evaluation criterion ($\mathtt{EV}_{\mathtt{i}}$) may be the maximum error (MAX_ERR) and $\mathtt{F}_{\mathtt{j}}^{\mathtt{i}}$ may be the fuzzy set HIGH.

The membership function $\mu_{\mathtt{HIGH}}(\mathtt{MAX\_ERR})$ may have the characteristics as shown in Figure 7.23.

The decision as to whether any rectification is necessary can then be based on a simple **IF-THEN-ELSE** formulation as follows:

IF HEALTH $\leq \gamma$, **THEN** STRATEGY=TRIGGER ALARM
ELSE STRATEGY=CONTINUE TO MONITOR.

γ can be seen as a threshold value. Suitable values for γ may be in the range $0.6 \leq \gamma \leq 0.9$.

Under this framework, it is relatively easy to include additional criteria for analysis and decision making on the system. The procedure will involve setting up the membership functions for the criterion, formulating the additional fuzzy rules required, and adjusting the scaling parameters—the αs in Equation (7.21)— to reflect the relative weight of the new criterion over existing ones.

In this way, under the monitoring mode, foreboding trends can often be spotted long before the vibration reaches a level that is detrimental to the machine.

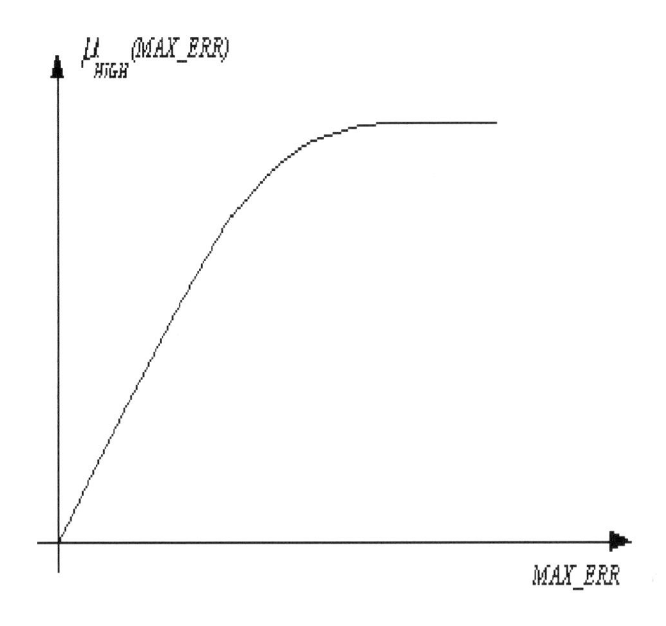

Fig. 7.23. Membership function for the the input MAX_ERR, $\mu_{\text{HIGH}}(MAX\_ERR)$

7.3.3 Diagnostic Mode

In the diagnostic mode, deliberate and explicit signals are applied as inputs to the machine and the vibration signal corresponding to each input signal is analysed against the associated signature, depending on the type of machine. Similar to the monitoring mode, there can be multiple evaluation criteria to be used in the diagnostic mode, so that the fusion technique described earlier is also applicable.

The input signals applied to the machine have to be designed carefully so as to yield as much information of the machine condition as possible in the operational regime of interest. Two important considerations are in the choice of amplitude and frequency.

Machines may have constraints in relation to the amount of travel that is possible. Too large an amplitude for the input signal may be not be viable for the machine due to the limit of travel or it may even damage the machine. Also, the frequency range of the input should be chosen so that it has most of its energy in the frequency bands that are important for the system. Where input signals cannot be applied to the system in the open-loop, the set-point signal will serve as the input for the closed-loop system, since it may not be possible to directly access the system under closed-loop control. Careful considerations of the mentioned issues will ensure that significant information can be obtained from the machine.

7.3.4 Experiments

A shaker table as shown in Figure 7.24 is used as the test platform for the experiments presented here.

Fig. 7.24. Test platform: the shaker table

The shaker table can be used to simulate machine vibrations and evaluate the performance of active mass dampers. This table is driven by a high torque direct drive motor. The maximum linear travel of the table is ± 2 cm. The shaker table is controlled *via* a DSP module implemented on a stand-alone mode, using the Texas Instruments' DSP emulator board (TMS320C24x model). The vibration analysis and monitoring program is coded in C24x assembly language. For the purpose of remote monitoring (to be described), the control can also be done using a general purpose data acquisition and control board.

The learning mode is first initiated to obtain the vibration signals with the shaker table operating under normal conditions. It is assumed here in the experiment that the normal condition corresponding to the input is the square wave signal. For the purpose of implementing the diagnostic mode, the vibration signals are also obtained for input signals of the sinusoidal and chirp type.

Input Variables - The Evaluation Criteria

Different types of EVs can be used as input variables for the determination of the machine condition. For this vibration analysis application, the input variables chosen for the computation of the HEALTH attribute and corresponding to the different operational modes are given below.

Monitoring Mode

$$\mathrm{EV}_1 = \frac{\sum_{q=1}^{N}(S_{sq,q} - S^*_{sq,q})^2}{N}, \tag{7.24}$$

$$\mathrm{EV}_2 = \frac{(\max(T_{sq,q}) - \max(T_{sq,q})^*)^2}{M}, \tag{7.25}$$

$$\mathrm{EV}_3 = \frac{\sum_{q=1}^{M}(T_{sq,q} - T^*_{sq,q})^2}{M}, \tag{7.26}$$

where $S_{sq,q}$ and $T_{sq,q}$ represent the vibration spectrum and the time-domain signal respectively corresponding to a square wave input. N relates to the number of frequencies for which the discrete spectrum, and M is the number of time series data points over an operational cycle. EV_1 thus refers to the mean-square deviation between the vibration spectrum and its signature, EV_2 refers to the square of the difference between the amplitude of the vibrational signal over one operational cycle compared to its signature, and EV_3 refers to the mean-square deviation between the vibration signal and its signature (time domain) over one operational cycle.

Diagnostic Mode

$$\mathrm{EV}_4 = \frac{\sum_{q=1}^{N}(S_{sq,q} - S^*_{sq,q})^2}{N}, \tag{7.27}$$

$$\mathrm{EV}_5 = \frac{\sum_{q=1}^{N}(S_{cp,q} - S^*_{cp,q})^2}{N}, \tag{7.28}$$

$$\mathrm{EV}_6 = \frac{\sum_{q=1}^{M}(S_{sn,q} - S^*_{sn,q})^2}{M}. \tag{7.29}$$

cp denotes a chirp input signal and sn denotes a sine input signal.

For the monitoring mode, the input attributes are related only to the square input, due to the assumption that the input signal, under normal operating conditions, is the square wave signal.

Evaluation Rules

The three rules for the computation of the HEALTH attribute are:

Monitoring Mode

IF EV_1 **IS** LOW, **THEN** $u=\mu_1$,

IF EV_2 **IS** SHORT, **THEN** $u=\mu_2$,

IF EV_3 **IS** LOW, **THEN** $u=\mu_3$.

The values of the scaling parameters, *i.e.*, αs in Equation (7.21), reflect the relative importance of the fuzzy rules in the determination of the HEALTH of the machine. The scaling values used are

$$\alpha_1 = 0.4,$$
$$\alpha_2 = 0.3,$$
$$\alpha_3 = 0.3.$$

The respective membership functions are

$$\mu_i(\mathtt{EV_i}) = e^{-n(\mathtt{EV_i})^\beta}, i = 1...3.$$

where n and β are scaling factors for normalisation of EV_i. In this application, they are selected as $n = 10$ and $\beta = 0.5$ respectively.

Diagnostic Mode

IF EV_4 **IS** LOW, **THEN** $u=\mu_4$,

IF EV_5 **IS** LOW, **THEN** $u=\mu_5$,

IF EV_6 **IS** LOW, **THEN** $u=\mu_6$.

The scaling values used are:

$$\alpha_4 = 0.4,$$
$$\alpha_5 = 0.4,$$
$$\alpha_6 = 0.2.$$

Similar membership functions are used as for the monitoring mode.

The machine condition attribute HEALTH is then computed as in Equation (7.22).

Tests

In what follows, real-time test results under the monitoring and diagnostic mode are provided which verify the operability of the device.

Monitoring Mode

In the monitoring mode, the normal input signal (*i.e.*, square wave) is input to the shaker table system. At $t = 5$, an additional sinusoidal signal with frequency $f = 5$ Hz is also input to the system to simulate a fault arising in the machine. The time domain signal of the machine (corresponding to square input) is as shown in Figure 7.25. The spectrum of the machine before and

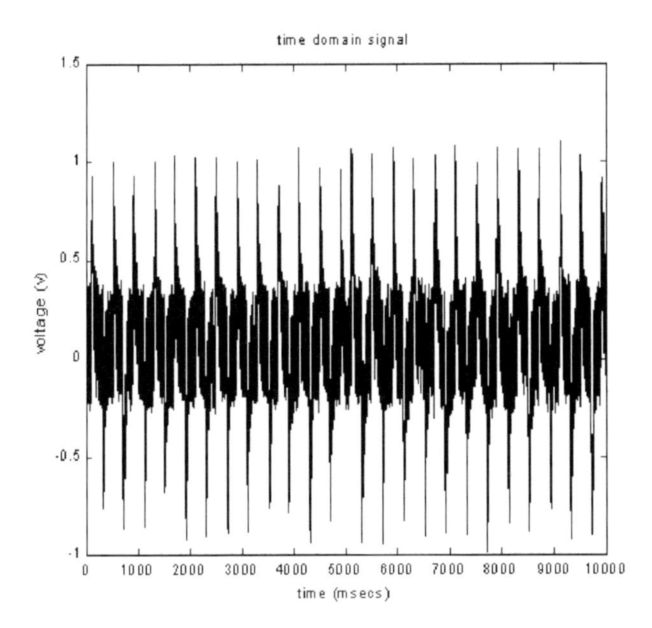

Fig. 7.25. Vibration signal corresponding to a square input signal (at $t = 5$, a fault is simulated)

after $t = 5$ are shown in Figures 7.26 and 7.27 respectively.

The vibration analysis algorithm is able to detect the fault in the machine. The `HEALTH` attribute of the shaker table falls to 0.2 which is below the threshold value (which is set at 0.6). The alarm is triggered.

Diagnostic Mode

In the diagnostic mode, different input signals (sine, square and chirp) may be used accordingly. To simulate a fault arising at $t = 5$, the input gain is

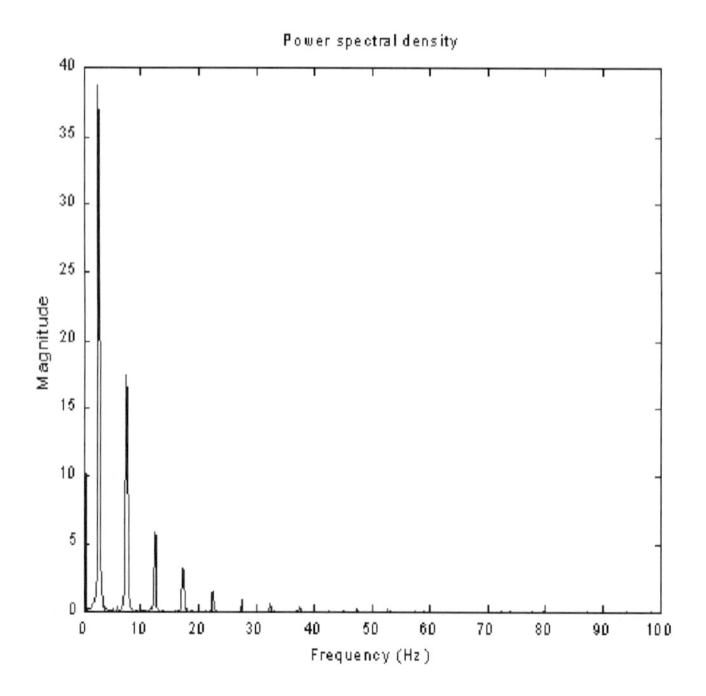

Fig. 7.26. Vibration signature corresponding to a square input signal

increased by a factor of two at t=5. The time domain vibration signal of the machine (corresponding to chirp signal) is as shown in Figure 7.28.

The spectrum (corresponding to chirp signal) of the machine before and after t=5 are shown in Figure 7.29 and Figure 7.30 respectively.

The time domain vibration signal of the machine (corresponding to sine wave input) is shown in Figure 7.31. The spectrum (corresponding to sine wave input) of the machine before and after t=5 are shown in Figures 7.32 and 7.33 respectively.

The vibration analysis algorithm is able to detect the fault in the machine. The `HEALTH` attribute of the shaker table falls to 0.1 which is below the threshold value (which is set at 0.6). The alarm is accordingly triggered.

7.3.5 Remote Monitoring

Efficient availability and organisation of production data directly from the shopfloor is highly critical to the responsiveness and competitiveness of a manufacturing company. This is even more critical today as the layout of an entire plant can be rather extensive, spreading across continents, in some cases. It is thus essential and highly useful to be able to enable remote monitoring and control, without requiring a physical presence which can be both

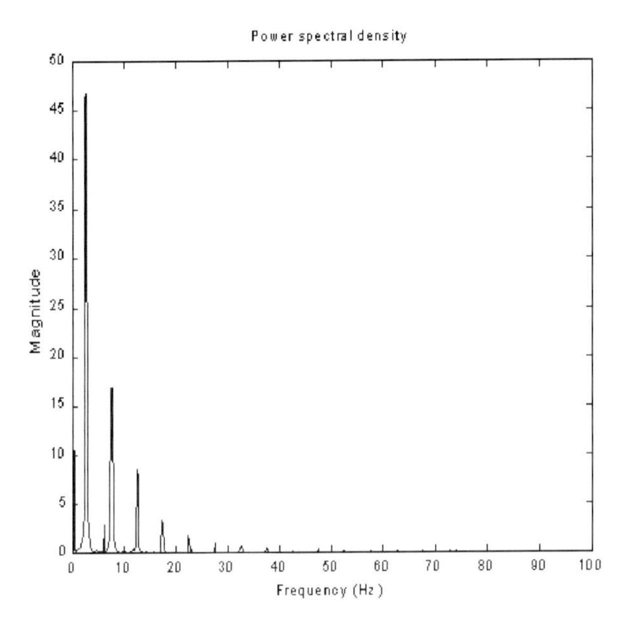

Fig. 7.27. Spectrum of machine corresponding to an input square signal after a fault has occurred

time consuming and economically inefficient. A possible approach towards remote monitoring of machines across affiliated production plants, using existing extensibe TCP/IP infrastructure will be highlighted in this section.

Hardware

The hardware, necessary and used, in the development of the remote monitoring at the server's end, consists of a data acquisition card installed onto a PC. As mentioned earlier, many types of data aquisition cards could be used. The acquisition card used here is one from National Instruments. The PC is used primarily for data acquisition and control of the remote machine. The hardware connection is depicted in Figure 7.34.

Software

The main idea behind the implementation of the remote monitoring session is to deliver data required to the client. On the client's end, it allows the client to monitor and control the remote machines and plants through the web browser *via* the CGI interface. Essentially, the distant remote monitoring application, as with other network applications, consists mainly of two parts: the client side and the server side. The client side is a simple web browser. At the server side, LabVIEW 5.0 supplemented with the Internet Developers Toolkit package are

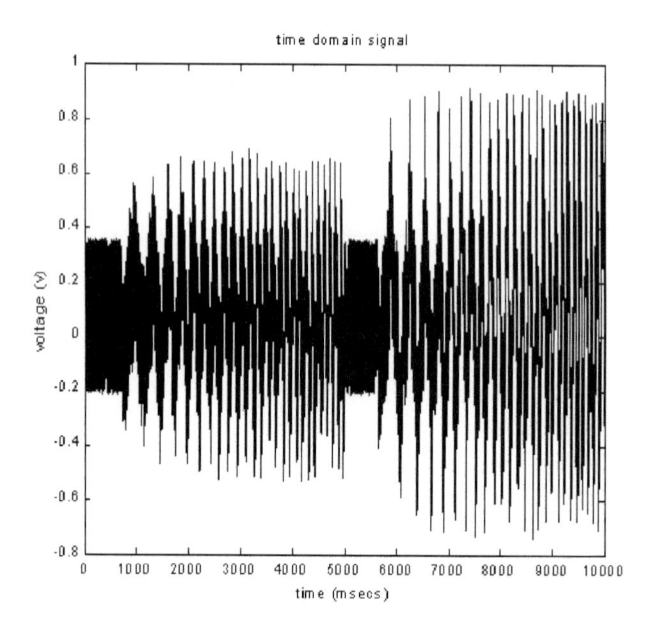

Fig. 7.28. Vibration signal corresponding to a chirp input signal (at $t = 5$, a fault is simulated)

the only essential software needed. The software requirements and interaction are depicted in Figure 7.35.

Operating Principles

The Internet Developers Toolkit package is an add-on component to LabVIEW. It is essentially a collection of libraries for converting Virtual Instruments (VIs) into Internet-enabled applications, such as electronic mails and FTP. First and foremost for WWW, the toolkit consists of the G Web Server. It is able to serve up the front panels of the VI as a picture to be viewed from the client's web browser. There are basically two modes of service. The first is the "snapshot" mode where a snapshot of the front panel image will be served. The second is the "monitor" mode. In this mode, the effect served to the user is an embedded animation of the panel image. This is achieved with Server-push technology. Successive "frames" of the panel's image will be pushed from the server to the client one after another without waiting for subsequent requests from the client. The rate at which the frames are pushed over can be set at the G-web Server configuration file. Presently, only Netscape's browser supports Server-push technology. With other web browsers, particularly Microsoft's Internet Explorer, it is also possible to create pseudo animation. This will cause the G-web Server to serve up the front panel image in the "monitor"

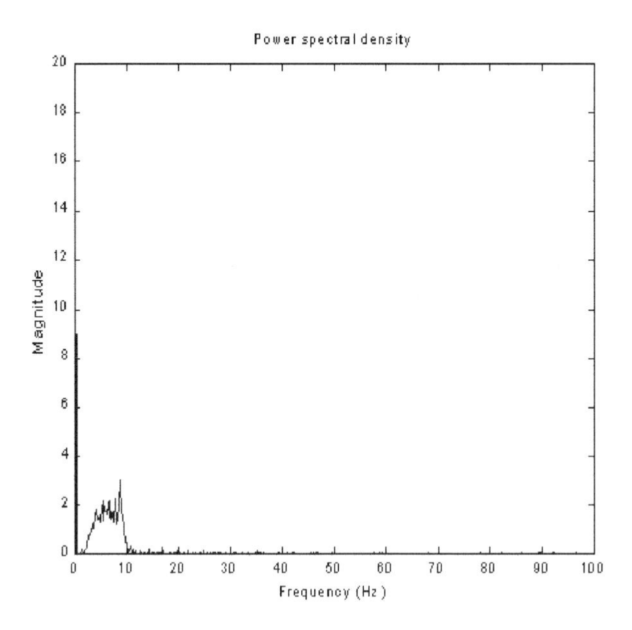

Fig. 7.29. Vibration signature corresponding to a chirp input signal

mode, animating the image to the client. Thus the G-web Server provides a convenient means of delivering the VI's front panel image across the Internet.

The package also includes a library for writing CGI (Common Gateway Interface) programs for use with the G Web Server. The CGI is the standard for interfacing external applications with information servers, such as HTTP or Web servers. Whenever the user request, for example by clicking on the hyperlink or image map, the URL (Uniform Resource Locator) corresponding to the CGI program, the CGI program will be executed in real-time on the server.

Access security is incorporated into the G Web Server. It is achieved *via* the Basic Access Authentication scheme as specified in HTTP/1.0. It is a simple challenge-response authentication mechanism that is used by a server to challenge a client request and by a client to provide authentication information. It is based on the model that the client must authenticate itself with a user-ID and a password for each realm. The server will service the request only if it can validate the user-ID and password for the protection space of the Request-URI. With the G Web Server, it is also possible to control access based on user name, password and user's IP address.

7.3.6 Implementation

On the client side, the operator may send commands over the internet to perfom remotely the vibration analysis of the machine. A typical session begins

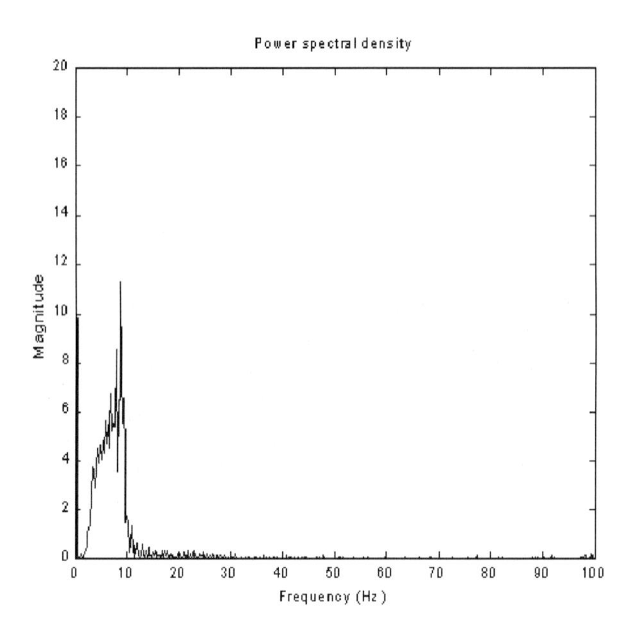

Fig. 7.30. Spectrum of machine corresponding to a chirp input signal after a fault has occurred

from the browser at the client end. Upon successful authentication of the user, the home page as shown in Figure 7.36 will appear. The user may initiate any mode of analysis by clicking on any of the hyperlinks. These hyperlinks actually refer to a CGI program. Whenever the server receives any request, this CGI program will request a session with the server side. Only one session may proceed at any point in time. Following these checks, the VI requested will be loaded into memory at the server. The front panel image will be served up and the results of the analysis will be shown. The membership functions, scaling factors and threshold necessary to implement the analysis can be modified remotely.

Clicking of the image maps or submission of standard HTML forms will send commands from the client to the server. These commands will be sent to the server *via* the Internet with the TCP/IP set of protocol. At the server side, the G-web Server will receive the client request through the designated socket and port number, usually 80 for HTTP. The server will process the client's request and serves up the front panel image to the client's browser. Snapshots of the browser executing the monitoring and diagnostic modes of the vibration analysis are given in Figures 7.37 and 7.38 respectively.

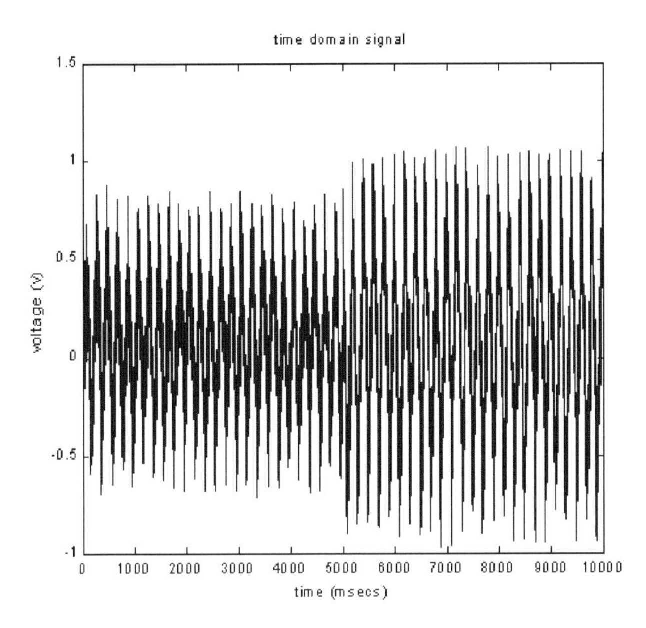

Fig. 7.31. Vibration signal corresponding to a sine input signal (at $t = 5$, a fault is simulated)

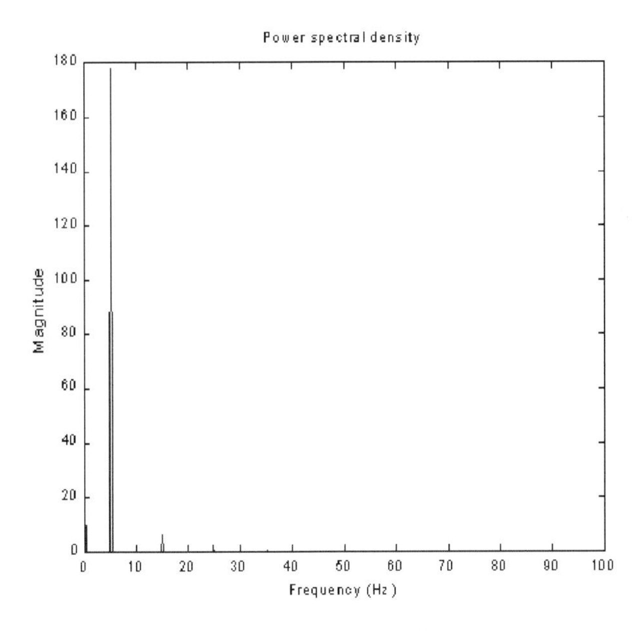

Fig. 7.32. Vibration signature corresponding to a sine input signal

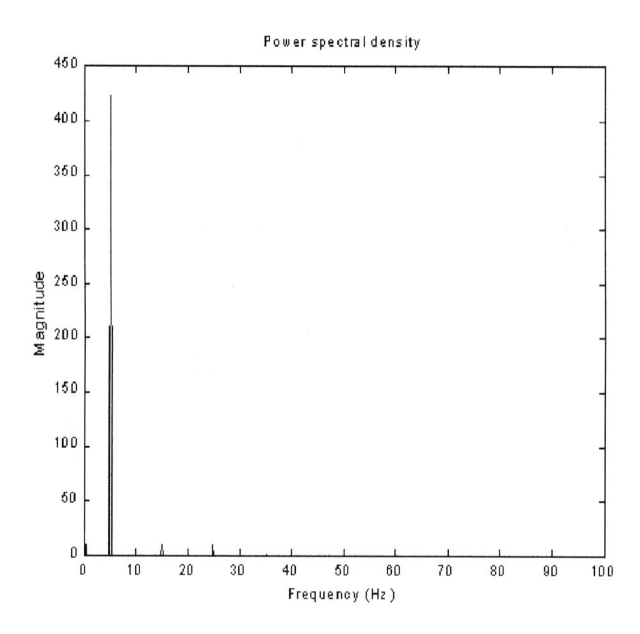

Fig. 7.33. Spectrum of machine corresponding to a sine input signal after a fault has occurred

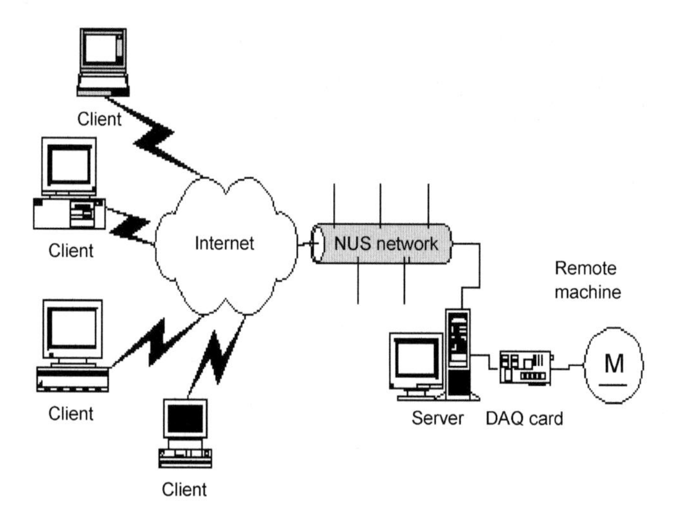

Fig. 7.34. Overview of the hardware configuration

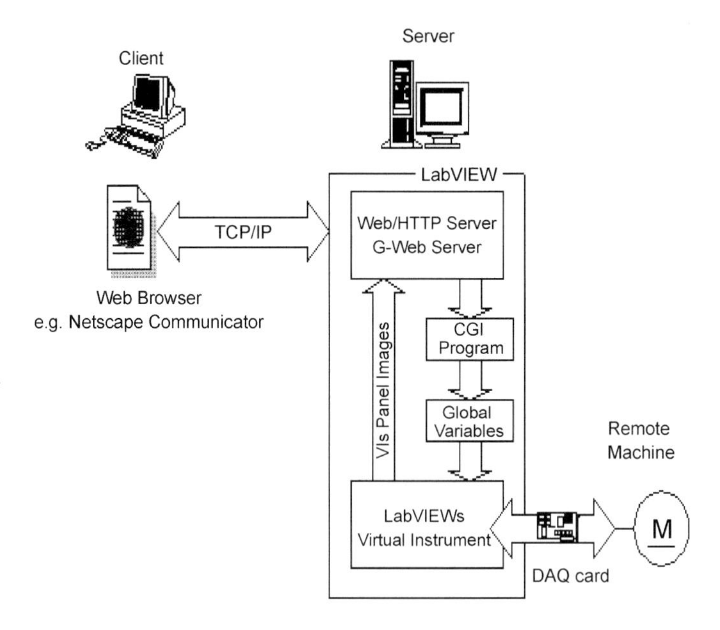

Fig. 7.35. Overall software configuration

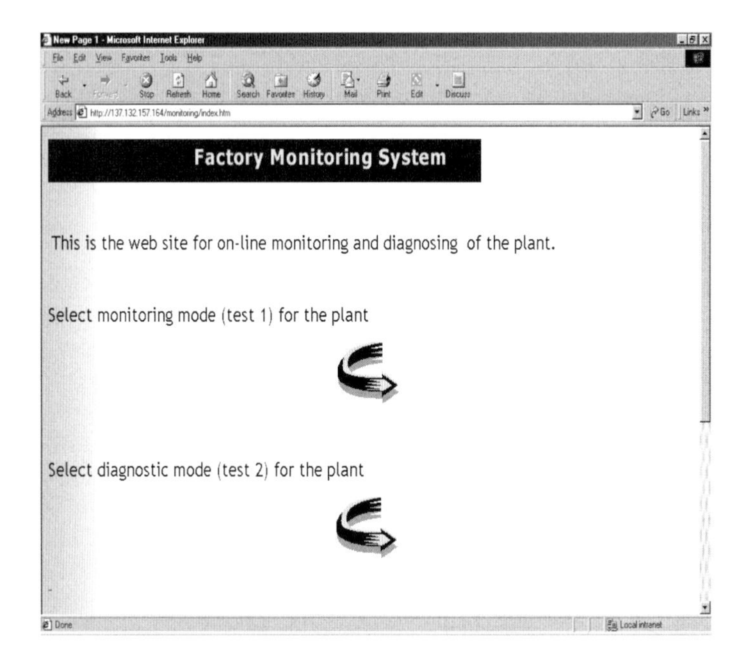

Fig. 7.36. The home page

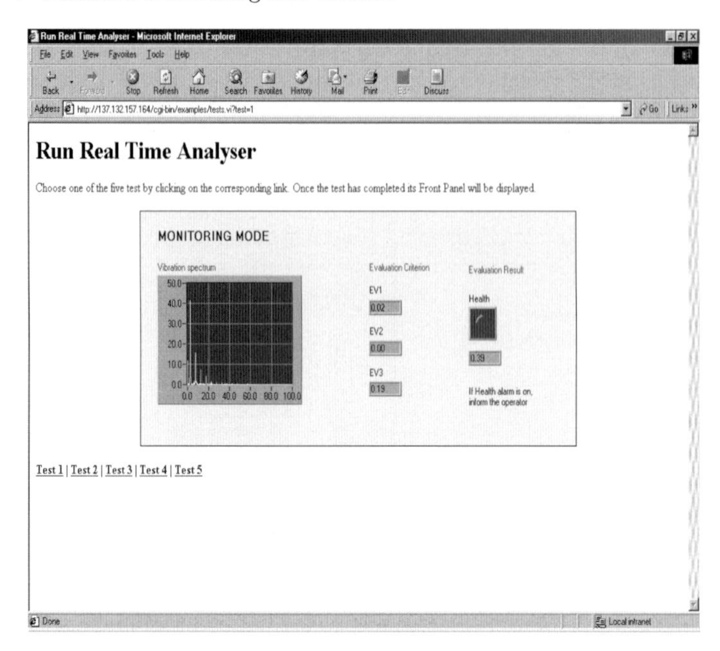

Fig. 7.37. Snapshot of the remote monitoring mode

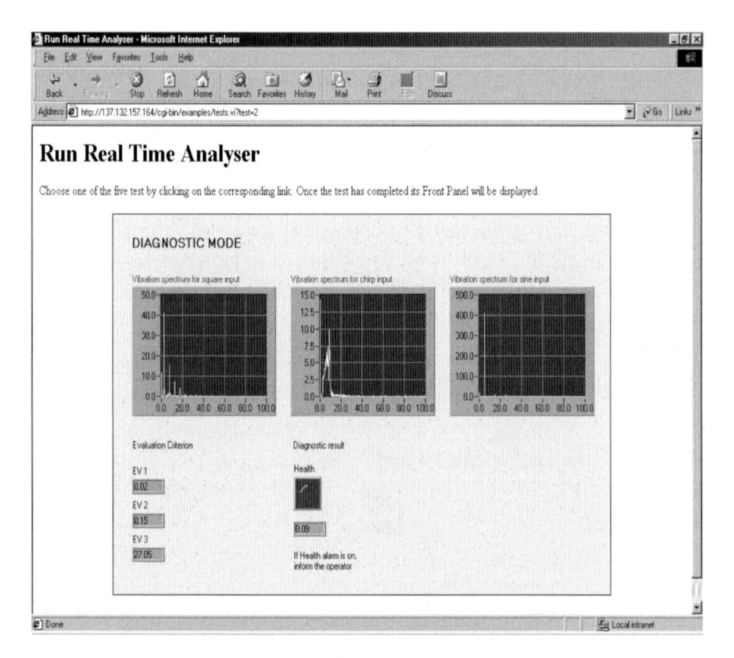

Fig. 7.38. Snapshot of the remote diagnostic mode

8

Other Engineering Aspects

A control system, based on the above developments, is developed and applied to a precision 3D Cartesian robotic system as shown in Fig. 8.1 with a travel of $250 \times 400 \times 50$ (dimensions are in mm). Other engineering aspects are signficant to the overall development effort, such as sizing and choice of components, hardware architecture, software development platform, user interface design and performance assessment. Details of these aspects are described in the ensuing sections.

Fig. 8.1. 3D Cartesian robotic system

8.1 Specifications

The control specifications are given in Table 8.1.

Table 8.1. Control specifications

Resolution R, (μm)	0.01
Repeatibility R_p, (μm)	0.05
Minimum velocity V_{min}, (μm/s)	2
Maximum velocity V_{max}, (m/s)	0.5
Step-settle time T_s, (s)	1.5

8.2 Selection of Motors and Drives

Linear electric motors manufactured by Anorad Corporation, U.S.A. are used for the construction of the robot. The LE series of high efficiency brushless linear servo motors are selected for the high continuous force specification of 78 N and a peak force of 191 N. The maximum acceleration thus depends on the moving mass and this peak force. In addition, the motion profile must be subject to the condition that the RMS motor force required over an operational cycle must be less than the continuous force. Assuming the RMS requirement is satisfied, a maximum acceleration of about 0.2 G is achievable if the maximum moving mass is less than 100 kg.

8.3 Selection of Encoders

Optical encoders are selected over laser interferometers mainly for the cost factor. In order to achieve a measurement resolution of $R = 0.01$ μm, Heidenhein linear encoder LIP481 (1Vpp type) is selected. The LIP481 has a signal period of $T_e = 2$ μm and it is accurate to ± 0.2 μm over 220 mm and accurate to ± 0.5 μm over 420 mm. The LIP581 can be used for a longer travel. It has a signal period of 4 μm and it is accurate to ± 1 μm over 1440 mm.

In order to yield a resolution of $R = 0.01$ μm, a minimum electronic interpolation of 400 and 200 are needed for LIP581 and LIP481 respectively. In order that the interpolation can work effectively, measurement noise must be suppressed to below 5 mV using proper shielding and grounding techniques to minimise undesirable interference effects from magnetic field.

8.4 Control Platform

The flexibility, quality, functionality and development time are crucial factors driving the selection of the hardware and software development platform for the control system. The dSPACE development platform is selected according to three main features and provisions: rapid control prototyping, automatic production code generation, and facilities for hardware-in-the-loop testing.

Rapid control prototyping implies that new and customised control concepts can be directly and quickly developed, and optimised on the real system via the rich set of standard design tools and function blocks available in commonly used software such as MATLAB®/Simulink®. Controllers can be directly and graphically designed in the form of functional block diagrams with little or no line programming necessary. Real-time code can be automatically generated from the functional block diagram and implemented on the machine through the automatic production code generation feature. The hardware-in-the-loop facilities further allow for a reliable and cost-effective method to perform system tests in a virtual environment. Peripheral components can be replaced by proven working mathematical models, while the actual physical components to be evaluated are inserted systematically into the loop. In addition to savings in time and costs, the modularity and reproducibililty associated with hardware-in-the-loop simulation greatly simplifies the entire development and test process.

8.4.1 Hardware Architecture

The overall system hardware architecture is shown in Fig. 8.2. To meet simultaneous high speed and high precision requirements, the control unit is configured with high speed processing modules. A dSPACE DS1004 DSP board is used together with a DS1003 DSP board. The DS1004 DSP board uses a DEC Alpha AXP 21164 processor capable of 600 MHz/1200 MFLOPS. This board is used to concentrate fully on the computationally intensive tasks associated with the execution of the control algorithms. The DS1003 DSP board uses the TMS320C40 DSP which is capable of 60 MFLOPS. It can deal effectively with all the necessary I/O tasks because of the high-speed connection to all I/O boards via the Peripheral High-speed (PHS) Bus. Both the boards are RTI (Real-time Interface) enabled, and they allow full programming from within Simulink®. This multiprocessor system is configured to give optimal performance via the decentralisation of computational and I/O tasks separately to the DS1004 and DS1003 boards respectively.

In addition to the processor boards, a DS2001 data acquisition board is used which has five parallel high-speed 16-bits A/D channels. The sampling and holding of signals along all channels can be executed simultaneously, with a short sampling time of 5 μs. A DS2102 high-resolution D/A board is used to drive the actuators. It has six parallel D/A channels, each with a 16-bits resolution. The typical settling time (full scale) is 1.3–2μs, and output voltage ranges (programmable) of ±5 V, ±10 V, or 0–10 V can be configured. The DS2102 Simulink® blocks and dialog boxes provided with RTI facilitate the configuration of user-defined voltage range for each channel, depending on the devices connected to the channel.

To achieve fine measurement resolution via analog incremental optical encoders, the DS3002 incremental encoder interface board with a maximum input frequency of 750 kHz is chosen. Sinusoidal encoder signals are captured through six channels in the DS3002, converted to 12-bit digital signals and then phase decoded by special highly optimised software functions to extract the relative position from these data. Along with the relative position, a search block will seek the encoder index lines and updates the corresponding counter when a new index is reported to give an absolute position information. Theoretically, in this way, an interpolation of 4096 can be achieved. This in turns implies that a measurement resolution of less than 1 nm can be achieved if the grating-line pitch is 4 μm. However, one should be cautious of the constraints in terms of interpolation errors associated with limited wordlength A/D operations, and imperfect analog encoder waveform with mean, phase offsets, noise as well as non-sinusoidal waveform distortion (Chapter 6).

A timer and digital I/O board, DS4001 with 32 in/out channels is used for status checking of the travel-limit switches and other safety enhancing digital devices. The 32 in/out channels can be divided into 8-bit groups.

Comprising the processor I/O boards as mentioned, the system hardware architecture is highly modular. The boards are installed into a 19" rack version expansion box which has a maximum of 20 full-size 16-bits ISA slots for 20 dSPACE boards. The expansion box includes a power supply, cooling fan and interface electronics. Connector panels are used in the control system to provide easy access to all input and output signals on the dSPACE I/O boards. Analog signals can be accessed via BNC connectors while most of the digital signals are accessed via Sub-D connectors. The Sub-D connectors on the connector panels are of a low density and they are grouped with respect to the I/O channels or functional units on the board.

The control platform configured will satisfy the velocity requirements as illustrated below.

Minimum velocity, V_{min}

The closed-loop bandwidth is estimated to be around 100Hz. Therefore, to avoid aliasing, the Shannon sampling theorem must be satisfied, requiring the sampling frequency to be $\omega_s \geq 2 \times 100$ Hz=200 Hz. This sampling frequency is easily supported by DS2001 which can achieve a *per* axis servo update interval of as low as 5 μs.

This further implies the encoder count frequency after interpolation, f_e, should satisfy

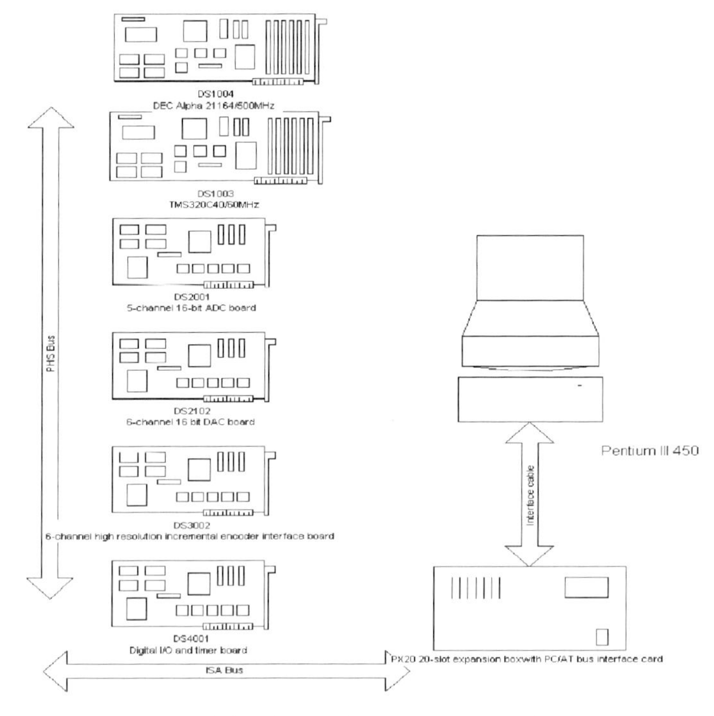

Fig. 8.2. Overall hardware architecture

$$f_e \geq 200Hz.$$

Therefore,

$$\frac{V_{min}}{R} \geq 200,$$
$$V_{min} \geq 2\mu\text{m/s}.$$

The minimum velocity of $V_{min} = 5\ \mu$m/s can be satisfied.

Maximum velocity, V_{max}

Maximum velocity achievable depends critically on the maximum bandwidth of the controller electronics. DS3002 supports a bandwidth of upto 750 kHz. Therefore,

$$V_{max}(1 \times 10^6)/T_e < 750(1000),$$
$$V_{max} < 1.5\text{m/s}.$$

The maximum velocity of $V_{max} = 0.5$ m/s can be satisfied.

8.4.2 Software Development Platform

The processor boards are well supported by popular software design and simulation tools, including MATLAB® and Simulink®, which offer a rich set of standard and modular design functions for both classical and modern control algorithms. The overall Simulink® control block diagram customised for the cartesian 3D gantry machine is shown in Fig. 8.3. The block diagram can be divided into three parts according to their functions:

- Control and automatic tuning,
- Geometric error calibration and compensation, and
- Safety features, such as emergency stops, limit switches, *etc.*

The control algorithms are included in the subsystem `x-ctrl`, `y-ctrl` and `z-ctrl`. Fig. 8.4 shows the Simulink® control block diagram for the x-axis. Apart from the PID feedback control which is fixed, the other advanced control schemes are configurable by the operator. An automatic tuning operation mode is also provided for the controllers. The operation modes (control or automatic tuning) can be selected through the switch blocks `X-Output-Switch`, `Y-Output-Switch` and `Z-Output-Switch`.

The geometric error calibration and compensation for the axes are integrated with the controllers via an S-function interface. These features are enabled through switches `Comp-x` and `Comp-y`, as shown in Fig. 8.3.

All the limit switch signals from the three axes are acquired through DS4001 board. These limit switch signals serve as the control input of the three switches shown in Fig. 8.3, to nullify the system control signal when the limit switch is activated. An operator emergency stop function is also provided in the overall Simulink® control block diagram.

A software component, running on MATLAB®/Simulink®, is written for the geometrical error compensation. Using this software, an S-function comprising Radial Basis Function (RBF) based error compensation can be automatically produced given the raw data set obtained from the calibration experiments, and simple user inputs on the RBF training requirements. Thus, little prior technical knowledge of RBF is required of the operator.

Upon a successful automatic code generation from the Simulink® control block diagram, the controller will run on the dSPACE hardware architecture configured. The user interface, designed using dSPACE CONTROLDESK, allows for user-friendly parameters tuning/changing and data logging during the operations. The control parameters can be changed on-line, while the motion along all axes can be observed simultaneously on the display.

8.4.3 User Interface

The user interface is designed as a virtual instrument panel based on the dSPACE CONTROLDESK instrumentation tool. CONTROLDESK is a com-

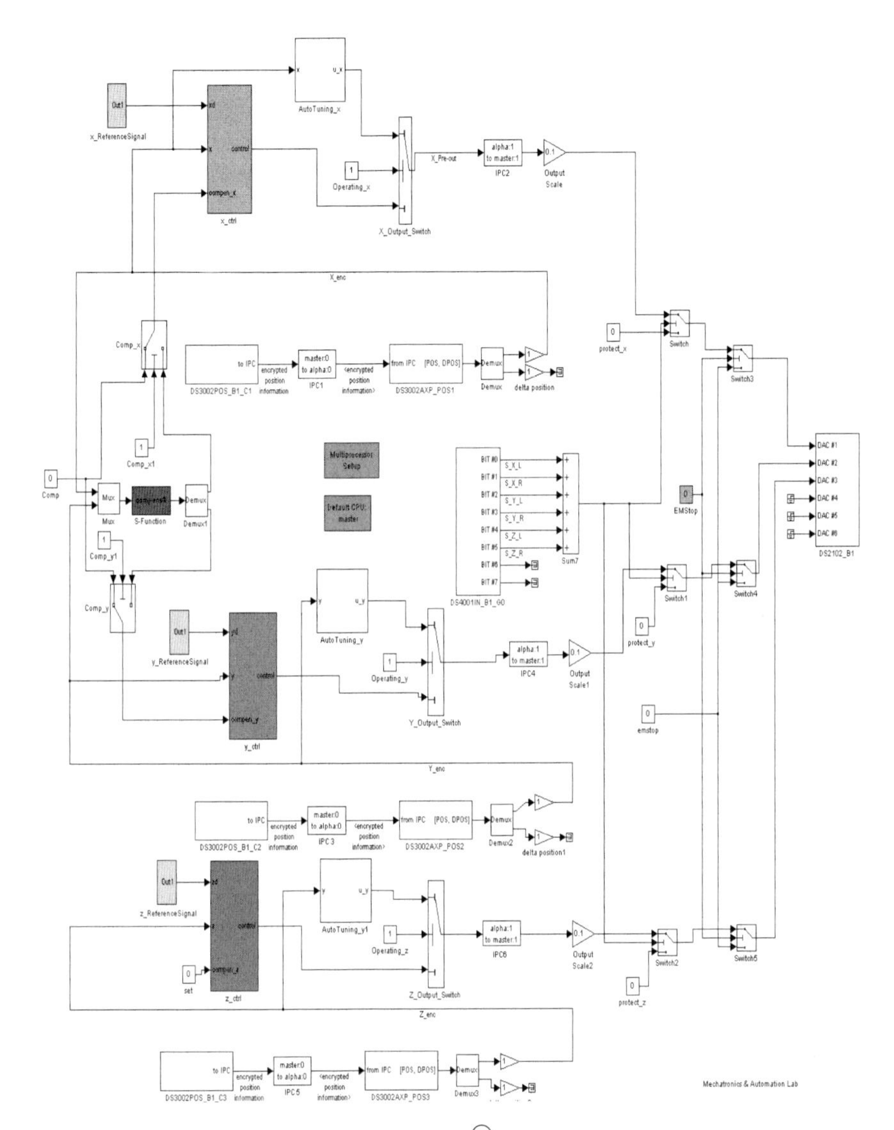

Fig. 8.3. Overall Simulink® control block diagram

prehensive design environment where designers can intuitively manage, instrument, and automate their experiments and operations. CONTROLDESK is seamlessly integrated within the dSPACE development platform. It can realise real-time data acquisition, online parameterisation and provide an easy access to all model variables without having to interrupt the running operations. The entire user interface design is achieved simply via drag and drop operations from the Instrument Selector provided. This greatly speeds up the

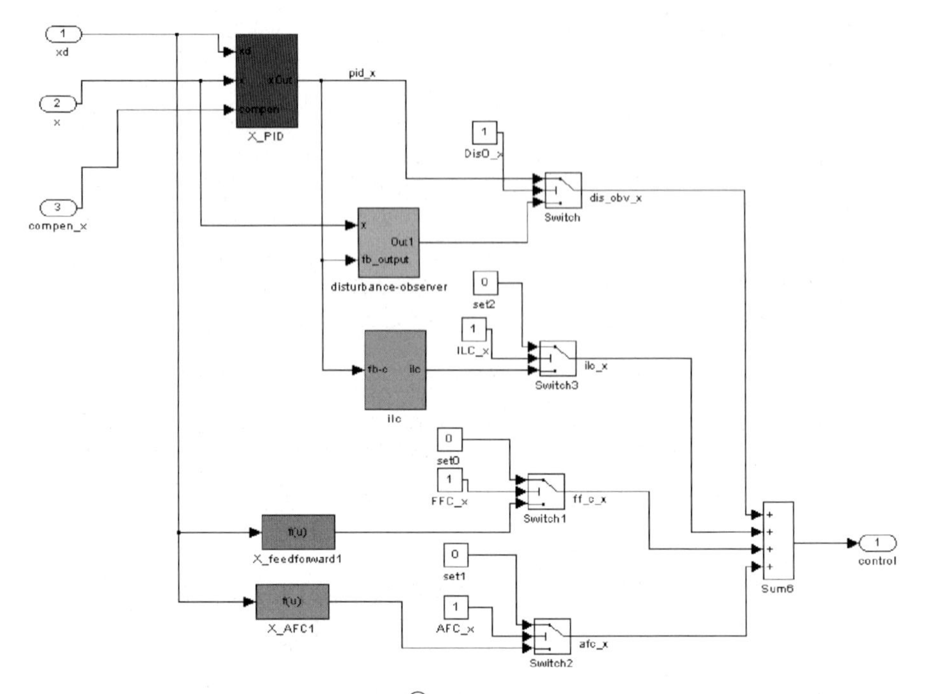

Fig. 8.4. The Simulink® control block diagram for X-axis

design process and help avoids standard design pitfalls associated with line programming.

Fig. 8.5 shows the user interface customised for the gantry motion system. The interface can be broadly split into six portions: `Operating Status`, `Control Buttons`, `Parameter Adjustments`, `Limit Switches`, `Reference Signals`, and `Results`.

The `Operating Status` part of the interface will reflect the present operating status of the system. An `EMStop` button is provided to abort the system from operation under emergency scenarios. The status of the system is indicated by the LEDs beside the buttons. The `Limit Switches` part of the interface will reflect the status of the limit switches, which are installed in the system to prevent travel outside allowable ranges. Under the `Control Button` part of the interface, the controllers for the individual axis can be configured accordingly. Multistate LEDs are used to indicate the controllers currently applied. The parameters for the controllers, such as control gains, can be adjusted online in the `Parameter Adjustments` area while the control results can be observed simultaneously in the `Results` area. The specific motion trajectories are configured in the `Reference Signal` area. Apart from the standard ones provided for demonstration purposes, other reference signals can also be added accordingly, depending on the specific applications.

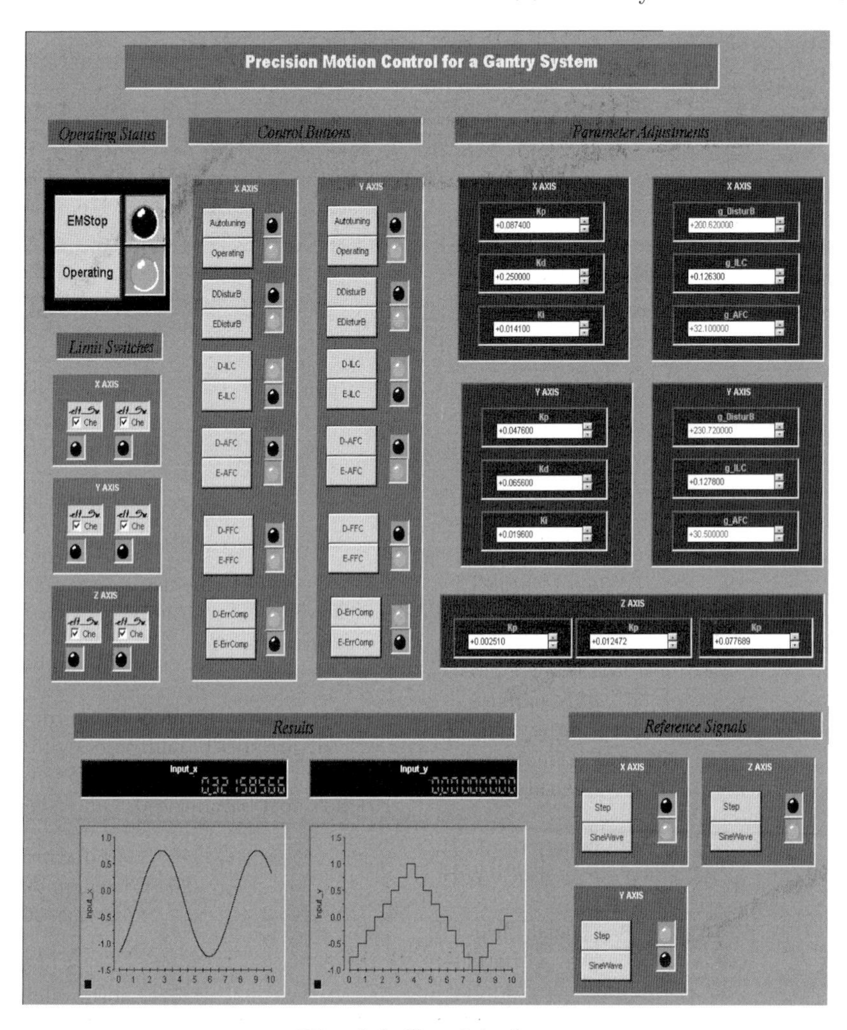

Fig. 8.5. User interface

8.5 Accuracy Assessment

The accuracy of a machine is usually measured according to prescribed procedures (*e.g.*, British Standards (1989)). By making some assumptions for the machine of concern, a first rough assessment of achievable accuracy can be established.

The following assumptions are made:

Assumption 8.1.

Repeatibility error is zero.

Assumption 8.2.

The machine is calibrated at points separated by one resolution.

Consider a 3D machine of the FXYZ type (Hocken 1980), where the F represents the machine frame, X,Y,Z represent the axes in the order of stacking (left to right order is from part to tool). The x, y, z components of the actual displacement with respect to the part frame can be written as

$$\begin{aligned}
\Delta x &= \delta_x(x) + \delta_x(y) + \delta_x(z) - y\alpha_{xy} - z\alpha_{zx} - y\epsilon_z(x) + \\
&\quad z[\epsilon_y(x) + \epsilon_y(y)] - y_t[\epsilon_z(x) + \epsilon_z(y) + \epsilon_z(z)] \\
&\quad + z_t[\epsilon_y(x) + \epsilon_y(y) + \epsilon_y(z)] + x_t, \\
\Delta y &= \delta_y(x) + \delta_y(y) + \delta_y(z) - z\alpha_{yz} - \\
&\quad z[\epsilon_x(x) + \epsilon_x(y)] + x_t[\epsilon_z(x) + \epsilon_z(y) + \epsilon_z(z)] \\
&\quad - z_t[\epsilon_x(x) + \epsilon_x(y) + \epsilon_x(z)] + y_t, \\
\Delta z &= \delta_z(x) + \delta_z(y) + \delta_z(z) + y\epsilon_x(x) - \\
&\quad x_t[\epsilon_y(x) + \epsilon_y(y) + \epsilon_y(z)] + y_t[\epsilon_x(x) + \epsilon_x(y) + \epsilon_x(z)] + z_t,
\end{aligned}$$

where x, y, z are the nominal values of the carriage positions, x_t, y_t, z_t are the x, y, z offsets of the tool tip, $\delta_u(v)$ is the translation error in the u-direction under v motion (the sign of the error is chosen as the sign of the u-direction), is the rotation about the u-axis under v motion. u and v are arbitrary arguments, which may be $x, y,$ or z. $\alpha_{xy}, \alpha_{xz}, \alpha_{yz}$ are the orthogonality errors in the xy, xz, yz planes respectively.

Suppose $|x| \leq 400, |y| \leq 250, |z| \leq 50$. Using a laser interferometry measurement system, linear and straightness errors can be measured accurate to 1 nm resolution, and the angular error (pitch,yaw) and squareness can be measured accurate to 0.002 arcsec resolution. Roll error cannot be directly measured using a laser interferometer. Instead an electronic level sensor may be used which is accurate only to 0.2arcsec. However, it may be possible to carry out interpolation to reach 0.002arcsec. Under Assumptions 8.1 and 8.2 , the errors may be estimated over the entire working volume, assuming $x_t = y_t = z_t = 0$:

$$\begin{aligned}
|\Delta x| &\leq 1 + 1 + 1 + 2.5 \times 10^8 \times 9.76 \times 10^{-9} + 0.5 \times 10^8 \times 9.76 \times 10^{-9} \\
&\quad + 2.5 \times 10^8 \times 9.76 \times 10^{-9} \\
&\quad + 0.5 \times 10^8 (9.76 \times 10^{-9} + 9.76 \times 10^{-9}) \\
&= 9.34\text{nm}, \\
|\Delta y| &\leq 1 + 1 + 1 + 0.5 \times 10^8 \times 9.76 \times 10^{-9} \\
&\quad + 0.5 \times 10^8 (9.76 \times 10^{-9} + 9.76 \times 10^{-9}) = 4.46\text{nm},
\end{aligned}$$

$$|\Delta z| \leq 1 + 1 + 1 + 2.5 \times 10^8 \times 9.76 \times 10^{-9} = 5.44\text{nm}.$$

The absolute diagonal error is given by

$$|\Delta d| = \sqrt{\Delta x^2 + \Delta y^2 + \Delta z^2} \leq 11.69\text{nm}.$$

Without interpolation of roll measurements, the errors increase to:

$$\begin{aligned} |\Delta x| &\leq 57.66\text{nm}, \\ |\Delta y| &\leq 52.78\text{nm}, \\ |\Delta z| &\leq 247.00\text{nm}, \\ |\Delta d| &\leq 259.07\text{nm}. \end{aligned}$$

If the repeatibility of the linear and straightness errors is λ nm, and angular error is φ times the above angular resolution, then the errors over working volume are given by

$$\begin{aligned} |\Delta x| &\leq 3\lambda + 6.34\varphi, \\ |\Delta y| &\leq 3\lambda + 1.46\varphi, \\ |\Delta z| &\leq 3\lambda + 2.44\varphi, \\ |\Delta d| &\leq \sqrt{27\lambda^2 + 48.28\varphi^2 + 61.44\lambda\varphi}. \end{aligned}$$

Using mechanical specifications of $\lambda = 2000$ nm, $\varphi = 100$ (angular repeatibility is 0.2 arcsec), it follows that

$$|\Delta d| \leq 10.99\mu\text{m}.$$

This is a conservative estimate since the error components are taken to cumulate in the worst possible manner.

8.6 Digital Communication Protocols

Preceding sections have focused specifically on sensors, actuators and the control platform. In order to integrate these components to work efficiently as a system, or to interconnect multiple control systems, signal communication protocols and standards will be necessary.

Fieldbus is a digital, bi-directional, serial bus communications network that links various instruments, transducers, controllers, final control elements, and other devices. It serves as a spinal column of distributed real time systems and tremendously simplifies the wiring among field devices.

In this section, the typical fieldbus protocol stack will be described and the common field protocols currently used in servo drives will be presented.

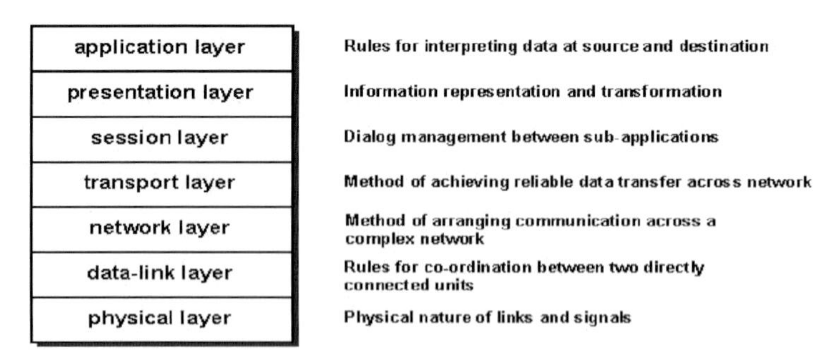

Fig. 8.6. OSI Model

8.6.1 Fieldbus Protocol Stack

The Open Systems Interconnect (OSI) model defines the main aspects of any communication system. The seven layers of the model are given in Figure 8.6.

This OSI model is a reference model towards standardization. It does not define the exact services or protocols, but just what each layer should do. The communication functions are implemented by the lowest three layers and the host functions are implemented by the top four layers. The lowest two layers deal with intermediate points. The remaining layers have end-to-end significance.

The OSI model is a general communication model. The protocol stacks for fieldbuses usually only involve the following layers:

- Physical layer
- Link layer
- Network layer
- Application layer

There may be a user layer above the application layer implementing high level user functions, or these functions may be implemented as part of the application layer. The remaining layers do not matter, except in more complex systems for the following reasons:

- Protocol overheads involved in traversing the full seven layers are not acceptable for a real-time response solution.
- Field communication requirements are not intensive, typically short to medium size messages, but protocol overheads must be low.
- Communication system at this level must be simple and easy to install and maintain, and it should be low cost.

In addition, the link and network layers are often closely tied together, with the network component being relatively simple. Some technical descriptions tend to treat such a combination of link/network layers, with the network part being only rudimentary, as a link layer.

Physical Layer

In this layer, the physical nature of data links and data signal are defined, *i.e.*, the type of data link used, the physical topology, the nature of signal on the link (voltage, current, frequency), the signaling rate, unit of data (bytes, words), the signal to logic translation, the type of connectors to be used.

Examples of topology are given in Figures 8.7 and 8.8.

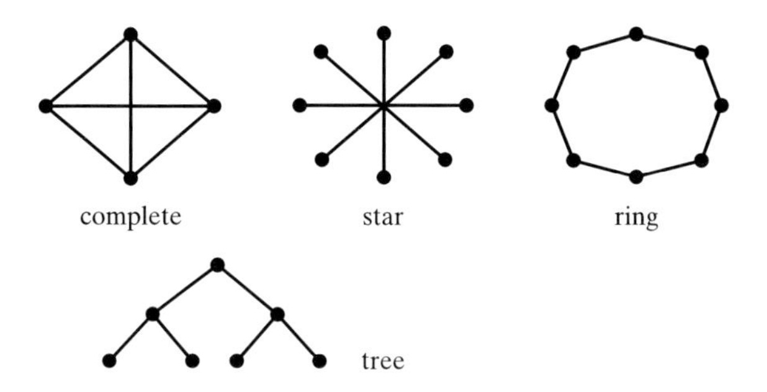

Fig. 8.7. Point-to-point topology

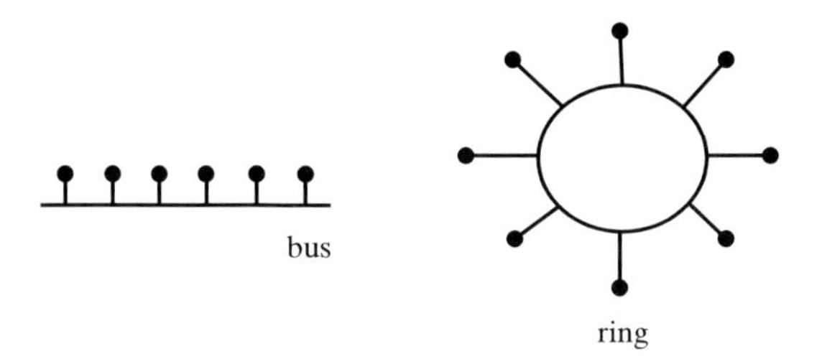

Fig. 8.8. Broadcast topology

Link Layer

The link layer defines rules of communication between a communicating entity (source or destination) and the element of the communication system that is directly connected to that entity.

The key elements involved at the link layer include:

- The establishment of a link between two points.

- Media-access control. *i.e.*, the rules of 'the right to speak' (flow control), is the link half-duplex or full duplex.
- Segregation of the link data and link commands.
- Error detection and correction techniques
- Indication of link failure and the methods of recovery

Media access control may be centered on popular schemes such as master-slave, token bus or carrier sense multiple access/collision detect (CSMA/CD).

Apart from media access control, error detection and correction are two important features that the link layer must be able to perform, since errors are ever present and happen in every system. Errors may be small (just one bit inverted) or larger (loss of whole command or response).

Error detection can be organized on the following levels:

- Byte level, which is a parity bit
- Data packet level, which is a checksum byte or a number of checksum bytes

If a whole message is lost, there is only one method to recover, which is re-transmission. Self-recovery protocols are risky and usually unreliable.

Network Layer

The network layer defines the rules for end-to-end communication. It takes care of the method of addressing in the network and the routing of data.

To provide orderly communication, all elements of a network must have their addresses assigned. Addressing is an important issue to consider because of the following reasons:

- Address space must be sufficient. If 8 bits are allocated, the network cannot have more than 256 members. If too many bits are used than necessary, network performance will be degraded.
- Address must be set in all network components, from sensors to controllers.
- Addresses must be unique.

Practically, all fieldbuses allow for messages that are not addressed to any specific device but instead are addressed to every device. Such messages are called broadcast messages.

Application Layer

The application layer defines the meaning of data transmitted and received. It is right at the access level for applications which require the network services. This layer will provide file, print, message, and application database services. It will identify and initiate the services necessary for a user's request. In a control system application, at this level, the code for various commands is defined with the format of response codes and data. This layer provides services for the various control functions.

User Layer

There may be a user layer which aims to implement high-level control functions more efficiently. It connects the individual plant areas and provides a high level environment for applications. Function blocks are now commonly used at this level to realize the control strategies. Each fieldbus device is described with a Device Description (DD). The DD can be viewed as the driver for the specific device. It includes all variables descriptions and corresponding operating procedures to use the device, thus making the DD truly interoperable. Any control system or host will be able to communicate with the device once it has the DD for that device.

Traversing the Stack

When a transmitting device sends a message (PDA - Packet Data Unit), the message will travel down through the layers to the physical medium at the physical level and subsequently move up the layers to the receiving device. When there is a request in the message, the receiving device will attend to it and responds in the reverse manner. While traversing down the stack, the original message from the transmitting device is added with a piece of information from each layer and the same information is stripped off in the corresponding layer of the receiving device.

8.6.2 Common Fieldbuses

The selection of field devices is driven by the supporting protocols and *vice versa*. Devices from a particular manufacturer are usually conforming to particular protocols, so that there is usually no real option what protocol and supporting devices to use. In this section, some common fieldbuses, with wider user bases, will be presented.

CANopen

Controller Area Network (CAN) is a serial bus network of microcontrollers that connects devices, sensors, and actuators in a system or sub-system for real-time control applications. CAN provides many powerful features, including multi-master functionality and the ability to broadcast or multicast telegrams. CAN offers many other advantages, among which are the low cost, high data reliability, short response time, and a huge user base. These strong points put CAN among the leaders in fieldbus technology, especially in the automotive and textile industries.

In this protocol, the message is broadcast to all nodes in the network using an identifier unique to the network. Based on the identifier, the individual node decides whether or not to process the message and also determines the priority of the message in terms of competition for bus access. This method allows for uninterrupted transmission when a collision is detected, unlike Ethernets that will stop transmission upon collision detection.

CANopen is a CAN-based higher layer protocol. It was developed as a standardized embedded network and designed for motion-oriented machine control networks, although it is now used in many other fields, such as medical equipment, off-road vehicles, maritime electronics, public transportation, building automation, *etc.*

CANopen provides a mechanism so that devices of different types can be integrated and can communicate in a standardized fashion. By making use of the device profile information, CANopen devices will ensure common operating functions. For example, two digital modules from two different manufacturers will have common functionality such as setting the outputs and reading the inputs. The profile specifies the functionality, which must be common for devices to be interoperable. With CANopen, the manufacturers are not constrained in making various features in their devices.

The fundamental part of device profile is object dictionary, consisting of a mixture of data objects, communication objects, and commands/actions. CANopen services give the user full access to the object dictionary, allowing reading and writing of data and commands. Data and commands are implemented using a 16-bit index addressing mechanism together with 8-bit sub-index, giving an address range from 0000H to FFFFH. Parts of the object dictionary are divided into different areas based on functionality. Index 6000H, for example, is reserved for reading.

CANopen provides an open protocol and allows direct data exchange between nodes on the network without participation of a bus master unit. It allows full broadcast/multicast features and a variety of communication modes designed to keep bus loading minimal and predictable. Therefore, CANopen is well suited to the concept of remote intelligence and is ideal for distributed control solutions.

Profibus

Profibus is an open digital communication system which is mainly used in factory automation. It is now one of the market leaders in data communication, with a 20% share and over 2000 products. Profibus is now widely applied in the food and pharmaceutical industries.

The protocol structure of Profibus comprises three layers: physical layer (layer 1), data link layer (layer 2), and application layer (layer 7).

Physical layer describes standard serial communication EIA-485 using twisted and shielded pair, with a bus topology tree expandable using repeaters and a digital transmission NRZ coded.

Data link layer defines a logical model of the network nodes. The network nodes consist of passive stations (which use LAN transmission medium only under an active station request) and active stations (which communicate among themselves and with passive stations). In the active stations, the medium access follows the token passing rule, *i.e.* a defined bit sequence that allows mutual communication between devices and grants the use of LAN. A logical ring connects active stations and guarantees defined transmission time and lack of collision with other data in the bus cable, as only the station with the token can transmit, while the others remain inhibited. Medium access control in passive stations follows master-slave methods, where the masters are the active stations. The data link layer is further divided into two sub-layers, one controlling medium access and the other giving higher levels interface with synchronous/asynchronous transmission services.

Application layer allows objects manipulation, with objects being variables, arrays, matrix, variable lists, program calls, subroutines, *etc.* It is divided into two sub-layers: FMS (Fieldbus Message Specification) and LLI (Lower Layer Interface)

Interconnecting drives and steering unit with Profibus yields the possibility to realize control loops through LAN. In an electric drive operation, for example, the active station handles path generation and position loops, while the passive station handles speed and torque loops.

Foundation Fieldbus

Foundation Fieldbus (FF) is a serial, two-way communications system that serves as the base-level network in factory automation. FF is typically implemented together with Ethernet as its hardware platform. FF defines two level networks, H1 and HSE, with 31.25 Kbps and 100 Mbps transfer rates, respectively.

The key concept of FF is a schedule timetable to ensure that all messages are transferred to correct destination nodes within a prescribed time. This contributes to FF capability of timely information access. The schedule timetable acts as a manager that control information traffic; it determines when a message is sent, what message is to be sent, where it should be sent, *etc.* Schedule timetable is in essence an algorithm of information execution, and the content can differ from one algorithm to another; however, the messages usually include periodic data (synchronous), request from user (asynchronous), and request from devices (asynchronous). All messages are transferred within one bus. Since there are several messages to be transferred, token mechanism is used to determine which should be transferred first, with priority given to synchronous messages.

The structure and characteristics of FF make it suitable for diagnostics purposes, for example in valve diagnostics in a hydraulic application, which requires timely information access. In this application, process information, such as pressure, temperature, and level, is monitored. This information is then processed with statistical tools and compared to a threshold value, so that the condition of the process can be diagnosed. For example, an exponential rise in the statistical distribution of temperature may indicate that the system is overheating.

Firewire

Firewire (also known as i.Link or IEEE 1394) is a personal computer and digital video serial bus interface standard offering high-speed communications and real-time data services. Firewire is a successor technology to SCSI Parallel Interface. It was developed by Apple computer and has been widely used in the computer and consumer electronics industries.

The advantages of Firewire are as follows:

- It is a low-cost, self-powered, high-speed, digital single cable serial bus suited for real time motion control applications.
- The bandwidth can be determined overhead.
- It is compatible with peer-to-peer communications.
- The speed ranges from 100 to 400 Mbps.
- It supports 63 devices in one bus.
- Up to 10 m cable length repeaters can be used to extend distance.
- It is compatible with digital plug-n-play setup with all parameters software driven.
- The interface has typically been included with every PC, with easy set-up and configuration.

The disadvantages of Firewire are as follows:

- There is no standard protocol.
- Firewire only addresses a small proportion of the real problems in developing an industry standard.

Sercos

Serial real-time communication system (Sercos) interface is a digital motion control bus that interconnects motion controls, drives, I/O and sensors. It is an open controller digital drive interface which is designed for high-speed serial communication of closed-loop data in real-time over a noise-immune, fiber optic cable. Sercos takes advantage of digital drive capabilities by not only replacing the standard 10 V analog standard interfaces, but also providing two way communications between control and drive.

The controls and drives use a standard medium for transmission, topology, connection techniques, signal levels, message structures, timing and data formats. This allows devices from different Sercos' manufacturers to communicate with each other in the same platform.

The advantages of Sercos are as follows:

- Data is exchanged between control and drives via fiber optic rings which eliminates the electromagnetic interference.
- The time taken to transmit command and actual values is very short which guarantees an exact synchronization with axes.

- It supports four operating modes including torque, velocity, position control and block mode.
- Sercos machines have plug-n-play capability, allowing easy adaptation to different applications.
- Implementation of Sercos improves system flexibility as one identical drive can be parameterized to handle multiple prime movers.
- The Sercos interface products from various vendors are interoperable.
- The master controls each ring, assigning timeslots to ensure proper transmission of data.

The disadvantages of Sercos are as follows:

- The speed of the Sercos interface has been pointed out as negative. The capability to transmit data 4 to 10 times faster than required does little to improve machine performance.
- The interface has 32,767 identification numbers for standard commands and it also has the capability to incorporate 32,767 identification numbers. This sometimes makes one SERCOS interface incompatible with the other.
- The cost of SERCOS interface may be higher than that of other standards.

Ethernet

Ethernet was initially mainly used for office automation purposes with the flexibility in layout and interoperability with a majority of office networks. The development of Ethernet has brought the present technology to communicate at 100 Mbps, allowing it to be used as a host for industrial networking.

Ethernet is a computer networking technique for local area network (LAN). The word itself comes from "ether", which was perceived by scientists to be the medium in outer space; Ethernet is in fact the hardware medium of communication of various devices in one network. Ethernet defines the connection in the physical layer and data link layer. Figure 8.9 presents the classification of Ethernet.

With Ethernet, devices with different protocols (but the same Ethernet platform) can exist in the same local area network without conflict, although no communication is established due to different protocols. This simplifies the electrical installation of the network. Furthermore, Ethernet addressing schemes further eliminate the possibility of conflict among devices.

The topology of Ethernet follows the physical layer topology. Signal attenuation due to the physics of the cable limits the length of the network and in this situation hubs and repeaters are required. Their role is essentially to refresh the signal, so that the signal can now be transmitted over a long distance. There is still, however, a limit imposed in the number of hubs and repeaters by the data transfer speed. For example, at a speed of 10 Mbit/s, the number of hubs and repeaters may not exceed four between two nodes (devices). Hubs and receivers broadcast a message to all ports so that each port only gets a part of the available bandwidth.

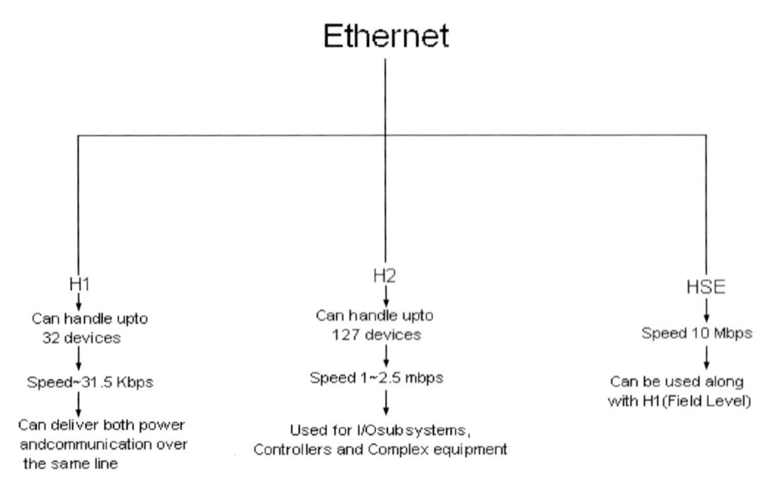

Fig. 8.9. Classification of Ethernet

In an industrial networking, it is very likely that the network grows bigger so that it becomes as if the entire system consists of several local networks. Another similar case occurs when there arises a merger among several small networks, each with their own Ethernet-based network, required to form a high performance system. Ethernet bridge allows connection of several networks to be accomplished, even though each may work at a different speed. Ethernet bridge has an ability to direct the message to the intended recipient by making use of the Ethernet addressing scheme. This is the difference between Ethernet bridge and hubs, as hubs do not filter a message.

The connection of devices into Ethernet is executed via Network Interface Card (NIC). NIC can automatically select the correct speed it should work on.

Ethernet, from being a standard for office level automation, is permeating into the industrial control environment. With an Industrial Ethernet backbone in place, diagnostics can easily be expanded and service functions will be available network-wide, and on a location-independent basis. Despite much development of Ethernet, the future of Ethernet is still full of challenges. Data capture is one of the many challenges which have to be overcome. It would be of great advantage if the plant data can be extracted from the field devices as well as local controllers. This requires a great deal of space in the controller memory. Current development in Ethernet technology includes the use of fiber optics and wireless communication.

Wireless Ethernet usually employs infrared or radio frequency communication and can be applied for both peer-to-peer and infrastructure communication. Although wireless Ethernet offers a great deal of flexibility, its effectiveness is significantly reduced by its incapability of transferring messages over a long distance (about 100 – 200 m). In addition, wireless Ethernet is more prone to message collision as some nodes may not be able to recognize another transmission due to physical obstruction (unlike the case of wire Ethernet). This problem is usually solved by virtual checking, whereby a transmitting node sends a request to all nodes and waits for permission prior to sending a message. This, however, is still an ongoing research area.

Wireless Ethernet can significantly boost the performance of motion control systems, as it can reach the parts of network unreachable by cables, for example when chemical conditions prevent cables/wires to be used in the network.

A

Laser Calibration Optics, Accessories and Set-up

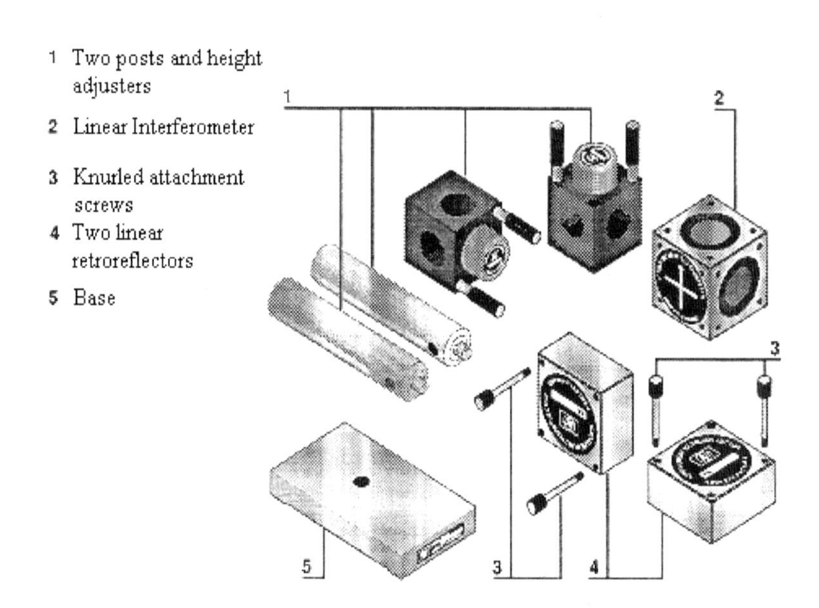

Fig. A.1. Optics and accessories for linear measurements

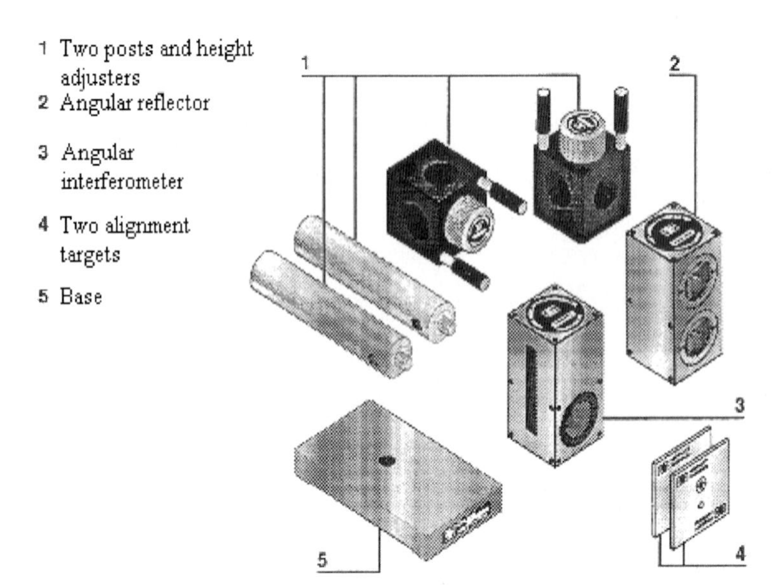

Fig. A.2. Optics and accessories for angular measurements

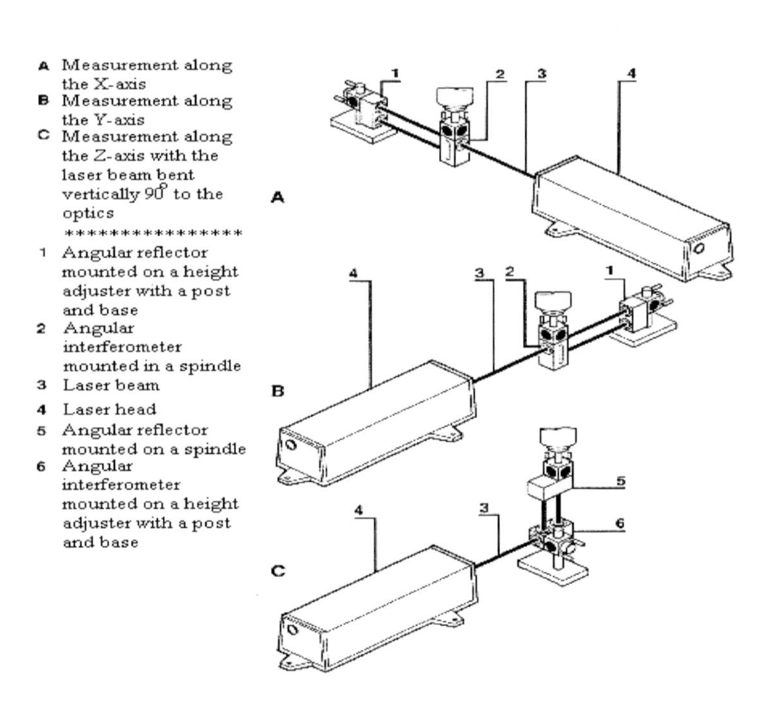

Fig. A.3. Set-up for pitch measurements

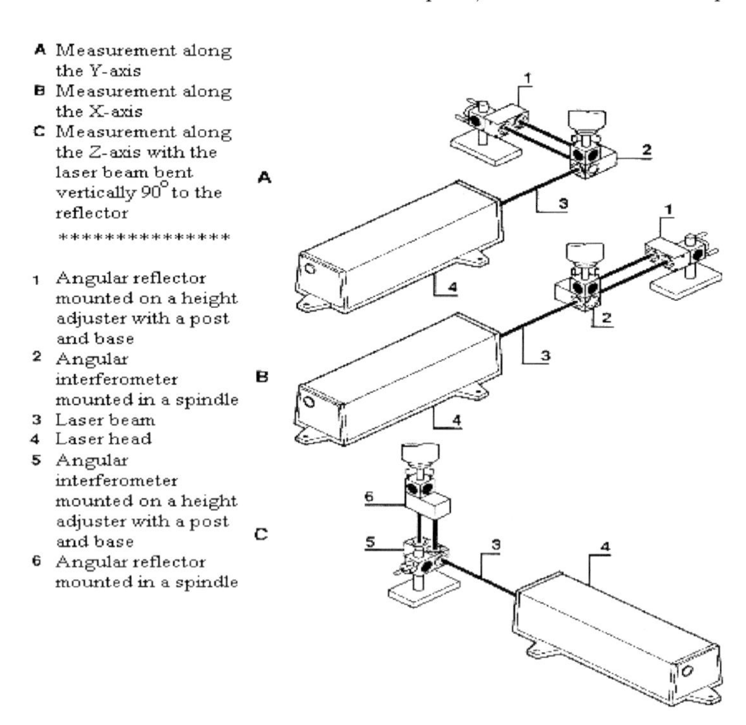

Fig. A.4. Set-up for yaw measurements

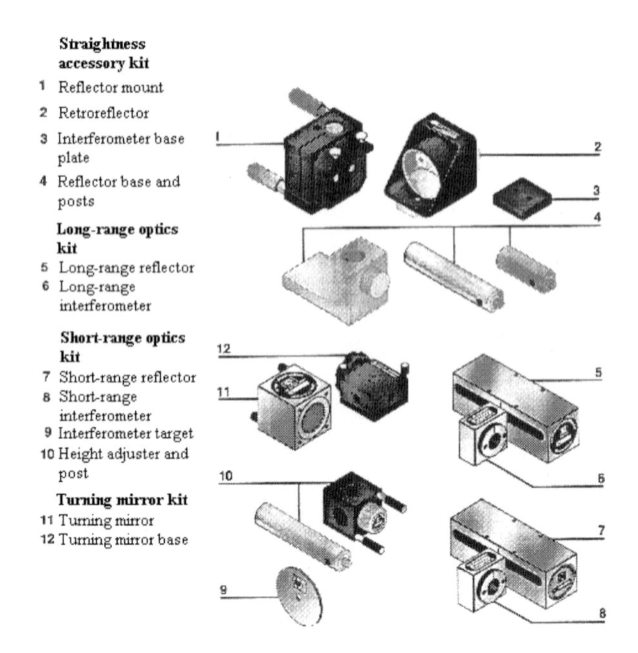

Fig. A.5. Optics and accessories for straightness measurements

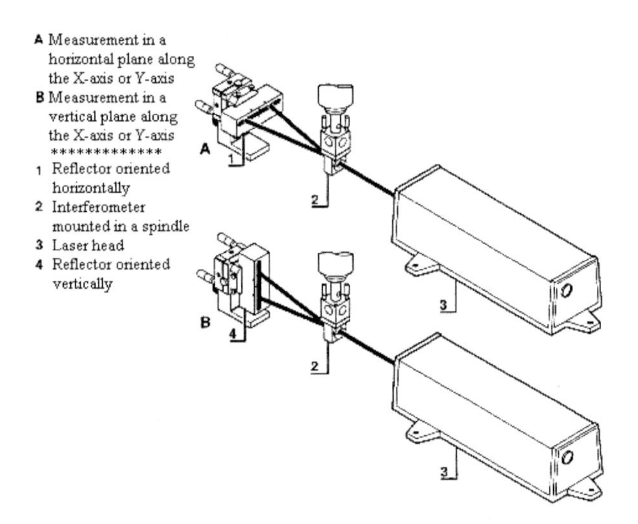

Fig. A.6. Set-up for X-axis and Y-axis straightness measurements

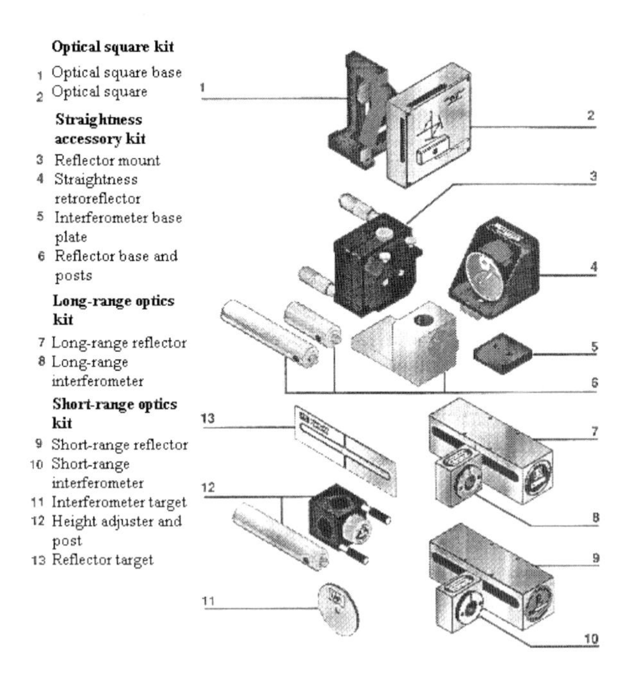

Fig. A.7. Optics and accessories for squareness measurements (horizontal plane)

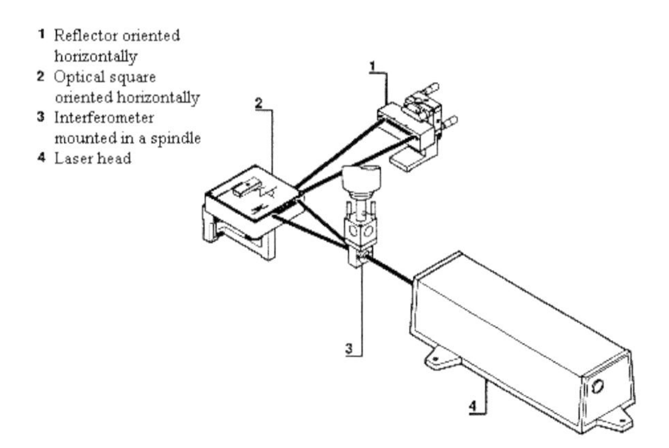

Fig. A.8. Squareness measurements - first axis (horizontal plane)

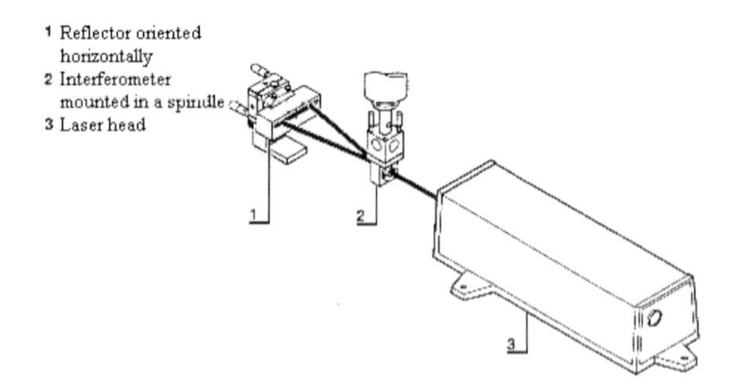

Fig. A.9. Squareness measurements - second axis (horizontal plane)

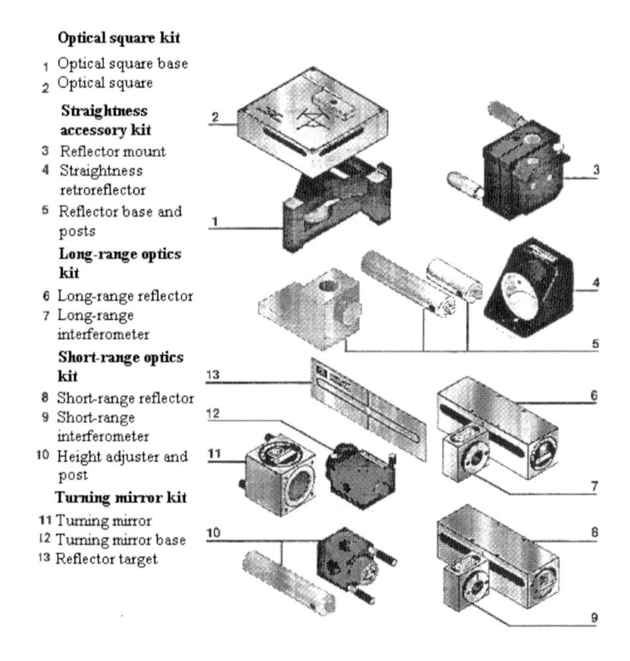

Fig. A.10. Optics and accessories for squareness measurements (vertical plane)

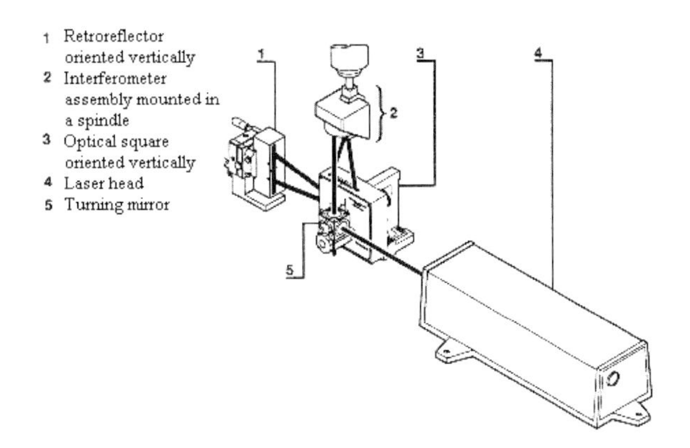

Fig. A.11. Squareness measurement - second axis (vertical plane)

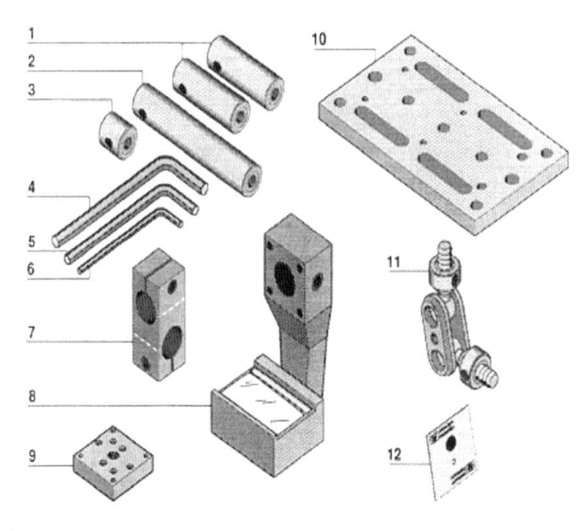

1 Post-medium
2 Post-long
3 Post-stub
4 Hex key 5mm
5 Hex key 4mm
6 Hex key 2.5mm
7 Right-angle clamp
8 Beam-steering assembly
9 Adapter plate
10 Base-large
11 Flexible ball-joint assembly
12 Alignment target

Fig. A.12. Diagonal measurement kit

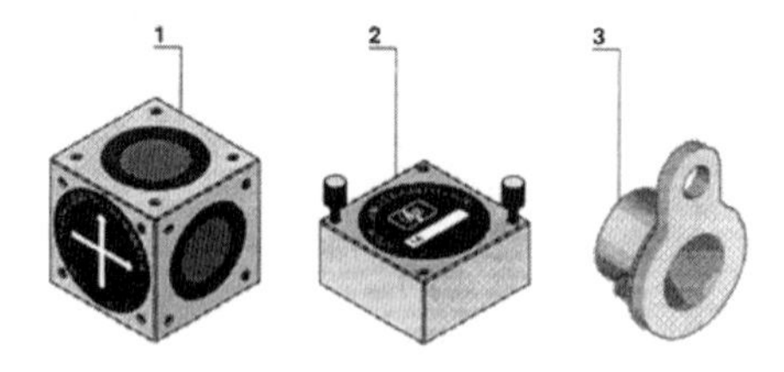

Fig. A.13. Optics and accessories for diagonal measurements

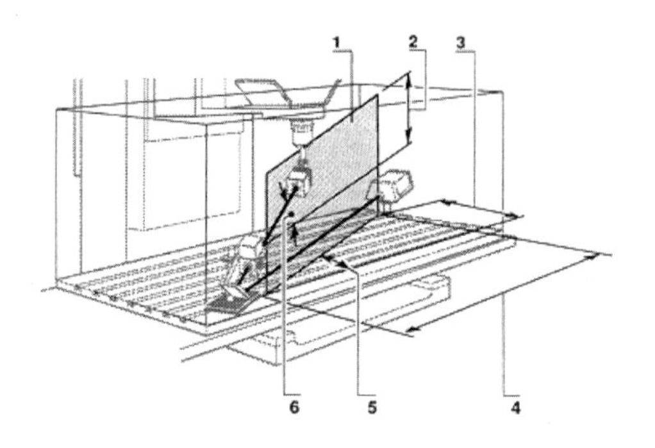

Fig. A.14. Typical diagonal measurement set-up

References

Ahlstrom, M.L. and W.J. Tompkins, Digital filters for real-time ECG signal processing using microprocessors, *IEEE Trans. on Biomedical Engineering*, vol. 32, pp. 708-713, 1985.

Alter, D.M. and T.C. Tsao, Control of linear motors for machine tool feed drives: design and implementation of H_∞ optimal feedback control, *ASME Journal of Dynamic Systems. Measurement and Control*, vol. 118, pp. 649-658, 1996.

Arimoto, S., S. Kawamura and F. Miyazaki, Bettering operation of robots by learning, *Journal of Robotic Systems*, vol. 1, no. 2, pp. 123–140, 1984.

Armstrong-Helouvry, B., P. Dupont and C.C. de Wit, A survey of models, analysis tools and compensation methods for the control of machines with friction, *Automatica*, vol. 30, no. 7, pp. 1083-1138, 1994.

Astrom, K.J. and T. Hagglund, *PID Controllers: Theory, Design, and Tuning*, 2nd Edition, Instrument Society of America, 1995.

Basak,A., *Permanent-magnet DC Linear Motors.* Monographs in Electrical and Electronic Engineering, Clarendon Press, Oxford, 1996.

Beards, C.F., *Structural Vibration Analysis: Modelling, Analysis and Damping of Vibrating Structures*, Ellis Horwood,Chichester, England, 1983.

Bertran, E. and G. Montoro, Adaptive suppression of narrow-band vibrations, 5th International Workshop on Advanced Motion Control, pp. 288-292, 1998.

Besancon-Voda and G. Besancon, Analysis of a class of two-relay systems, with application to Coulomb friction identification, *Automatica*, vol. 35, no. 8, pp. 1391-1399, 1999.

Birch, K.P., Optical fringe subdivision with nanometric accuracy, *Precision Engineering*, vol. 12, no. 4, October 1990.

Blanding, D.L., *Exact Constraint: Machine Design Using Kinematic Principles*, ASME Press, New York, 1999.

Braembussche, P.V., J. Swevers, H. Van Brussel and P. Vanherck, Accurate tracking control of linear synchronous motor machine tool axes, *Mechatronics*, vol.6, no. 5, pp. 507-521, 1996.

Brian, A.H., D. Pierre and C.D.W. Carlos, A survey of models, analysis tools and compensation methods for the control of machines with friction, *Automatica*, vol.30, pp. 1083-1138, 1994.

BritishStandard: Coordinate Measuring Machines, Part 3. Code of practice, BS 6808,1989.

Bryan, J., International status of thermal error research, *Annals of the CIRP*, vol.39(2), pp. 645-656, 1990.

Bush, K., H. Kunzmann and F.Waldele, Numerical error correction of Coordinate measuring machines, Proc.Int.Symp.on Metrology for Quality Control in Production, Tokyo, pp. 270-282, 1984.

Canudas-de-Wit, C., H. Olsson, K. Astrom, and P. Lischinsky, A new model for control of systems with friction, *IEEE Trans. on Automatic Control*, vol. 40(3), pp. 419-425, 1995.

Chang, S.B., S.H. Wu and Y.C. Hu, Submicrometer overshoot control of rapid and precise positioning, *Journal of the American Society for Precision Engineering*, vol.20, pp. 161-170, 1997.

Chen, J.S. and C. C. Liang, Improving the machine accuracy through machine tool metrology and error correction, *International Journal on Advanced Manufacturing Technology*, vol. 11, 1996.

de Jager, Bram, Acceleration assisted tracking control. *IEEE Control Systems Magazine*, vol. 14, no. 10, pp. 20–27, 1994.

Dorato, P., C.Abdallah and V.Cerone, *Linear-Quadratic Control: An Introduction*, Prentice Hall, Englewood Cliffs, New Jersey, 1995.

Dou, H., Z. Zhou, Y. Chen, J.-X. Xu and J. Abbas, Robust motion control of electrically stimulated human limb via discrete-time high-order iterative learning scheme, *Proceedings of 1996 Int. Conf. on Automation, Robotics and Computer Vision (ICARCV'96).* Singapore. pp. 1087–91, 1996.

Duffie, N.A. and S.J. Maimberg, Error diagnosis and compensation using kinematic models and position error data, *Annals of the CIRP*, vol.36, no.1, pp. 355-358, 1987.

Evans, C.J., Precision Engineering:an evolutionary view, Cranfield Press, Cranfield, UK, 1989.

Ferdjallah, M. and R.E. Barr, Adaptive digital notch filter design on the unit circle for the removal powerline noise from biomedical signals, *IEEE Trans. on Biomedical Engineering*, vol. 41, no. 6, pp. 529-536, 1994.

Fleming, J.F., *Analysis of Structural Systems*, Prentice-Hall, New Jersey, 1997.

Friman, M. and K.V. Waller, A Two-Channel Relay for Autotuning, *Ind. Eng.Chem.Res.*, vol. 36, pp. 2662-2671, 1997.

Fujimoto, Y. and A. Kawamura, Robust servo-system based on two-degree-of-freedom control with sliding mode, *IEEE Trans.on Industrial Electronics*, vol. 42, no. 3, pp. 272-280. 1995.

Gelb, A. and W.E. Vander Velde, *Multiple-input describing functions and nonlinear system design*, McGraw-Hill Book Company: USA. 1968.

Gilles, R., D. Dragan, and S. Nava, Separation of nonlinear and friction-like contributions to the piezoelectric hysteresis, pp. 699-702, 2001.

Glover, J.R., Comments on digital filters for real-time ECG signal processing using microprocessors, *IEEE Trans. on Biomedical Engineering*, vol. 34, pp. 962-963, 1987.

Goldfarb, M., and N. Celanovic, Modeling piezoelectric stack actuators for control of micromanipulation,*IEEE Transactions on Control Systems Magazine*, vol. 17 (3), pp. 69-79, 1997.

Goto, S. and M. Nakamura, Accurate contour control of mechatronics servo system using Gaussian Networks. *IEEE Trans. on Industrial Electronics*, vol. 43, pp. 469–476, 1996.

Hagiwara, N. and H. Murase, A method of improving the resolution and accuracy of rotary encoders using a code compensation technique, *IEEE Trans. on Instrumentation and Measurement*, vol. 41, no. 1, February 1992.

Hayati, S.A. Robot Arm Geometric Link Parameter Estimation, *Proceedings of the 22nd IEEE Conference on Decision and Control,* pp. 1477-1483, 1983.

Haykin, S., *Neural Networks: A Comprehensive Foundation*, Prentice Hall International, Inc., 1994.

Heydemann, P.L.M., Determination and correction of quadrature fringe measurement errors in interferometers, *Applied Optics*, vol. 20, no. 19, October 1981.

Hiroyuki, A., Linear motor system for high speed and high accuracy position seek, *Proceedings of the Second International Symposium on Linear Drives for Industry Applications*, Tokyo, Japan, pp. 461-464, 1995.

Hocken, R., J. Simpson, B. Borchardt, J. Lazar and P. Stein, Three dimensional metrology, *Annals of the CIRP*, vol.26, no. 2, pp. 403-408, 1977.

Hocken, R., Machine Tool Accuracy, in Report of Technology of Machine Tools, vol.5, Lawrence Livermore Laboratory, University of California, 1980.

Hornik, K., M.Stinchcombe,and H.White, Multiilayer feedforward networks are universal approximators, Neural Networks, vol.2, pp. 359-366, 1989.

Inonnou, P.A., and J. Sun, *Robust adaptive control, Prentice-Hall International*, Inc., 1996.

Jeon, J-Y., S-W Lee, H-K Chae and J-H Kim, Low velocity friction identification and compensation using accelerated evolutionary programming.

Proceedings of IEEE International Conference on Evolutionary Computation, vol. 41, no. 1, pp. 372–376, 1996.

Kim, S., B. Chu, D. Hong, H.K. Park, J.M. Park and T.Y. Cho, Synchronizing dual-drive gantry of chip mounter with LQR approach. *Proceedings of IEEE international conference on Advanced Intelligent Mechatronics*, pp. 838-843, 2003.

Krstic, M., I. Kanellakopoulos and P. Kokotovic, *Nonlinear and Adaptive Control Design*, John Wiley & Sons, 1995.

Kwan, T. and K. Martin, Adaptive detection and enhancement of multiple sinusoids using a cascade IIR filter, *IEEE Trans. on Circuits and Systems*, vol.36, no. 7, pp. 937-947, 1989.

Le-Huy, H., R. Perret and R. Feuiliet, Minimization of torque ripple in brushless DC motor drives, *IEEE Trans. on Industrial Applications*, vol. 22, no. 4, pp. 748-755, 1986.

Li, G.,A stable and efficient adaptive notch filter for direct frequency estimation, *IEEE Trans. on Singal Processing*, vol.45, no. 8, pp. 2001-2009, 1997.

Ljung, L. *System identification theory for the user*, 2nd-Edition, Prentice-Hall, Inc., Englewood Cliffs, New Jersey, 1997.

Longman, R.W. *Iterative Learning Control - Analysis, Design, Integration and Application.* Chapter in Designing Iterative Learning and Repetitive Controllers, Kluwer Academic Publishers, pp. 107–145, 1998.

Love, W.J. and A. J. Scarr, The determination of the volumetric accuracy of multi axis machines, 14th MTDR, pp. 307-315, 1973.

Mayer, J.R.R., High resolution of rotary encoder analog quadrature signals, *IEEE Trans. on Instrumentation and Measurement*, vol. 43, no. 3, June 1994.

McInroy, John E. and George N. Saridis, Acceleration and torque feedback for robotic control: experimental results. *Journal of Robotic Systems* vol. 7(6),pp. 813–832, 1990.

McKeown, P., Geometric error measurement and compensation of machines, *Annals of the CIRP*, vol.44, no. 2, pp. 599-609, 1996.

McKeown, P., Reduction and compensation of thermal errors in machine tools, *Annals of the CIRP*, vol.44, no. 2, pp. 589-598, 1996.

Moore, K.L., Iterative learning control - an expository overview, *Applied & Computational Controls, Signal Processing, and Circuits*, vol.1, no.1, pp. 425-488, 1998.

Ni, J., CNC machine accuracy enhancement through real time error compensation, *Journal of Manufacturing Science and Engineering*, vol 119, November 1997.

Otten, G., J.A.de Vries, J.van Amerongen, A.M.Rankers and E.W.Gaal, Linear motor motion control using a learning forward controller, *IEEE Trans. on Mechatronics*, vol.2, no. 3, pp. 179-187, 1997.

Pritschow, G. and W. Philipp, Research on the efficience of feedforward controllers in direct drives, *Annals of CIRP*, vol. 41, no. 1, pp. 411–415, 1992.

Ramirez, R.W. *The FFT Fundamentals and Concepts*, Prentice-Hall Inc, Englewood Cliffs, 1985.

Regalia, P.A., S.K. Mitra and P.P. Vaidyanathan, The digital all-pass filter: a versatile signal processing building block, *Proc. of IEEE*, vol.76, no. 1, pp.19-37, 1988.

Robet, P.P., M. Gautier, and C. Bergmann, A frequency approach for current loop modeling with a PWM converter, *IEEE Trans. on Industry Applications*, vol.34, no. 5, pp. 1000-1014, 1998.

Saab, S.S. Discrete-time learning control algorithm for a class of nonlinear system, *In Proceeding of American Control Conference*, Seattle, Washington, USA, pp. 2739-2743, 1995.

Satori, S., G. Colonnetti and Zhang, G.X., Geometric error measurement and compensation of machines, Annals of the CIRP, vol.44, no.1, pp. 599-609, 1995.

Shen, Y.L., Comparison of combinational rules for machine error budgets, *Annals of the CIRP*, vol.42, no. 1, pp. 619-622, 1993.

Shiro, U., K. Takuya, S. Takayuki and H. Hironobu, High performance positioning system with linear DC motor under self-tuning fuzzy control,

Proceedings of the Second International Symposium on Linear Drives for Industry Applications, Tokyo, Japan,1995, pp. 295-298, 1995.

Slotine, J.J.E. and J.A. Coetsee, Adaptive sliding controller synthesis for nonlinear systems, *International Journal of Control*, vol.43, no. 6, pp. 1631-1651, 1986.

Taghirad, H.D. and P.R.Bélanger, Robust friction compensator for harmonic drive transmission, *Proc. of 1998 IEEE Int'l Conf. on Control Applications*, pp. 547-551, 1998.

Takagi, T. and M. Sugeno, Fuzzy identification of systems and its applications to modelling and control, *IEEE Trans. on Systems, Man, and Cybernetics,* vol. 15, pp. 116-132, 1985.

Tan, K.K., H. F. Dou, Y. Q. Chen and T. H. Lee, High precision linear motor control via artificial relay tuning and zero-phase filtering based iterative learning, *IEEE Trans. on Control Systems Technology*, vol. 9, pp. 244-253, 2001.

Tan, K.K., T. H. Lee, H. F. Dou and S. J. Chin, PWM modelling and application to disturbance observer-based precision motion control, *PowerCon 2000*, Perth, Australia, 2000.

Tan, K.K., S.N. Huang, H.F. Dou, S.Y. Lim and S.J. Chin, Adaptive robust motion control for precise trajectory tracking applications, *ISA Trans.*, vol. 40, pp. 57-71, 2001.

Tan, K. K., T.H.Lee, S.N.Huang, H.L.Seet, H.F.Dou, and K.Y.Lim, Probabilistic approach towards error compensation for precision machines, *Proceedings of the Third International ICSC Symposia on Intelligent Industrial Automation II'A99 and Soft Computing SOCO'99*, Italy, pp. 293-299, 1999.

Tan, K. K., S. N. Huang, and H. L. Seet, Geometrical error compensation of precision motion systems using radial basis functions, *IEEE Trans. on Instrumentation and Measurement*,vol.49, no.5, pp. 984-991, 2000.

Tan, K. K., S. N. Huang, and T.H. Lee, Geometrical error compensation of precision motion systems using neural network approximations, *IEEE Trans. on Instrumentation and Measurement*, vol. 40, pp. 57-71, 2001.

Tan, K. K., T. H. Lee, H. F. Dou and S. Y. Lim, An adaptive ripple suppression/compensation apparatus for permanent magnet linear motors,

Technical Report MA99-05, Department of ECE, National University of Singapore, 2000.

Tan, K.K., T.H.Lee, S.N.Huang and S.Y. Lim, Adaptive control of DC permanent magnet linear motor for ultra-precision applications, *Proceedings of the International Conference on Mechatronic Technology*, Taiwan, pp. 243-246, 1998.

Tan, K.K.,T.H.Lee, S.Y.Lim and H.F.Dou, Learning enhanced motion control of permanent magnet linear motor, *Proc. of the third IFAC International Workshop on Motion Control*, Grenoble, France, pp. 397-402, 1998.

Tan, K.K.,T.H.Lee and F.M.Leu, Automatic tuning of 2 DOF control for D.C.Servo Motor Systems,*Intelligent automation and soft computing*, vol.6, pp. 281-290, 2000.

Tan, K.K.,T. H. Lee, S. Huang and X. Jiang, Friction modeling and adaptive compensation using a relay feedback approach, *IEEE Trans. on Industrial Electronics*, vol. 48, pp. 169-176, 2001.

Tan, K.K., H. X. Zhou and T. H. Lee, New interpolation method for quadrature encoder signals, *IEEE Trans. on Instrumentation and Measurement*, vol. 51, pp. 1073-1079, 2002.

Tan, K.K., S.Y. Lim, S. N. Huang, H. Dou and T.S. Giam, Co-ordinated motion control of moving gantry stages for precision applications based on an observer-augmented composite controller. *IEEE Trans. on Control Systems Technology*, vol. 12, pp. 984-991, 2002.

Tan, K.K., S. Y. Lim, T. H. Lee and H. F. Dou, High precision control of linear actuators incorporating acceleration sensing, *Robotics and Computer-Integrated Manufacturing*, vol. 16, pp.295-305, 2000.

Tang, K. Z., K. K. Tan, C. W. de Silva, T. H. Lee, and S. J. Chin, Monitoring and Suppression of Vibration in Precision Machines, *Journal of Intelligent and Fuzzy Systems*, vol. 11, pp. 33-52, 2000.

Tymerski, R., V. Vorperian, Fred C.Y. Lee and W.T. Baumann, Nonlinear modeling of the PWM switch, *IEEE Trans. on Power Electronics*, vol.4, no.2, pp.225-233, April 1989.

Veitschnegger, W.K. and C.H. Wu, Robot accuracy analysis based on kinematics, *IEEE Journal of Robotics and Automation,* vol. RA-2/3, pp. 171-179, 1986.

Vierck, R.K., *Vibration Analysis*, Crowell, New York, 1979.

Watanabe, K. and H. Yokote, A microstep controller of a DC servomotor, *IEEE Trans. on Instrumentation and Measurement*, vol. 39, no. 6, December 1990.

Weekers, W.G. and P.H.J. Schellekens, Assessment of dynamic errors of CMMs for fast probing, *Annals of the CIRP,* vol.44, no.1, pp. 469-474, 1995.

White, M. T. and M. Tomizuka, Increased disturbance rejection in magnetic disk drives by acceleration feedforward control and parameter adaptation, *Control Engineering Practice*, vol. 5, no. 6, pp. 741–751, 1997.

Wu, S.M.and J. Ni, Precision machining without precise machinery, *Annals of the CIRP*, vol.38, no. 1, pp. 533-536, 1989.

Xu, J.X.and Z. Bien. *Iterative learning control - analysis, design, integration and application.* Chapter in The Frontiers of Iterative Learning Control, Kluwer Academic Publishers. pp. 9–35, 1998.

Xu, L.and B. Yao, Adaptive robust precision motion control of linear motors with ripple force compensations: theory and experiments, *Proceedings of the 2000 IEEE International Conference on Control Applications*, pp. 373-378, 2000.

Yamada, K., S. Komada, M. Ishida and T. Hori, Analysis of servo system realized by disturbance observer, *Proc. IEEE International Workshop on Advanced Motion Control*, pp. 338-343, 1996.

Yamada, K., S. Komada, M. Ishida and T. Hori, Analysis and classical control design of servo system using high order disturbance observer, *23rd International Conference on Industrial Electronics, Control and Instrumentation IECON 97*, vol.1, pp.4-9, 1997.

Yao, B. and L. Xu, Adaptive robust control of linear motors for precision manufacturing, *Proceedings of the 14th Triennial World Congress*, Beijing, pp. 25-30, 1999.

Yeh, S-S. and P-L Hsu, Analysis and design of the integrated controller for precise motion systems. *IEEE Trans. on Control Systems Technology*, vol. 7, no. 6, pp. 706–717, 1999.

Yokote, H. and K. Watanabe, A hybrid digital and analog controller for DC and brushless servomotors, *IEEE Trans. on Instrumentation and Measurement*, vol. 39, no. 1, February 1990.

Zadeh, L. A., Outline of a new approach to the analysis of complex systems and decision process, *IEEE Trans. on Systems, Man, and Cybernetics*, vol. 3, pp. 28-44, 1973.

Zhang, G., R. Veale, T. Charlton, R. Hocken,B. Borchardt, Error compensation of coordinate measuring machines, Annals of the CIRP, vol.34, pp. 445-448, 1985.

Zhu, Z.Q., Z.P.Xia, D.Howe and P.H.Mallor, Reduction of cogging force in slotless linear permanent magnet motors, *IEE Proc. Electrical Power Applications*, vol.144, no. 4, pp. 277-282, 1997.

Index